Mesozoic Sea Dragons

MESOZOIC SEA DRAGONS

Triassic Marine Life from the Ancient Tropical Lagoon of Monte San Giorgio

OLIVIER RIEPPEL

RECONSTRUCTIONS BY **BEAT SCHEFFOLD**

Indiana University Press

This book is a publication of

Indiana University Press
Office of Scholarly Publishing
Herman B Wells Library 350
1320 East 10th Street
Bloomington, Indiana 47405 USA

iupress.indiana.edu

Manufactured in the
United States of America

Cataloging information is available
from the Library of Congress.

ISBN 978-0-253-04011-4 (hdbk.)
ISBN 978-0-253-04013-8 (web PDF)

1 2 3 4 5 24 23 22 21 20 19

Contents

C

vii   **Acknowledgments**

3   **1   The Dragon Mountain**

31   **2   Fishes**

73   **3   A Sketch of Reptile Evolution**

91   **4   Ichthyosaurs**

105   **5   *Helveticosaurus*, *Eusaurosphargis*, and the Placodonts**

123   **6   Pachypleurosaurs**

139   **7   Lariosaurs and Nothosaurs**

153   **8   Thalattosaurs**

169   **9   Protorosaurs**

191   **10   A Dinosaur Lookalike from Monte San Giorgio**

199   **11   The Tethys Sea: Connections from East to West**

215   **Epilogue—In the Shadow of the Chinese Dragon**

219   **Literature Cited**

239   **Index**

Acknowledgments

I am very grateful to the Series Editor, James O. Farlow, as well as Gary Dunham, Peggy Solic, and Mary Jo Rhodes from Indiana University Press, who helped to see this book through publication. Christian Klug, and Torsten Scheyer from the Paleontological Institute and Museum of the University of Zürich, as well as Tony Bürgin (St. Gallen), Markus Felber (Morbio Inferiore, Ticino), Nicholas C. Fraser (Edinburgh), Heinz Furrer (Zürich), Hartmut Haubold (Halle/Saale), Heinz Lanz (Winterthur), Hans Rieber (Zürich), Christopher R. Scotese (Evanston, IL), Giorgio Teruzzi (Milan), Karl Tschanz (Zürich), the Archivio Sommaruga, the Commissione Scientifico Transnationale Monte San Giorgio, and the Fotostiftung Schweiz all helped and supported the book project in various ways, especially by providing illustrations. John Weinstein (photographer) and Marlene Donnelly (scientific illustrator), both at the Field Museum, improved the quality of the illustrations. Sincere thanks to all of these friends and colleagues.

Mesozoic Sea Dragons

1.1. The field crew at the Acqua del Ghiffo site in 1928. From *left* to *right*: Emil Kuhn, Giuseppe Buzzi, "Sergente" Buzzi, Vittorio Marogna, "Bergum" (a citizen of Meride), T. Bresciani, B. Bianchi (digital rendition Heinz Lanz; Photo © Max P. Linck / Fotostiftung Schweiz).

The Dragon Mountain

It had been a hot day, in spite of the surrounding trees, which offered speckled shade. But the team of paleontologists had managed to clear another slab of Triassic fossiliferous sedimentary rock from the overburden. In the oblique light of the late afternoon, the contours of several promising fossils could be made out—some fish, several other small, lizard-like pachypleurosaurs. No larger fossils were found that day. They had circled the fossils with white chalk lines and planned to cut them out the next day. Then they would further expand their dig in the following days and weeks, when they would hit it big! The find would be a complete skeleton of a new lariosaur genus and species, 104 cm in total length (small by comparison to other, later finds of the same species), which Peyer christened *Ceresiosaurus calcagnii*, in honor of Commendatore Emilio Calcagni (Peyer, 1931a). Calcagni was the landowner who had graciously allowed excavations to proceed since the spring of 1927, when Peyer first found fossils at this locality. Peyer derived the genus name, *Ceresiosaurus*, from the local name for Lake Lugano, which embraces the eastern, northern, and western flanks of the northward-jutting pyramidal mountain; the team was working on the western slope of this mountain, some way above the Italian lakefront town of Porto Ceresio. Satisfied with the dig's progress, Peyer poured himself a strong coffee, espresso really, from his Thermos flask. To complete what he called the trimming of a fossil find, the coffee was to be accompanied by a shot of Grappa del Ticino, a spirit distilled from pomace of Merlot, the signature grape of the Canton Ticino, Switzerland.

Peyer sat down and proceeded to stuff his pipe, looking at his crew of workers through his rimless spectacles. Absent-mindedly, he picked up a chestnut, one of the first to have ripened this season. The Ticinesi, the local inhabitants of the Canton Ticino, collect them in the fall to roast over fire or glowing charcoal. Some would travel to northern cities in Switzerland—Lucerne or Basel—to sell their freshly roasted *Marroni* on the streets. With all the excitement of fossil hunting, Peyer had not realized how hungry he was. He looked forward to ordering braised rabbit with polenta for dinner at the local *Grotto*, the Ristorante Alpino in Serpiano, paired with a bottle of the local Merlot, and followed of course by another trimming—an espresso con grappa (Kuhn-Schnyder, 1968)! It was late summer of 1928, and his team was fossil hunting at the Acqua del Ghiffo locality near Crocifisso (fig. 1.1), on Monte San Giorgio, the latter located in southern Switzerland just across the border from Italy.

Bernhard Peyer

Peyer had a lean, wiry physique and was wearing baggy trousers stuck in rubber boots. During the heat of the day he had taken his jacket off and put it aside, but being the only academic on the site, he had thought it proper to keep his vest on. His hair, currently disheveled, was cut short and combed to one side. As Peyer stroked his moustache, which partly obscured his thin lips, he thought it needed trimming. Bernhard was born in the Swiss town Schaffhausen on July 25, 1885, son of the textile manufacturer Johann Bernhard Peyer and his wife, née Sophie Frey (H. Fischer, 1963; H. C. Peyer, 1963). The parents guided their son Bernhard through the Schaffhausen school system toward graduation in a classical humanistic education. Bernhard displayed a mastery of foreign languages—and these included not just French, English, and Italian but also Latin and classic Greek. One of his preferred leisure-time activities became reading and reciting Homer, author of the *Iliad* and the *Odyssey*, in the original. But Bernhard felt equally at ease in the outdoors and developed an early interest in natural history, paleontology, and geology. He obtained his first formal training in zoology and comparative anatomy as a student of Arnold Lang (1855–1914), a former student, then assistant, and eventually colleague of the famous Jena zoologist and evolutionist Ernst Haeckel (1834–1919), also known as the "German Darwin." Of Swiss extraction, Lang joined the University of Zurich in 1889, where he pursued a stellar career until his death in November 1914 (Haeckel, Hescheler, and Eisig, 1916). Peyer (fig. 1.2) studied further at the University of Munich, where he heard the famous zoologist Richard Hertwig (1850–1937), another Haeckel student, and forged relations that would evolve into longtime friendships with the paleontologists Ferdinand Broili (1874–1946) and Ernst Stromer von Reichenbach (1871–1952).

Back at the University of Zurich, Peyer obtained his PhD under Arnold Lang with a dissertation on the embryonic development of the skull of the asp viper, *Vipera aspis*, a venomous snake native to central and southern Europe, a species first described by Linnaeus in 1758. Karl Hescheler (1868–1940), former student and eventual successor of Arnold Lang as professor of zoology and comparative anatomy at the University of Zurich, encouraged Peyer to apply for the *venia legendi*—the honor and duty to teach at the university level—with the submission of his work on the fin-spines of catfish as a *Habilitation* thesis. Hescheler is considered the initiator of Swiss paleontology, not only through his research on fossil mammals but also as a founding member of the Swiss Archeological and Paleontological Societies. A bachelor throughout his life, Hescheler bequeathed his estate to the University of Zurich, thus establishing the Karl Hescheler endowment. The latter would support Peyer's paleontological excavations in important ways in the years to come and continues to support the Zoological and Paleontological Museums of the University of Zurich to the present day.

On the excavation in 1928, Peyer was still a lecturer at the University of Zurich, but in 1930 he was promoted to associate professor. In 1943 he was voted full professor of paleontology and comparative anatomy, and

1.2. Bernhard Peyer (1885–1963), date unknown (Photographer unknown; digital rendition Heinz Lanz).

director of the Zoological Museum of the University of Zurich. During his long and extraordinary career, Peyer published extensively, producing numerous voluminous monographs on the Triassic reptiles from Monte San Giorgio and also on fossil remains of sharks, bony fishes, reptiles from other localities and time horizons, and important papers on fossil mammals, most notably the Late Triassic haramiyids from Hallau near his home town Schaffhausen (Peyer, 1956). He also published on the development and histology of vertebrate hard tissues, especially teeth. But additionally he contributed publications in the history of science, such as comments on the biological writings of Aristotle; a biography of his famous forefather, the medic Johann Conrad Peyer (1653–1712); an account of the biological writings of the medic Johannes von Muralt (1645–1733); a portrait of the senior town physician and polymath Johann Jakob Scheuchzer (1672–1733) from Zurich; an *aperçu* of the founding father of stratigraphy, Nicolaus Steno (1638–1686); and an overview of Johann Wolfgang von Goethe's (1749–1832) vertebral theory of the skull. When he happened to observe the exotic reproductive behavior of the large land slug *Limax cinereoniger* during fieldwork at Monte San Giorgio in the years 1927 and 1928, he published that too, together with his field assistant, the student Emil Kuhn.

1.3. (top) and **1.4.** (bottom) The little town of Meride in 2000 (Photo © Heinz Furrer/Paleontological Institute and Museum, University of Zurich); Via Bernardo Peyer, Meride, March 18, 2001 (Photo © Heinz Lanz/Paleontological Institute and Museum, University of Zurich).

Near the site of Peyer's excavation was Meride, a small, historic municipality located at the southern foot of Monte San Giorgio (fig. 1.3). The houses are grouped along the main street that stretches from east to west along the side of the mountain, forming the backbone of the village. In its center stands the church San Rocco, dating from the seventeenth century. On slightly elevated grounds west of the village is the church San Silvestro, dating from the sixteenth century. Corn (maize) fields stretch out to the south of the hamlet, mingled with orchards and the occasional wheat field. The vineyards creep up the lower reaches of Monte San Giorgio behind it. Today, Meride boasts a refurbished paleontological museum, designed by star architect Mario Botta; the new museum opened in 2012 with Triassic fossils from Monte San Giorgio on display. In 1967, four years after his death, Peyer was named honorary citizen of Meride, and the street on which the museum is located was named after him, the Via Bernardo Peyer (fig. 1.4). Even today, Meride is not easy to reach from Zurich. An intercity train takes passengers to Lugano, a regional train continues on to Mendrisio, where they have to board the PostBus—the *auto da posta*—that goes to Meride. Since 1955, fieldwork at Monte San Giorgio has been organized out of Meride. The dig in 1928 at Acqua del Ghiffo targeted limestone deposits, the so-called Cava Superiore layers of the lower Meridekalke (Meride Limestone) of Ladinian age, approximately 238 million years old. These were not the deposits, however, that originally attracted Peyer's attention or the interest of other paleontologists. Munich paleontology professor and later friend Ferdinand Broili first pointed Peyer in that direction. The slightly older layers of a different nature, approximately 241 to 245 million years old, had first been recognized for their potential to yield a rich variety of fossils: these were the bituminous black shales and dolomites of the so-called Grenzbitumenzone that straddles the Anisian–Ladinian boundary of the Middle Triassic.

Since the mid-eighteenth century, the government of Lombardy, based in Milan, had been concerned about maintaining a sufficient energy supply for the city (the present account of the industrial exploitation of the mid-Triassic bituminous layers of Lombardy and Switzerland follows the account by Furrer, 2003:31–34). One source of combustible fuel that could potentially be exploited was the Middle Triassic bituminous layers of rock rich in carbon and oil situated above Besano, a village located just southwest of Porto Ceresio in northwestern Lombardy, near the Italian-Swiss border. Today, these layers are called the Besano Formation. Equivalent outcrops of the same layers on the slopes of Monte San Giorgio were discovered in 1856. These bituminous black shales and dolomites on the Swiss side are called the Grenzbitumenzone. The term *Grenzbitumenzone*, which translates into "bituminous boundary zone," was introduced by Albert Frauenfelder (1916:264), as he believed it to form the upper boundary of the *Paraceratites trinodosus* biozone of Anisian age (268).

Of Oil and Fossils

P. Brack and H. Rieber draw the Anisian-Ladnian boundary at the base of the *Eoprotrachyceras curionii* biozone, which places the Anisian-Ladinian boundary in the upper part of the Grenzbitumenzone (Brack and Rieber, 1993; see also Brack et al., 2005; see below for further discussion). In the more recent literature, the term *Besano Formation* has also been applied to the Monte San Giorgio localities because the sediments at both localities were deposited in the same marine basin (Furrer, 2003:37). In this book, I use the terms *Besano Formation* for the Besano locality and *Grenzbitumenzone* for the Monte San Giorgio localities; this choice is not for geological reasons but to keep the geography of these fossiliferous deposits from becoming confusing.

The discovery of these bituminous shales marked the beginning of a mining history that involved both Lombardian and Swiss interests. It soon became clear, however, that the carbon content of these bituminous rocks was insufficient for them to serve as an efficient source of energy. What instead became the target of industrial exploitation was their oil content. An oily substance called *Saurol* (sometimes also referred to as *Ichthyol*) could be extracted from these black shales through pyrolysis; this became the raw material from which the pharmaceutical industry in Milan and Basel produced anti-inflammatory ointments. The company that headed these efforts, beginning in 1908, was the Società Anonima Miniere Scisti Bituminosi di Meride e Besano, founded in 1906 (Rieber and Lanz, 1999; Felber and Tintori, 2000). The name of the primary substance, *Saurol*, indicates the occurrence of fossils of saurians in the bituminous shales that this company exploited. The first saurian to be described from the surroundings of Besano was a pachypleurosaur, called *Pachypleura Edwardsii* by Emilio Cornalia (1824–1882), conservator and later director of the Museo Civico di Storia Naturale di Milano, in a publication dating to 1854 (Cornalia, 1854; the valid name of the species today is *Neusticosaurus edwardsii*: Sander, 1989a). The earliest paleontological fieldwork conducted at the Besano site dates to the years 1863 and 1878. It was organized by the Milan Natural History Museum under the direction of Cornalia, and the Italian Association for Natural Sciences under the leadership of the abbot Antonio Stoppani (1824–1891), who succeeded Cornalia as the director of the Milan Natural History Museum in 1882 (Stoppani, 1863; Pinna and Teruzzi, 1991:5; Rieber and Lanz, 1999:78; a preliminary account of the results of those fossil excavations was published by Francesco Bassani in 1886).

For its 1919 annual meeting, the Swiss Academy of Sciences (at that time called the Schweizerische Naturforschende Gesellschaft) chose Lugano in the Canton Ticino for its venue. Bernhard Peyer attended the meeting and used the opportunity to visit the plant of Miniere Scisti Bituminosi di Meride e Besano, the Fabbrica di Olio at Spinirolo near Meride (fig. 1.5). The mining company granted him permission to search for fossils in a pile of bituminous rock that awaited industrial processing. And sure enough, he came up not only with the remains of fossil fishes but also with the well-preserved fore fin of a smallish ichthyosaur. Peyer

1.5. The *Fabbrica di Olio Spinirolo* near Meride around 1940 (Photographer unknown, digital rendition Heinz Lanz; © Archivio Sommaruga, Fondazione del Monte San Giorgio-CH).

extended his search to the float of bituminous shales at the Cava Tre Fontane site above the village of Serpiano on the western slope of Monte San Giorgio, where he found additional fossil material. Due to the industrial methods of mining, including blasting, these fossil remains were badly broken and incomplete and hence useless for science, but they indicated where more could be found and collected observing scientific standards (Kuhn-Schnyder, 1974:10ff.).

Peyer explained the great potential and importance of these discoveries to his supervisor, Karl Hescheler, who immediately saw their significance and in the years to come would generously support Peyer in his activities at Monte San Giorgio. With the initial support of a grant from a Swiss endowment, the Georges und Antoine Claraz Schenkung, Peyer started his first field campaign in 1924 (Peyer would later name a fossil from Monte San Giorgio for that family: Peyer, 1936a). Peyer approached the mining company and obtained permission to try to collect fossils using minor blasting in the gallery they had dug at the Cava Tre Fontane site on the western shoulder of the mountain right above the village of Serpiano. Dissatisfied with the results, Peyer approached the management again requesting to be allowed to dig for fossils at a locality called Val Porina, a valley running down the southern slope of Monte San Giorgio. There, the Miniere Scisti Bituminosi di Meride e Besano was operating an opencast site for the extraction of the black shales. Permission was granted, but as Peyer complained, "against a not inconsiderable compensation"

1.6. Relaxing at the table in front of the *Knappenhaus* near Cava Tre Fontane in September 1929. From *left* to *right*: Karl Hescheler (1868–1940), Jean Strohl (1886–1942; professor of physiology at the University of Zurich), Bernhard Peyer, and Mr. Waldisbühl (digital rendition Heinz Lanz; Photo © Max P. Linck/Fotostiftung Schweiz).

(Peyer, 1931b:5). On the plus side, the company assigned two workers to Peyer's project and allowed the crew to stay in the house it owned near the entry to the gallery they had mined at Cava Tre Fontane—"a highly estimable base of operations" thought Peyer, who dubbed the lodging the *"Knappenhaus"* (squire's house; fig. 1.6) (5).

Results were much improved at the Val Porina site (fig. 1.7). A sizable portion of the bedding planes of the black shales was cleared and leveled down layer by layer. Fossil fishes, complete skeletons of the ichthyosaur *Mixosaurus*, and a spectacular find of the armored placodont *Cyamodus* richly rewarded the back-breaking effort (Rieber and Lanz, 1999:79). The success was spectacular enough to motivate Peyer to return to the site the following year, 1925, this time accompanied by an undergraduate student tutored by Hescheler, Emil Kuhn (1905–1994; after his marriage, he would sign his publications Emil Kuhn-Schnyder), who would eventually become Peyer's successor at the University of Zurich. Over an area of roughly 100 square meters, the black shales were again dug through layer by layer, an eminently successful approach, which Kuhn-Schnyder once likened to leafing through an illustrated book on evolution (Kuhn-Schnyder, 1968). The phenomenal finds from 1924 and 1925 at the Val Porina site made the search for fossils at Monte San Giorgio a permanent item on Peyer's agenda. Peyer and his crew would prospect for fossiliferous outcrops beyond the sites operated by the mining company, locating fossils not only in the Grenzbitumenzone, but also in three horizons in the overlying lower Meride Limestone (Meridekalke). Fieldwork at Monte San Giorgio became an annual affair for Peyer from 1927 through

1.7. The Val Porina field site, probably 1929. *Left*: Guiseppe Buzzi; *right*: Vittorio Marogna (Photographer unknown; digital rendition Heinz Lanz).

1933, and again in 1937 and 1938, with Peyer eventually buying an abandoned farmhouse near Crocifisso to serve as his headquarters. Crocifisso is a wayside cross along the street from Serpiano to Meride, close to the Acqua di Ghiffo localities, which in Peyer's use lent its name to the nearby building he had acquired.

Starting in 1931, Peyer presented his findings in a series of voluminous monographs under the general title "The Triassic Fauna of the Ticino Limestone Alps" (Die Triasfauna der Tessiner Kalkalpen). He had originally planned to run a few excavations and then present the findings collectively and in systematic order. But the fossil record proved so rich that he had to change his approach. He decided to present each taxon in a monograph of its own, in the order they were collected and prepared. He was concerned that this would result in a somewhat confusing arrangement of the material, but he had no other choice: "The available material is so copious that its description will probably run through a greater number of volumes of our memoirs" (i.e., the Swiss Paleontological Memoirs: Peyer, 1931b:1). In fact, they continue to the present day. Peyer did, however, publish an overview of the results he obtained between 1924 and 1944 (Peyer, 1944).

The Mountain of the Dragonslayer

The Monte San Giorgio where Peyer was excavating is named after the martyr Saint George, who, according to legend, killed a dragon that guarded a water well near an ancient oriental city. To gather water, the citizen had to offer the dragon an oblation, a sheep or, if unavailable, a maiden. Passing by on his travels, Saint George killed the dragon, that way sparing the life of a princess. The Bay of Beirut is also called the Saint Georg Bay, as it is believed to have been the location near which St. George killed the dragon. Saint George is regarded as the patron saint of knights, horsemen, and wanderers, fitting for a mountain that not only lies alongside a major north-south trade route but also yields an abundance of fossil dragons, discovered and collected by Peyer, and dug up in abundance by Kuhn-Schnyder.

Emil Kuhn-Schnyder was born on April 29, 1905, in Zurich, the son of a railroad worker (fig. 1.8). He would later in life paint himself as a proud blue-collar democrat, who consciously eschewed academic aspirations. As a teenager, he realized, however, that the desire had blossomed in him to become a natural scientist. He managed to pass the federal qualifying exam that allowed him to study at the university. Given his childhood interest in the Neolithic lake dwellings in Switzerland, it was natural that he would be attracted to Karl Hescheler's courses and research. In the spring of 1925, Hescheler allowed Kuhn to volunteer in the zoological preparation laboratories, where he met Peyer, who was cataloging the fossils he had collected the year before at Monte San Giorgio. Kuhn immediately became involved in the project, and later that same year accompanied Peyer to the site of the ongoing fieldwork. Having obtained his diploma in 1927, Kuhn started his professional career as a high school teacher, working in his free time on his PhD thesis on mammal remains from Neolithic lake dwellings under the supervision of Hescheler. He successfully defended his thesis in 1932, and all the while teaching school, he pursued his interests in paleontology, accompanying Peyer on his nearly annual digs after 1927. Eventually, however, Kuhn switched career paths, accepting a position as a research assistant under Peyer at the Zoological Museum of the University of Zurich. He defended his *Habilitation* thesis in 1947—the necessary precondition for employment at the professorial level. Upon Peyer's retirement, Kuhn became his successor in 1955, first as associate, and as of 1962 as full professor. In 1956, the newly founded Paleontological Institute and Museum of

1.8. Emil Kuhn (later Emil Kuhn-Schnyder; 1905–1994), around 1940 (digital rendition Heinz Lanz; Photo © Max P. Linck/Fotostiftung Schweiz).

the University of Zurich was inaugurated, with Kuhn as its director (the biographic data for E. Kuhn-Schnyder derive from Kuhn-Schnyder, 1968; Ziegler, 1975; and Rieber, 1995).

Kuhn-Schnyder was a brilliant administrator and networker, who built up the Paleontological Institute and Museum of the University of Zurich and its treasures of Monte San Giorgio fossils. Throughout his career, he pursued interests as broad as those of his one-time supervisor Peyer. Vertebrate paleontology, and in particular the Triassic fauna of the Monte San Giorgio, were the focus of his research, which he paired with studies in the history of biology and earth sciences. Georges Cuvier (1769–1832), Lorenz Oken (1779–1851), Karl Ernst von Baer (1792–1876), and Louis Agassiz (1807–1873) were among the towering figures in comparative anatomy, embryology, and paleontology whose life and work Kuhn researched and wrote about. His appreciation of humanities led him to become a member of the German chapter of the Teilhard de Chardin Society with headquarters in Munich, which dedicated the proceedings of its 1975 meeting on the evolution of language to Kuhn-Schnyder on the occasion of his 70th birthday. Kuhn-Schnyder had served as president of the German chapter of the society since 1968.

Kuhn-Schnyder's major claim to fame was the herculean excavations at Survey Point 902 (902 meters above sea level; coordinates: 716 325 / 085 475) near Mirigioli on Monte San Giorgio, an annual affair that ran

1.9. Excavation at Mirigioli, Point 902. Kuhn-Schnyder (*center*) in discussion with Heinz Lanz, summer 1963 (Photo © Hans Rieber/ Paleontological Institute and Museum, University of Zurich).

from 1950 through 1968 (fig. 1.9). The campaign at Point 902 was driven by Kuhn-Schnyder's unflinching conviction that only fossils could provide direct evidence of evolutionary relationships. The strength of such evidence increased with the number of complete, well-preserved fossils collected. Across an area of initially 240 square meters, the bituminous black shales and intercalated dolomitic deposits were quarried layer by layer through 16 meters, the fossils in each layer identified and calibrated, which allowed insights into the paleoecology and taphonomy of this Middle Triassic biota (Furrer, 2003:20; Ziegler, 1975, writes of 400 square meters). The enormous collections of fossils that were amassed during these years were deposited in the Paleontological Institute and Museum of the University of Zurich (PIMUZ), where the material was successively prepared and described. In 1955, Kuhn-Schnyder moved the headquarters for his excavations to Meride, where his wife would cook marvelous dinners for the field crew that returned tired from their hard day's work. Merlot flowed freely. One of the preparators once told me: "If a professor drinks wine, he's called a connoisseur; if a blue-collar preparator drinks wine, he's called a drunkard." In 1967, Kuhn-Schnyder—together with Peyer—was named honorary citizen of Meride.

A famous formative episode in Kuhn-Schnyder's early scientific career that forcefully brought home the importance of the search for complete, well-preserved fossils concerned a signature fossil found in the Middle Triassic of Monte San Giorgio and Besano, *Tanystropheus longobardicus*. The species (or rather one of the putatively two species in the genus found at Monte San Giorgio) can grow up to six meters in length, with a neck that is as long as the trunk and tail combined; even in a juvenile specimen, the neck equals three times the length of the trunk. And yet, the neck is made up of only 13 vertebrae, which from the third to the eleventh are extremely elongated. As mentioned above, the Milan Natural History Museum organized fossil digs at the Besano locality in 1863 and 1878. In 1886, Francesco Bassani published a preliminary account of the fossils thus obtained, promising that a more detailed, monographic treatment of the finds would follow later (a project Bassani never completed). Among the specimens recovered was the curled-up skeleton of a small reptile, comprising the skull sporting jaws furnished with tricuspid teeth; some vertebrae, limb, and girdle elements; and gastral ribs along with strange, slender, and much elongated elements. Bassani interpreted the skeleton as that of a pterosaur, a new genus and species he named *Tribelesodon longobardicus* for its tricuspid teeth and its geographical provenance (Bassani, 1886:25). Since he never fulfilled his promise to deliver a monographic treatise, the Besano fossils remained rather poorly known, except to the Austria-Hungarian paleontologist Baron Franz von Nopcsa (1877–1933), who regularly visited the Milan Natural History Museum and its collections. In his preliminary account, Bassani did not illustrate the putative Besano pterosaur; the first illustration of *Tribelesodon* was a photograph taken by Nopcsa in 1902 and published by the Viennese paleontologist Gustav von Arthaber (1864–1943) in 1922 (Arthaber, 1922:6, fig. 3a; the photograph shows the fossil not at a slightly reduced size as indicated but slightly enlarged). In 1923, Nopcsa followed up on Arthaber's publication with a monographic treatment of *Tribelesodon*, where he commented on the circumstances under which the photograph was taken, and where he also quickly dismissed Arthaber's reconstruction of the skull of the beast (Nopcsa, 1923). The significance of the fossil was that it was then by far the oldest pterosaur known, or so people thought. Pterosaurs are known for their hollow limb bones, and the interpretation of the Besano fossil as a pterosaur seemed confirmed not only by its limb bones but also by those strange, slender and much-elongated elements,

1.10. Bernhard Peyer (*left*) and Ferdinand Broili (*right*) in the Val Serrata, 1929 (Photographer unknown; digital rendition Heinz Lanz).

which were also found to be hollow. It was thus only logical to consider that these elements represented limb bones as well, and Nopcsa accordingly interpreted them as the much-elongated phalanges in the fourth digit of the hand, that is, the digit that in pterosaurs supports the leading edge of the wing membrane (patagium). It wasn't until 1929, when Peyer, accompanied by Kuhn-Schnyder, recovered the complete and articulated skeleton of a marine reptile at the Val Porina locality at Monte San Giorgio that the true nature of these elements was recognized: they are the much-elongated neck vertebrae of a saurian with an extremely long neck. Peyer immediately recognized the similarity of these neck vertebrae to isolated cervical vertebrae from the Middle Triassic Muschelkalk deposits near Beyreuth, Germany, which had been described in 1852 under the name of *Tanystropheus conspicuus* by the Frankfurt paleontologist Hermann von Meyer (1801–1869) (Meyer, 1847–1855). The genus name *Tanystropheus* thus took priority over the name *Tribelesodon*. And, of course, *Tanystropheus* was no longer recognized as a pterosaur, but, as Peyer contended, possibly related to the origin of modern lizards (squamates). Peyer could not wait to notify his friend Ferdinand Broili from Munich of that sensational find. Broili immediately boarded the train and rushed to visit Peyer in his field quarters (the *Knappenhaus*) to examine the spectacular fossil (fig. 1.10).

The taxonomy and relationships of the genus *Tanystropheus* will require a detailed discussion later on. But as far as Kuhn-Schnyder was concerned, he supported squamate relationships of *Tanystropheus* throughout his career, as Peyer had done in the past and his own graduate student would continue do later (Kuhn-Schnyder, 1947, 1959a; Wild, 1973). Kuhn-Schnyder's own research on *Tanystropheus*, but even more so his investigations of *Macrocnemus*, another reptile from Mont San Giorgio considered to be closely related to *Tanystropheus* and consequently again placed close to the origin of squamates (lizards), eventually led to a bitter feud between him and his erstwhile mentor, Bernhard Peyer. It was in 1953 that the editor of the highly respected science magazine *Endeavour* approached Kuhn-Schnyder with the suggestion that he write an article of about 2,500 words on Monte San Giorgio fossils for his journal. Kuhn-Schnyder suggested a contribution that dealt with the origin of lizards (squamates). By the time the paper was published in 1954, Peyer and Kuhn-Schnyder were no longer on friendly terms (Kuhn-Schnyder, 1954a, b). As he read through the draft manuscript before submission, Peyer got increasingly incensed. He felt that his own work was being slighted in Kuhn-Schnyder's account. In his manuscript, Kuhn-Schnyder argued that three fossil forms from Monte San Giorgio, the protorosaur *Tanystropheus*, the related *Macrocnemus*, and the unrelated thalattosaur *Askeptosaurus* (more of those three taxa later), were all close to the origin of lizards—a view no longer entertained today. Peyer was furious and requested that Kuhn-Schnyder insert into the flow of his narrative a sentence clarifying that he, Peyer, had already pointed to the close relationships of these forms to the squamates and rhynchocephalians, the latter a group that includes the modern lizard-like tuatara (*Sphenodon punctatus*). Peyer found the concluding paragraph of the paper the most insulting one and requested that it be dropped and replaced by more neutral language. Following is the passage that so infuriated Peyer: "Progress in paleontology depends on the acquisition of new and ever better material. Progress thus depends on original production. It may well be that under the influence of powerful theories, preconceived ideas may guide research in important ways. But this will only go so far. What must be added to theory, what infuses its pallor with blood and life, are the fossils themselves."[1]

Peyer had himself published monographs on *Macrocnemus* and *Tanystropheus* in 1931 and 1937, respectively (Peyer, 1931c, 1937). Kuhn-Schnyder in turn had submitted his monograph on *Askeptosaurus*, worked up under Peyer's supervision, as *Habilitation*-thesis in 1947, but its publication was delayed until 1952 (Kuhn-Schnyder, 1952). However, Kuhn-Schnyder had also described new material of *Tanystropheus* found after 1931, and published a paper on its skull in 1947 (Kuhn-Schnyder, 1947). And in 1952, he presented an as yet unpublished new reconstruction of the skull of *Macrocnemus* at the annual meeting of the Swiss Academy

of Sciences (then Schweizerische Naturforschende Gesellschaft), again based on a specimen that was collected in 1938, that is, after the publication of Peyer's monograph on the genus. This new reconstruction of the skull of *Macrocnemus* Kuhn-Schnyder first published in his contentious 1954 paper in *Endeavour* (see also Kuhn-Schnyder, 1962a). What Kuhn-Schnyder's passage thus implied was that yes, Peyer may have had an intuition, called a "preconceived idea" by Kuhn-Schnyder, that *Askeptosaurus*, *Macrocnemus*, and *Tanystropheus* were close to the origin of squamates. But it was only his own work, based on later, newly discovered fossils that validated that hypothesis, indeed provided the (putative) proof that these three genera from Monte San Giorgio stood at the root of the lizard lineage.

The feud between the two protagonists of vertebrate paleontology at the University of Zurich, the senior and his junior, continued on into 1958, when Peyer requested a clarifying discussion with Kuhn-Schnyder that took place on December 3, 1958, in the attendance of the president of the university and an independent witness. Following that meeting, Kuhn-Schnyder wrote up an elaborate defense of himself against any possible charge of plagiarism, of which he sent copies to his peers.[2] Nothing much came of it. Already in 1956—the same year that the Paleontological Institute was founded with Kuhn-Schnyder as its director—Peyer had abandoned the institute, never to return.

But it remains to the merit of both, Bernhard Peyer (fig. 1.11) and Emil Kuhn-Schnyder, to have recognized the importance of Monte San Giorgio for vertebrate paleontology; to have conducted large-scale excavations and thus to have amassed an invaluable collection of Middle Triassic marine fishes and reptiles that generated a large body of research, and continues to do so. Through their efforts, the Monte San Giorgio was recognized as containing exceptional conservation deposits of international significance, offering a unique window into the rich tropical marine life of the Middle Triassic. On the initiative of Markus Felber, then at the Natural History Museum of Lugano (Ticino), and his supporters, such as Andrea Tintori from the Università degli Studi di Milano, the World Heritage Committee of UNESCO inscribed the Monte San Giorgio on the World Heritage List in 2003, a designation that in 2010 was extended to the Besano locality in Lombardy, Italy (Felber and Tintori, 2004).

The Fossil Deposits

The Monte San Giorgio is a pyramidal mountain jutting northward and rising to a total height of 1,097 meters above sea level. It lies at the southern tip of Switzerland, its northern, eastern, and western flanks descending into Lake Lugano (fig. 1.12). The historic town of Riva San Vitale, with its medieval Baptistery of San Giovanni, is located at the base of its steeply descending eastern flank, just south of the southern tip of the eastern leg of Lake Lugano. At the southern tip of the western leg of lake Lugano lies Porto Ceresio, already on Italian territory (Province of Varese, Italy). Meride is a historic village sitting on the more gently descending southern slope of Monte San Giorgio. Serpiano is a village sitting on the western flank of Monte San Giorgio; the Via Serpiano, winding around Monte San Giorgio, connects it with Meride. Today, the Monte San Giorgio is a tourist attraction, allowing wonderful outlooks over neighboring mountains, lakes, valleys and rivers. There is a nature trail of about seven kilometers in length, starting near Meride and circling the mountain, with signposts that explain plant and animal life as well as the geology and the fossils found in the Triassic sediments.

Monte San Giorgio is part of the southern Alps, which form a tectonic unit of Gondwanan origin. The somewhat boomerang-shaped supercontinent Pangea was still intact during the Middle Triassic, encompassing the southern Gondwanaland and the northern Laurasian continental

components. Pangea's eastern boundary formed a vast, deep embayment, bounded at one end by the South China Block, at the other by what is Australia today. The sea that filled this embayment is called the Tethys, or Tethys Ocean. The sea that extended east of the South China Block and of the "Australian tip of Gondwana" all the way around the globe to the opposing coast of Pangea is called Panthalassa, or the Panthalassic Ocean, the latter essentially an early version of the Pacific. The "Alpine-Mediterranean" Triassic of the southern Alps stretching along the northwestern coast of the Tethys Ocean from southernmost Switzerland and northern Italy eastward to what today are the eastern Alps of the Grisons and Austria encompasses marginal marine deposits, lagoonal deposits, and intraplatform basins, the latter particularly rich in fossil fishes and reptiles.

The Monte San Giorgio sits on a pedestal of pre-Permian and Permian bedrock of brownish, purple, or reddish hue. The earliest Triassic sediments at Monte San Giorgio overlap unconformably Permian rocks of volcanic origin, which indicates strong erosional activity during the uppermost Permian and the Lower Triassic, a time span of approximately 35 million years. These basal Triassic sediments of Anisian (early Middle Triassic) age, called the Bellano Formation, were deposited along coastal stretches of a shallow sea. In general, the depositional environment of sedimentary layers of Monte San Giorgio was located, during the Middle Triassic, along the western margin of the Tethys Ocean, "in the tropical belt near the equator" (Etter, 2002:228). Such a location made for a climate "influenced by monsoonal circulation" (Renesto and Stockar, 2015:95; see also Preto et al., 2010).

In the late Anisian, a marine transgression spread westward across an area that today corresponds to the western Lombardian Alps, initiating the deposition of "shallow water carbonate sediments" (Furrer, 1995:843f). As was observed by the longtime curator of the Paleontological Institute

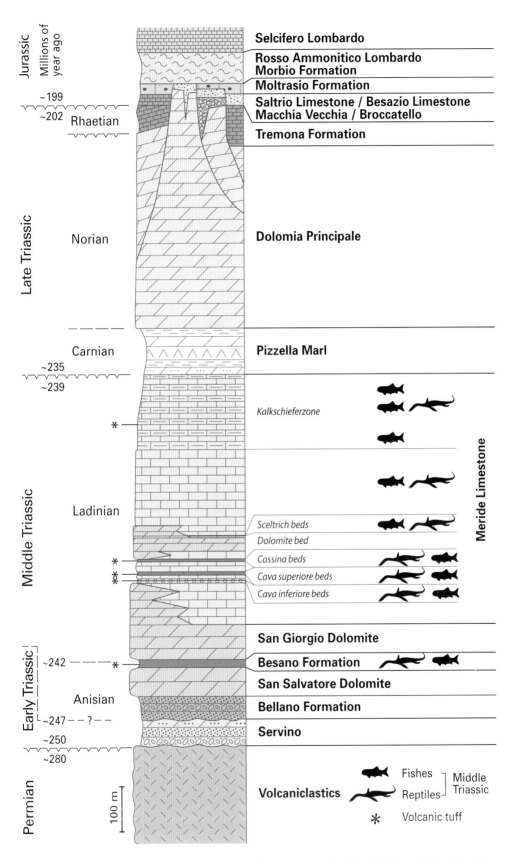

1.13. Stratigraphy of the fossiliferous beds yielding vertebrates at Monte San Giorgio (© Commissione Scientifico Transnazionale Monte San Giorgio, 2014).

and Museum, Heinz Furrer, "Tectonic activity coupled with extensive volcanism during the latest Anisian and Ladinian led to the formation of several basins separated by carbonate platforms" (844). It is in one of these basins that the Grenzbitumenzone and later fossiliferous sedimentary layers were deposited that crop out on the slopes of Monte San Giorgio and surrounding localities on Italian territory (Besano, Pogliana, Ca' del Frate: Province Varese, Lombardy) (fig. 1.13). The Grenzbitumenzone itself straddles the Anisian-Ladinian boundary, the latter located in its upper part (Rieber, 1973a; Brack and Rieber, 1993). Younger fossiliferous layers of Monte San Giorgio and its surrounding area are of Ladinian age.

Vertebrate fossils have been collected at five different horizons at Monte San Giorgio and its surroundings, with deposits that are from 230 to 245 million years old. The most basal, that is, oldest horizon is the Grenzbitumenzone, up to 16 meters thick, which on the Italian side also crops out near Besano and Pogliana (Besano Formation) (Furrer, 2003, 1999; see also Rieber and Lanz, 1999). Of all the fossiliferous horizons at Monte San Giorgio, the Grenzbitumenzone contains the taxonomically most diverse and most densely packed fossil record (Furrer, 1995:830). It consists of a sequence of interbedded layers of thin, finely laminated and highly bituminous black shales and thicker, less bituminous dolomites. The organic matter in the black shales is predominantly of bacterial origin. Given their slower sedimentation rates, the bituminous black shales are richer in fossil content, but the fossils are strongly compressed as a consequence of sediment compaction. Due to their more rapid sedimentation rate, vertebrate fossils are less common in the dolomite layers but retain a near three-dimensional structure. The occurrence of well-preserved invertebrates such as cephalopods (ammonoids, belemnites) and bivalves (*Daonella*) is likewise restricted to the weakly bituminous dolomite layers. The closeness of coastal waters is indicated by the rare occurrence of plant remains, such as calcareous algae and remains of the conifer *Voltzia*. The Grenzbitumenzone is richest in oil and fossil content in its middle section of approximately 3.9 meters thickness (Furrer, 2003:39) (fig. 1.14). It is this commercially exploited section of alternating layers of bituminous dolomites and black shales that was first targeted by Peyer in 1924. As he found the collecting of complete fossils in the abandoned narrow mine of Tre Fontane difficult, Peyer moved on to opencast mining at Val Porina, where his efforts were rewarded with several specimens of the small ichthyosaur *Mixosaurus* and a new species of an armored placodont. A much larger ichthyosaur was collected at the Cava Tre Fontane site in 1927, and the strange morphology of *Tanystropheus* was clarified when an articulated specimen comprising the skull and the neck was collected in 1929 at the Val Porina locality along with other taxa (an unarmored placodont, and a thalattosaur: Furrer, 2003:18). Peyer continued to dig in the Grenzbitumenzone until 1933, while Kuhn-Schnyder pursued his large scale opencast mining operations at P. 902 / Mirigioli from 1950 through 1968.

1.14. The Grenzbitumenzone (Besano Formation) at Mirigioli, Point 902, Monte San Giorgio (Photo © Hans Rieber/ Paleontological Institute and Museum, University of Zurich).

1.15. Excavations in the Alla Cascina beds in 1933. From *left* to *right*: Giuseppe Buzzi, B. Bianchi, E. Ponti, Vittorio Marogna, "Sergente" Buzzi; in upper right corner, Fritz Buchser (digital rendition Heinz Lanz; Photo © Max P. Linck/ Fotostiftung Schweiz).

It was in the fall of 1927 that two men from Meride, Gaetano Fossati and A. Zappa, pointed Peyer to younger fossiliferous horizons in the Middle Triassic of Monte San Giorgio (Furrer, 2003:20). There are three such horizons, all part of the lower Meride Limestone of Ladinian age. The lowest and oldest one is called the Cava Inferiore beds, 1.5 meters combined thickness, the lower ones yielding small actinopterygian fishes, the higher ones packed with specimens of a pachypleurosaur species (*Neusticosaurus pusillus*) (43). The intermediate fossiliferous horizon of the lower Meride Limestone is called the Cava Superiore beds, of 10 meters thickness. It is in these deposits that Peyer collected in the fall of 1928 three specimens of the large lariosaur *Ceresiosaurus calcagnii*, along with many smaller pachypleurosaurs (44). The third, and youngest, fossiliferous horizon in the lower Meride Limestone is called the Cascina beds, of about three meters thickness, cropping out near the Cascina Chapel located above Meride (fig. 1.15). First discovered by the fossil preparator Fritz Buchser from Meride in 1933, the deposits again yielded larger pachypleurosaurs as well as the lariosaur *Ceresiosaurus*, and the protorosaurs *Macrocnemus* and *Tanystropheus* (46).

The uppermost Meride Limestone forms the Kalkschieferzone, which once had been considered to be of Carnian (Upper Triassic) age; based on new paleobotanical evidence it is now referred to the late

Ladinian (Furrer, 2003:46). It crops out at Monte San Giorgio and a neighboring locality in Italian Territory, Ca' del Frate, and is well known for its small actinopterygian fishes as well as the occasional reptile. In the words of Heinz Furrer (1995), "The Kalkschieferzone represents a late stage evolution of intraplatform basin, beginning with open marine influence in late Anisian time (Grenzbitumenzone) and increasing restriction by growing carbonate platforms during early Ladinian (lower Meride Limestone). In late Ladinian time, the basin was filled progressively" (827). This, of course, means that starting with the Grenzbitumenzone, the successive fossiliferous horizons of Monte San Giorgio and its surroundings were deposited in the same basin, which during the late Anisian and Ladinian underwent an evolution of progressive restriction after an initial connection with the open sea.

Invertebrates known from the Grenzbitumenzone comprise bivalves, gastropods, cephalopods, echinoderms, brachiopods, a single occurrence of a shrimp (crustacean), and conodonts—the latter of controversial chordate affinities (Turner et al., 2010). Among these, the bivalves and cephalopods are by far the most frequently found faunal elements. Predominant among the bivalves is the thin-shelled genus *Daonella*, of which several species have been described, some of them

Box 1.1. The invertebrate fauna from Monte San Giorgio

© Beat Scheffold

thought to represent a discrete evolutionary lineage (Rieber, 1968, 1969; Schatz, 2001). Among the cephalopods, the ceratitid (Ceratitidae) ammonites are the most frequently found, of which 10 different species in five genera have been described. Nautiloid cephalopods are comparatively rare, as are the straight conical shells or phragmocones of coleoideid (Coleoidea) belemnites. Shown in these reconstructions are ceratitid ammonites with a strongly sculptured, spiny shell. The degree of sculpturing of the shell among the Monte San Giorgio ammonites was found to be subject to extensive variation across the different species, hence representing a poor taxonomic character. More in the background is shown a group of nautiloids (*Michelinoceras*) with a straight conical shell (orthocone). All cephalopods are featured as pelagic organisms roaming the open water column; the bottom water in the Grenzbitumenzone basin is thought to have been oxygen depleted. The extant cephalopod *Nautilus* from the Indian Ocean represents somewhat of a living "model" of an extinct ammonite, whereas squids and cuttlefish are believed to be related to extinct belemnites.

The Grenzbitumenzone basin comprises the outcrops both at Monte San Giorgio and at Besano-Pogliana (hence called the Besano Formation by Italian workers: Nosotti and Teruzzi, 2008). The basin is estimated to have had a diameter of minimally 10 kilometers, and a depth of 100 to 150 meters (Furrer, 1995:845). According to W. Etter (2002), the east-west extension of the basin is 20 kilometers, whereas the "south-west extension cannot be constrained because of the thick Cenozoic cover in the south" (227). Again according to Etter, the water depth during the deposition of the Grenzbitumenzone was only between 30 and 100 meters (228). The basin was partially cut off from the open sea by a shallow water belt composed of reef complexes and sandbanks in a lagoonal setting not unlike today's Bahamas (Furrer, 1999:102). The richness and taxonomic diversity of its fossil content clearly mark the Grenzbitumenzone as a conservation deposit, a so-called fossil *Konservat Lagerstätte*, defined as "rock bodies unusually rich in palaeontological information, either in a quantitative or qualitative sense" (Seilacher and Westphal, 1985:5). Vertebrate fossils in the bituminous black shales of the Grenzbitumenzone are usually exquisitely preserved, complete and fully articulated skeletons, yet strongly compressed. Such compression is less marked in fossils from the bituminous dolomitic layers, which were subject to lesser compaction. Among the various types of fossil Lagerstätten that have been recognized, the orthodox interpretation of the Grenzbitumenzone follows the model of a stagnation deposit (Seilacher et al., 1985:10ff, and fig. 1; see also Peyer, 1944; Rieber, 1973b, 1975, 1982; Rieber and Sorbini, 1983; Bernasconi, 1994; Furrer, 1995, 2003). The accumulation of that many fossils, complete skeletons in articulation, has been explained by an extremely slow sedimentation rate of 1 to 5 mm per thousand years under anoxic conditions at a depth below storm wave base (Furrer, 1995:832; according to Etter, 2002:227, the deposition of the entire Grenzbitumenzone lasted through 2 to 3 million years). The sedimentation rate is slower for the black shales than for the bituminous dolomite layers, however, which explains the greater fossil content of the black shales. There is "absolutely no sign of bioturbation or physical reworking" of the sediment (225). The orthodox model claims that there is also no trace of a benthic invertebrate fauna. During the deposition of the Grenzbitumenzone, the basin in which it formed was located between carbonate platforms and reef complexes. The fauna it contains indicates an unhindered exchange of nektonic and planktonic organisms with the Tethys Ocean. In order for the carcasses of these organisms to be buried in sediment under anoxic conditions requires a stagnant water column reaching to a depth below storm wave base. To stabilize such stratification of the water column, it has been stipulated that "aerobic photosynthetic (Cyanobacteria) and anaerobic chemoautotrophic bacteria formed a bacterial plate at the oxic-anoxic interface" (Etter, 2002:228; Bernasconi, 1994; Bernasconi and Riva, 1993).

But an alternative interpretation seems to gain more currency lately. Investigating the taphonomy of small actinopterygian fishes with "skeletons made up of a great number of elements . . . very sensible [sensitive] to different depositional conditions" in several intraplatform basins in the Alpine Triassic (Ladinian through Carnian), the paleoichthyologist Andrea Tintori from the Università degli Studi di Milano reached different conclusions (1992:396). With regard to the Grenzbitumenzone (Besano Formation) in particular, Tintori found that although many fishes remained fully articulated, "several show unimodal [unidirectional] dispersal of skull bones and distal fin elements" along the antero-posterior axis of the fish body (397). Using field sketches from Kuhn-Schnyder's excavation campaign at Point 902, he further found that small fishes are preserved in parallel orientation, their skulls pointing in the same direction (see also Kuhn-Schnyder, 1974, fig. 8). This to him constituted "evidence of bottom currents" that would carry at least small amounts of oxygen (Tintori, 1992:398). Additional specimens figured in the literature show the displacement of vertebral centra of considerable weight, indicating "relatively swift currents," which again would carry at least some oxygen along the water/sediment interface (398; see also Bürgin et al., 1989; Tintori, 1992:404). These conclusions of Andrea Tintori were rejected, however, by Heinz Furrer, who found them to be "based on wrong assumptions or not convincing arguments" (Furrer, 1995:844). In support of this criticism, Furrer cited the description of the only decapod specimen ever collected in the Grenzbitumenzone (a new species of shrimp in the genus *Antrimpos*) as a faunal element that—given anoxic bottom conditions—must have been washed into the basin from near-shore environments, as is also assumed to have been the case for the rare conifer remains in the genus *Voltzia*. In the description of that unique decapod, the author Walter Etter, a former graduate student at the Paleontological Institute in Zurich, admitted the possibility of weak bottom currents but denied their correlation with "oxygenation events in the proper basin" (1994:227). He would later change his mind, however.

In the reconstruction of the depositional environment of the Grenzbitumenzone, a lot seems to hinge on the interpretation of the lifestyle of the most frequently found invertebrates, which are thin-shelled bivalves in the genus *Daonella*. The monographic treatment of these bivalves recognized the presence of seven species of *Daonella* in the Grenzbitumenzone, of which five chronospecies succeed one another through a sequence of layers four meters thick (Rieber, 1968, 1969; Kuhn-Schnyder, 1974:93). Given the thinness of their shells, and his belief in anoxic bottom waters, Rieber stipulated a pseudoplanktonic life style for the *Daonella* of the Grenzbitumenzone. This was challenged by Etter, who takes *Daonella* species to have been benthic organisms. There can be no question that the bottom waters of the Grenzbitumen basin were "severely oxygen depleted . . . but not anoxic," allowing the highly tolerant bivalves, but not other, less tolerant benthic invertebrates, a bottom-dwelling existence (2002:229). This alternative interpretation also allows for the bacterial mat, including Cyanobacteria and chemoautotrophic bacteria, to develop on the sediment surface rather than at the oxic-anoxic interface in a stagnant water column. The upshot of this alternative model is that it "circumvents the problem of maintaining stable water stratification during several million years" (229). Such microbial mats developing on the sediment surface could furthermore have sealed the carcasses of dead vertebrates, thus leading "to the observed preservational patterns" (232).

Toward the upper limit of the Grenzbitumenzone, the connection of the basin with the open Tethys Ocean became increasingly restricted, which resulted in the deposition of superimposed sediments, the San Giorgio Dolomite, which is devoid of fossils. At the time of deposition of the lower Meride Limestone, a sea level rise had renewed a large basin of stagnant water surrounded by sandbanks and flat islands. Levels of salinity probably fluctuated in the superficial water layers, which required a greater salinity tolerance of its inhabitants. Toward the uppermost Meride Limestone, evidence of the influence of alternating periods of tropical rain and drought precedes the ultimate silting up of the basin (Furrer, 1999:102).

Notes

1. *"Der Fortschritt in der Paläontologie ruht auf der Beschaffung von neuem und immer besserem Material. Der Fortschritt ruht auf der Urproduktion. Wohl können unter dem Einfluss mächtiger Theorien vorgefasste Meinungen die Forschung bedeutend fördern. Sie bringen sie jedoch nur bis zu einem gewissen Punkte. Was zur Theorie hinzukommen muss, was ihrer Blässe erst Blut und Leben verleiht, das sind die Fossilien selbst."*

2. A copy of this document served as the basis of the above account of the discord between Peyer and Kuhn-Schnyder; courtesy of Heinz Lanz, Winterthur, Switzerland.

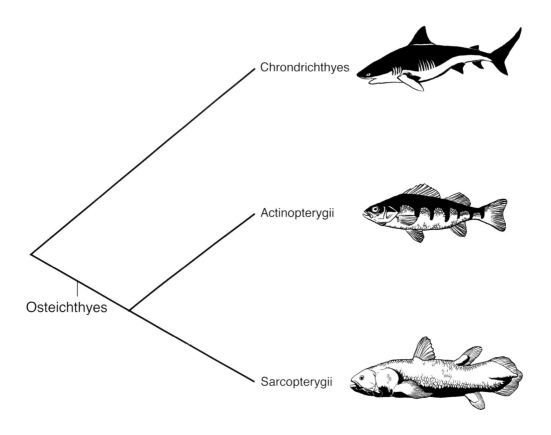

Chrondrichthyes

Actinopterygii

Osteichthyes

Sarcopterygii

2.1. Interrelationships of major groups of fishes represented in the Middle Triassic of Monte San Giorgio (artwork by Marlene Donnelly, the Field Museum, Chicago).

Fishes

2

A report published in 1839 tells of a certain Lodovicio Trotti who roamed the Grigna mountains that rise up from the eastern shore of the eastern leg of Lake Como, locally also known as Lake Lecco. His goal was the canyon that the Esino River had cut into the mountainside south of Perledo, a small village located above the lakeside town Varenna. The Grigna mountains were known for their fossils, but Lodovicio Trotti was particularly interested in the Calcare di Perledo, the Perledo-Varenna Limestone that was quarried in the area around Perledo. And indeed, he struck it lucky! He found the fossil remains of a sizable lizard-like reptile, along with two fishes, material that he turned over to Guiseppe Balsamo-Crivelli (1800–1874), professor of mineralogy and zoology at the University of Pavia, for description (Balsamo-Crivelli, 1839). Balsamo-Crivelli recognized the reptile as a distant relative of *Plesiosaurus dolichodeirus* and its kin, marine reptiles from the Lower Jurassic of southern England well known since the 1820s (Taylor, 1997). Balsamo-Crivelli refrained from formally naming the Perledo saurian, however, referring it instead simply to the family of *Paleosauri*. The better preserved of the two fish fossils he did name after its discoverer, *Lepidotus Trotti* (today also known as *Furo trottii*), while the second, less well-preserved fish he left unnamed again (1839:427). Vertebrate fossils are found in the upper part of the *Calcare di Perledo*, corresponding to the lower Kalkschieferzone and hence of late Ladinian age; they are thus younger than the fossils retrieved from the Besano Formation (Furrer, 1995:848; the *Calcare di Perledo* might be lowermost Carnian in its uppermost part: Tintori, Muscio et al., 1985:199).

The second saurian from Perledo that had come to light was described by Giulio Curioni (1796–1878), geologist in the service of the city of Milan, in a publication dated 1847 (Curioni, 1847). In that paper, Curioni described the new fossil under the name of *Macromirosaurus Plinj*, and christened the species represented by the Balsamo-Crivelli specimen *Lariosaurus balsami*. He noted, however, that "more species of fishes and of other reptiles are found in the bituminous layers of Perledo," and boasted of his collection that included ichthyosaur remains, fishes, and invertebrate fossils such as ammonites (Curioni, 1847:165). Johann Jakob Heckel (1790–1857), from the Natural History Museum in Vienna, visited Curioni's collection and erroneously identified one of the fishes from Perledo as *Palaeoniscus*, which would suggest a Paleozoic age for these sediments (1849:500). The first fossil fish remains from the bituminous black shales of Besano were published by Emilio Cornalia in 1854, who mentioned and illustrated, but neither described nor discussed, two dorsal fin spines of hybodontiform sharks (1854:56, n.1, pl. II, fig. 3). A more comprehensive study of Triassic fishes from Lombardia was published in 1857 by Cristoforo Bellotti (1823–1919), an honorary associate of the Museo Civico di Storia Naturale in Milan (Bellotti, 1857). The fishes from Perledo figured most prominently in that account, with 16 species in five genera described. Noteworthy is the first record of an actinistian from the Perledo-Varenna Limestone, *Heptonema paradoxa* (Bellotti, 1857:435). Among the fishes described were also two species from Besano, an actinopterygian (*Ichthyorhynchus* [*Saurichthys*] *curionii*), and the shark *Leptacanthus cornaliae*. The latter species description was based on the two dorsal fin spines that had been identified by Cornalia in 1854 (de Alessandri, 1910:17). Later, Francesco Bassani mentioned an unpublished manuscript penned by Bellotti, a *Catalogo dei pesci fossili del Museo Civico di Milano* (1873), recognizing two hybodontiform sharks and four actinopterygian species from Besano (Bassani, 1886:18; see also Kuhn-Schnyder, 1945:664).

The bituminous shales of Besano, their geology, geochemistry, and industrial exploitation were commented upon by Curioni (1863), in a paper that also made passing reference to fossil fish. It was not until 1886 that a more comprehensive treatment of the vertebrate fauna from the black shales of Besano was delivered by Bassani. He recognized five genera and species of sharks and 17 species of actinopterygians, but no actinistian, the latter a group by that time already known from Perledo. The Perledo fishes were the subject of a critical revision only three years later by Wilhelm Deecke (1862–1934), at that time associate professor for geology and paleontology at the University of Greifswald, Germany. He dismissed Bellotti's account of 1857 as "pretty useless" because no illustrations accompanied the systematic account of the fossil fish fauna from Perledo (Deecke, 1889:110). With his monograph, Deecke laid the groundwork for a modern understanding of the Perledo fish fauna. The next milestone in the understanding of the Triassic fishes of Lombardy was Giulio de Alessandri's monograph of 1910, based on the rich material that had by then accumulated in the collections of the Museo Civico di Storia Naturale in Milan, with specimens from Besano as well as from the Cava Tre Fontane locality on Monte San Giorgio (de Alessandri, 1910). He identified the works of Bellotti (1857), Bassani (1886), and Deecke (1889) as the classic sources on the subject. Alessandri compiled tables listing separately all known species collected at four localities strung out from north to south along the mountain slopes east of Porto Ceresio. The northernmost locality is Cava Tre Fontane above Serpiano, in Swiss territory. Moving south from there, Triassic fossil fishes were collected at Monte Grumello, Monte Nave, and finally Cà del Frate, the latter located above Besano and of considerably younger age (Kalkschieferzone) than the Besano Formation. For the Monte San Giorgio-Cava Tre Fontane locality Alessandri listed six species of actinopterygian fishes in five genera. This list was found to be incomplete by Carl Wiman (1867–1944), professor of paleontology at the University of Uppsala. Although his primary interest was to compare the ichthyosaurs he had collected in the Triassic of Spitsbergen (Svalbard) with *Mixosaurus cornalianus* from Besano, he noted in passing that one species was missing from Alessandri's list of the Besano actinopterygians (Wiman, 1910). Six years later, in 1916, Erik Andersson (Erik A:son Stensiö, 1891–1984) described in detail the "not insignificant" collection of fossil fishes that C. Wiman had obtained from Besano and, more important, from Cava Tre Fontane for the Geological Institution of the University of Uppsala (Stensiö, 1916:13). In his taxonomic account, Stensiö recognized 12 species present at these localities, one actinistian, and 11 actinopterygians in 10 genera.

The last major monograph on the Besano fish fauna to be mentioned in this historical review was published in 1939 by James Brough (1904?–1988), then lecturer at the University of Edinburgh, later professor of zoology at the University College Cardiff (Brough, 1939). In this monograph, Brough described the so-called Bender-collection, which the British Museum (Natural History) had bought from the German private collector and fossil dealer Carl Bender from Waiblingen, a town neighboring Stuttgart. Bender claimed that his fossils had come from Besano, when in fact some of the specimens described by Brough were collected in Monte San Giorgio outcrops (Bürgin, 1992:9). In his treatise on Permian actinopterygians from East Greenland, Hermann Aldinger (1902–1993) mentions *Gyrolepis* specimens from the Cava Tre Fontane locality as being part of Bender's collection (Aldinger, 1937:241, n. 1). More recently, Toni Bürgin identified specimens of the genus *Prohalecites* as part of the British Museum (Natural History) Bender collection, a genus otherwise known in the southern Alpine Triassic only from the Cá del Frate locality of distinctly younger age (Kalkschieferzone). He thus concluded that the commercially acquired collection described by Brough in 1939 is both geographically as well as stratigraphically heterogeneous (Bürgin, 1992:9).

A Note on Fish Phylogeny and Classification

A natural, that is, monophyletic group must include its ancestor and all, and only, its descendants. By virtue of the fact that one group of "fishes" that is nested within the lobe-finned fishes or Crossopterygii gave rise to land-dwelling tetrapods in the distant past, tetrapods are descendants of fishes. To render the group of fishes monophyletic would thus require the inclusion of tetrapods within "fishes." This is not, however, how the term "fishes" is commonly understood, as it is generally contrasted with the term "tetrapods." Talk of "the fishes" from Monte San Giorgio hence uses the term in a non-technical sense. Or, to put it differently, used in that sense the term "fishes" refers to a paraphyletic group, that is, to a group that comprises its ancestor and some, but not all, of its descendants. Representatives of three major clades of fishes have been found in the Middle Triassic marine deposits of Monte San Giorgio and Besano (fig. 2.1): hybodontiform sharks, actinopterygians (ray-finned fishes), and actinistians, extinct relatives of the extant coelacanth.

Sharks and their relatives constitute a clade called Chondrichthyes, also known as "cartilaginous fishes." This name refers to the fact that their cartilaginous endoskeleton may—in part—calcify ("encrusted with prismatic calcifications"), but never ossifies (cartilage being replaced by bone) (Miles, 1971:162; see also Enault et al., 2016). This contrasts with the Osteichthyes, the "bony fishes." The endoskeleton of all vertebrates is initially, during embryonic development, formed of cartilage. In osteichthyans, parts of the endoskeleton undergo ossification, a process during which cartilage breaks down and is replaced by bone. Ray-finned fishes and actinistians are representatives of two major clades of osteichthyans, the Actinopterygii and their sister group, the Sarcopterygii, respectively. Whereas chondrichthyans or cartilaginous fishes show no sign of ossification in their endoskeleton, this does not mean that they do not possess bone. Chondrichthyans are characterized by the presence of placoid scales, minute denticles embedded in the skin and providing a dense cover over the entire body. Each of these denticles morphologically resembles a miniature tooth: a crown composed of dentine and covered by enamel that sits on a base made up of bone. Another chondrichthyan characteristic is their peripheral fin support, which is made up of ceratotrichia, thin yet densely packed and flexible, thread-like support structures made up of an elastic protein resembling keratin. This contrasts with the peripheral fin support structures of osteichthyans, which are complex bony rods of exoskeletal derivation called lepidotrichia. Chondrichthyans and osteichthyans also show important differences in the structure of their jaws, and the position of the teeth. The primary jaws of vertebrates are derived from anterior gill arch components and hence are endoskeletal in nature, as seen in chondrichthyans, where the jaws are made up of calcified cartilage. The upper part (epibranchial) of the first gill arch (mandibular arch) forms the upper jaw in chondrichthyans, which is called the palatoquadrate. The lower part (ceratobranchial) of the mandibular arch forms the lower jaw, which is called Meckel's cartilage. The tooth-bearing elements in chondrichthyans are thus the primary jaws of endoskeletal origin. In osteichthyans, dermal bones that ossify directly in the deep (dermis) layer of the skin without going through a cartilaginous precursor stage, and hence are of exoskeletal origin, are superimposed on the primary jaws—among others, the maxilla in the upper jaw and the dentary in the lower jaw. In osteichthyans, therefore, the position of the teeth has shifted from the endoskeletal primary jaws to exoskeletal (dermal) elements that have become superimposed on the primary jaws. Other osteichthyan characteristics include the at least partial ossification of the endoskeleton, the exoskeletal body armor composed of large bony plates in the skull and shoulder girdle, and the scales covering the body. The split between the chondrichthyan and osteichthyan lineages is dated back to the Paleozoic, "at least 423 million years ago" (Brazeau and Friedmann, 2015:490).

Of the five genera and species of sharks known from the Middle Triassic of Monte San Giorgio, four belong to the Hybodontiformes, the latter a Mesozoic clade that is the sister taxon to the Neoselachii, or the "modern-level" sharks (Dick, 1978). The split between hybodontiforms and neoselachians has been dated back to some 300 to 350 million years ago, although their fossil record remains scant until it starts to pick up at about 200 million years ago (Maisey, 2011:422). This renders the hybodontiform and neoselachian clades the only ones among a number of Paleozoic shark lineages that not only survived the end-Permian mass extinction but subsequently also diversified. The hybodonts became the dominant shark lineage during the Triassic and Jurassic, but they disappeared during the Cretaceous. Only the neoselachians persisted into the Cenozoic and to the present day (the extant genus *Heterodontus* is no longer considered to represent a surviving hybodont: Maisey, 1982:34ff, 2012:928; Maisey et al., 2004:45). Hybodonts comprise sharks of very large to small body size, whose fossil remains suggest a great diversity of feeding habits in both marine and non-marine environments. Their extinction was probably caused by competition with neoselachians (Maisey, 2012:945). Among the characters that unite hybodonts as a monophyletic clade are "fin spines with a convex posterior wall and a series of retrorse denticles along the posterior midline" that support the leading edges of the two dorsal fins (Maisey et al., 2004:18; see also Lane, 2010). One of the synapomorphies that unite hybodontiform sharks with neoselachians are the placoid scales described as "'non-growing' (synchronomorial) denticles" (Maisey et al., 2004:23). These are also present in the fifth genus and species of shark known from the Middle Triassic of Monte San Giorgio, which otherwise shows a curious mixture of hybodontiform and neoselachian characteristics that renders its classification difficult. This highlights the fact that much still remains to be learned about the phylogeny and classification of Mesozoic sharks (Maisey, 2011).

The Osteichthyes again comprise two monophyletic sister groups, the Actinopterygii and the Sarcopterygii. Among many other characteristics, it is the fin anatomy that most readily identifies these two clades. The name of the actinopterygians derives from the actinotrichia, structures that support the very periphery of the fins. The paired (pectoral and pelvic) fins of actinopterygians, such as trout, salmon, gar, or bowfin, are fan-shaped and supported by bony fin-rays, or lepidotrichia. Proximally the lepidotrichia articulate with the cartilaginous (endoskeletal), rod-shaped radials that form the basal support structures of the fins. In the primitive condition there is more than one lepidotrichium per radial; in the derived condition, a 1 to 1 ratio has been established. Each of these lepidotrichia is hollow, indeed formed from two halves, or hemitrichia, the two combining to form the hollow lepidotrichium. From within the distal end of the lepidotrichia delicate, thread-like structures or fibrils, the actinotrichia, project into the margins of the fins. Lepidotrichia and actinotrichia also support the single dorsal, caudal and anal, that is, the unpaired fins of actinopterygians. The caudal fin is primitively asymmetrical (heterocercal) in actinopterygians but becomes symmetrical (homocercal) in advanced forms (teleosts).

The classification of the Middle Triassic actinopterygians from Monte San Giorgio is complex and controversial, which again indicates how much still remains to be learned about their phylogeny. Among extant actinopterygians, it is the sturgeons, paddlefish, and their immediate fossil relatives that are considered to form the most basal clade. They retain a heterocercal tail and thick, enamel-covered scales, unless these are reduced as in the modern representatives. The endoskeleton is subject to a limited degree of ossification, hence the name Chondrostei that is applied to this monophyletic clade (Grande and Bemis, 1991). Another major monophyletic clade among the actinopterygians, the putative sister group to chondrosteans, is the Neopterygii, so named after the fact that the endoskeletal support structures (radials) in the dorsal and anal fins are of the same number as the bony fin-rays (lepidotrichia) that they support (J. S. Nelson, 2006:90; see also G. J. Nelson, 1969; Patterson, 1973). In the Neopterygii the thickness of the scales is reduced, the heterocercal tail is abbreviated and becomes symmetrical (homocercal) in teleosts, and the endoskeleton (vertebral centra) undergoes significant ossification. Some of the lineages represented by Triassic fishes from Monte San Giorgio are classified as neopterygians, with one putative teleostean among them (the classification of the Pholidophoridae as basal teleosts takes the origin of the monophyletic teleostean lineage down into the Triassic: Arratia, 2013). The other, non-neopterygian fish lineages from the Middle Triassic of Monte San Giorgio are commonly relegated to the chondrosteans. Whereas the clade comprising the extant sturgeons and paddlefish and their immediate fossil relatives is considered to be monophyletic, the inclusion of additional fossil taxa renders the Chondrostei non-monophyletic: "The classification of this group is very insecure. It is a group of great structural diversity, and evidence is lacking for monophyly not only for this subclass, but also for most of the groups herein recognized" (J. S. Nelson, 2006:90).

Sarcopterygians have two dorsal fins and primitively a heterocercal caudal fin. They differ from actinopterygians primarily in the structure of the paired fins, the narrow, fleshy (muscular) base of which gave the group its name. The endoskeleton of the paired appendages of sarcopterygians shows an axial organization, one called archipterygium. As already mentioned, the paired fins of actinopterygians are fan-shaped, multiple radials radiating out from the scapulocoracoid in the shoulder girdle to articulate with the peripheral lepidotrichia. In the archipterygium, there is a series of central radials that form the axis of the paired appendages; a single proximal radial of that axis articulates with the scapulocoracoid of the pectoral girdle. In the biserial archipterygium, shorter radials are arranged along the length of the axis in two series, a preaxial and a postaxial one. In the uniserial archipterygium, only a single series of peripheral radials is arranged alongside the axis. The Sarcopterygians had classically been thought to divide into two monophyletic sister clades, the Crossopterygii, and the Dipnoi (lungfish). Of the latter group, three living species survive in the southern continents (Australia, Africa, and South

America). The crossopterygians comprise a more diverse array of clades, the principal ones being the Actinistia, comprising the fossil relatives of the extant coelacanth, and the Rhipidistia, which include among other clades the ancestor of tetrapods and all of its descendants. Classically, the rhipidistians had been subdivided into two major clades again, the osteolepiforms (including the tetrapods) and the porolepiforms. However, in modern cladistic analyses the dipnoans have been found to be more closely related to porolepiforms than to any other sarcopterygians. As a consequence, the Dipnoi are nested inside sarcopterygians close to the porolepiforms, and the distinction of Crossopterygii versus Dipnoi has become moot (Cloutier and Ahlberg, 1996).

All three sarcopterygians known from the Middle Triassic of Monte San Giorgio are actinistians, a group that ranges from the Middle Triassic to the Upper Cretaceous, with the two species in the coelacanth genus *Latimeria* (a West Indian Ocean and an Indonesian one) the sole survivors to the present day. Actinistians are readily identified on the basis of their diphycercal three-lobed caudal fin, and the forward position of the anterior dorsal fin. The air-bladder (lung) of fossil actinistians is calcified, and that of *Latimeria* is filled with fat; it thus could not and cannot function in respiration or buoyancy control, as it does in rhipidistians. Again, unlike rhipidistians, there are therefore also no internal nostrils (choanae) in actinistians (Carroll, 1988:148). This places *Latimeria* and its fossil relatives rather far from the tetrapod root. Contrary to earlier expectations, *Latimeria* was not observed to "walk" on the sea floor, and indeed avoided substrate contact when studied in its natural environment. But it does show an alternate coordination of pectoral and pelvic fins during swimming, a locomotor pattern that is also observed in tetrapod walking (Fricke et al., 1987; Fricke and Hissmann, 1992; see also discussion in Forey, 1998).

Four genera and species of hybodontiform sharks have been described from Monte San Giorgio: *Acrodus georgii*, *Asteracanthus* cf. *reticulatus*, *Hybodus* cf. *plicatilis*, and *Paleobates angustissimus*. All hybodont material collected at various localities on Monte San Giorgio comes from the Grenzbitumenzone. The most frequently found shark fossils are those of *Acrodus*, a genus first described by Louis Agassiz in 1838, who recognized several species in the genus (Agassiz, 1838, vol. 3, p. 139; pp. 73–140 of Agassiz, *Recherches sur les Poissons Fossiles* (1833–1844) were published in 1838: Brown, 1890:xxvii). Louis Agassiz (1807–1873) was a Swiss-born paleontologist and geologist, most famous perhaps for his ice age theory and his work on fossil fishes. Working at the University of Neuchâtel in Switzerland, Agassiz engaged in the publication of his monumental five-volume treatise on fossil fishes, the *Recherches sur les Poissons Fossiles*, the different parts of which were published in a loose and rather chaotic sequence through the years 1833 to 1844 (Brown, 1890). The compendium was printed and beautifully illustrated with stunning

The Sharks from the Middle Triassic of Monte San Giorgio

lithographs at the expense of the author himself. In 1846 Agassiz fled the debt he had incurred through this ambitious project to the United States, where he founded—gifted fundraiser that he was—the Museum of Comparative Zoology of Harvard University. Later in life, Agassiz became one of the most prominent critics of Darwin's theory of evolution (on Agassiz see, among others, Winsor, 1976; Lurie, 1988).

Box 2.1. The Hybodont Shark *Acrodus georgii*

Hybodontiform sharks were common during the Triassic and Jurassic; their fossil record overlaps with the earliest occurrence of modern sharks (neoselachians). The most common hybodonts are species in the genera *Acrodus* and *Hybodus*, the latter distinguished only by their dentition. *Acrodus* typically has a crushing dentition indicating durophagous habits; *Hybodus* is characterized by piercing multicuspid ("cladodont") teeth. Species in the two genera could reach a total length of up to three meters. They preyed on invertebrates and fishes. A *Hybodus* from the Lower Jurassic of Holzmaden (southern Germany) reveals a stomach content comprising approximately 200 belemnite phragmocones (Hauff and Hauff, 1981:58). A diagnostic feature of hybodonts are the head spines (cephalic spines) located above and behind the eyes. They may have played a role in mating behavior. The leading edge of the two dorsal fins is strengthened by prominent fin spines. The caudal fin is typically asymmetrical (heterocercal). The broad-based pectoral fins are inserted low on the body. Together with the asymmetrical caudal fin they provide lift during locomotion for the negatively buoyant shark. The jaw suspension was primitive (amphistylic), which did not allow for much protrusion of the jaws when the jaws opened to bite.

© Beat Scheffold / PIMUZ

Acrodus remains have been found at several outcrops of the Grenz-bitumenzone: Point 902, Cava Tre Fontane, Val Porina, and Valle Stelle (Kuhn-Schnyder, 1945; Rieppel, 1981). The Acrodus fossils from Monte San Giorgio have been referred to a separate species not known from other localities, A. georgii, named after the locality of their provenance (Mutter, 1998). Other than isolated teeth and fin spines, there are also more complete specimens, which show teeth, jaw fragments, fin spines, and head spines in association (fig. 2.2). The presence of head spines, a pair of them located behind and above each orbit—as demonstrated by a complete and articulated specimen of Hybodus from the Liassic (Lower Jurassic) of Holzmaden, southern Germany—is indeed diagnostic of hybodontiform sharks (Maisey, 1982:41; Fraas, 1889). They exhibit a characteristic tri-radiate base that carries a recurved spine from the tip of which projects a single, ventrally pointing barb. If used in mating behavior rather than as a defensive device, they might indicate internal fertilization for hybodonts.

As is indicated by its dentition, Acrodus was a benthic durophagous feeder, which could grow to considerable size. A large anterior fin spine

2.2. Associated skeletal remains of the hybodontiform shark *Acrodus georgii* (Paleontological Institute and Museum, University of Zurich T3926) from the Cava Tre Fontane. The original width of the fossil-bearing slab is 290 mm (Photo © Heinz Lanz/ Paleontological Institute and Museum, University of Zurich).

2.3. An articulated quadrant of the dentition of *Acrodus georgii* (Paleontological Institute and Museum, University of Zurich T3814) from the Val Porina. The width of the fossil-bearing slab is 160 mm (Photo © Heinz Lanz/ Paleontological Institute and Museum, University of Zurich).

of 311 mm total length, associated with *Acrodus* teeth, led Kuhn-Schnyder to estimate a total length of the shark of two to three meters (Kuhn-Schnyder, 1945). *Acrodus* would thus have grown to a similar body size as *Hybodus*, the latter an active predator of fast-moving prey with high-crowned, multicuspid teeth (Maisey, 2012:945). The teeth of *Acrodus georgii*, in contrast, form a typical crushing dentition (fig. 2.3). The teeth are low-crowned, transversely elongated, bearing a single, low and blunt main cusp located in a central or slightly distal position. Along the length of the crown runs a weakly expressed longitudinal crest, from either side of which striations radiate toward the labial and lingual margins of the crown. The crown itself is made up of osteodentine that encloses distinct vascular canals branching in a tree-like fashion. The osteodentine is capped by a layer of single-crystallite enamel. A dolomitic layer in the Grenzbitumenzone at the Val Porina locality yielded one of the very rare articulated dentitions of *Acrodus* (fig. 2.3); a full quadrant is perfectly preserved (Kuhn-Schnyder, 1945, 1974:87, fig. 68; Rieppel, 1981:329, fig. 2).

In his study of the dentition of the extant species in the genus *Heterodontus*, of which the Port Jackson shark (*Heterodontus portusjacksoni*) is a widely known example, Wolf-Ernst Reif from the University of Tübingen drew a comparison with the *Acrodus* dentition from Monte San Giorgio (Reif, 1976:30, and fig. 39a). In fact, hybodont sharks had once been classified, together with *Heterodontus* and other, putatively related fossil forms, in an order called Heterodontiformes (Andrews et al., 1967:667). The hallmark of that group (no longer recognized as monophyletic) is the heterodont dentition, where tooth morphology changes from the symphysis outward along the jaw. *Heterodontus* shows some of the most striking changes in morphology between the pointed anterior (symphyseal) teeth and the flattened lateral crushing teeth. Morphological change of tooth shape is, however, quite distinctly expressed in *Acrodus* as well, as also in other sharks. In view of such a heterodont dentition encountered in *Acrodus*, Otto Jaekel as early as 1889 drew attention to the fact that isolated teeth from the Middle Triassic of the Lorraine region (France), each described as a different species in the genus *Acrodus*, and even as species of different genera, might in fact represent different tooth positions (tooth families) in the jaw of a single species, namely, *Acrodus lateralis* (Jaekel, 1889:314). Jaekel based his conclusion on a complete, articulated dentition (left and right quadrant) of *Acrodus anningiae*, described and figured by E. C. H. Day in 1864, and again in Arthur Smith Woodward's *Catalogue of the Fossil Fishes in the British Museum (Natural History)* of 1889 (Day, 1864:57, pl. III; Woodward, 1889, fig. 10). The status of the species *Acrodus anningiae* is problematic. Agassiz figured a series of teeth mimicking the dentition of a quadrant, but most probably artificially arranged, from the Lyme Regis (Lower Jurassic) of West Dorset, England (Woodward, 1889:289). That figure, published in 1843, Agassiz labeled as *Acrodus anningiae*, in honor of Mary Anning (1799–1847), a pioneer explorer of the Lower Liassic Lyme Regis vertebrate fossils (Brown, 1890:xxvii). Agassiz never formally described this species, however (Day, 1864:60, footnote †). In the accompanying text, previously published in 1839, Agassiz referred to the same figure under the name *Acrodus undulatus*, a name that thus takes priority over *A. anningiae* (Agassiz, 1833–1844, vol. 3, 144, pl. 22, fig. 4). The specimen described and illustrated by E. C. H. Day in 1864, under the name of *Acrodus anningiae*, was not Agassiz's specimen, however, but a second one from the same locality comprising the dentition of the left and right quadrant.[1] Jaekel's use of such articulated material in his analysis of species-level taxonomy based on isolated teeth of *Acrodus* highlights the importance of *Konservatlagerstätten* for an improved understanding of the fossil record. The Middle Triassic of Monte San Giorgio plays a key role in that regard as well, as it yielded a second articulated dentition of *Acrodus*, in this case *A. georgii*. Even more important in that respect proved to be articulated shark remains from Monte San Giorgio to be discussed below.

The fin spines are identical in their morphology and histology in the genera *Acrodus* and *Hybodus* (fig. 2.4). Sharks are characterized by two dorsal fins, the leading edge of which is supported by a fin spine in hybodonts. The anterior one is generally taller than the posterior one. Hybodont fin spine structure and development was extensively studied by John Maisey from the American Museum of Natural History (Maisey, 1978). In *Acrodus* and *Hybodus*, the mantle shows a distinctive costate ornamentation, ribs and grooves running along the spine from the tip down to the "root." The most basal portion of the spine, its "root," is devoid of ornamentation as it was inserted in the epaxial muscles of the

2.4. Isolated fin spine of *Acrodus* or *Hybodus* (Paleontological Institute and Museum, University of Zurich T3815; specimen *c* in Kuhn, 1945) from the Valle Stelle. The original width of the fossil-bearing slab is 250 mm (Photo © Heinz Lanz/ Paleontological Institute and Museum, University of Zurich).

trunk. The posterior surface of the spine is somewhat concave and carries a double row of posteroventrally projecting denticles. Toward the apex of the spine, as it gets narrower, the two rows of denticles merge into a single row.

In contrast to *Acrodus*, teeth of *Hybodus* are much rarer in the Middle Triassic of Monte San Giorgio. Only a couple of isolated teeth have ever been collected, in spite of the massive excavation efforts at Point 902. One of those two teeth was collected in 1924, during Peyer's first field season at the Cava Tre Fontane locality. The teeth of *Hybodus* are those of an active predator, high-crowned with a prominent, pointed central cusp and a variable number of smaller, accessory cusps on either side of it (fig. 2.5). The root of the teeth is deep and may show multiple nutritive foramina. It was again Otto Jaekel, who in his monograph of 1889 commented on the intraramal tooth variation in the genus *Hybodus*, noting issues of synonymy and thus highlighting the difficulty of assigning isolated teeth to a particular species (Jaekel, 1889:294). The teeth from Monte San Giorgio compare most closely to those of *Hybodus plicatilis*, first described by Louis Agassiz in 1839, to which they have been tentatively referred (Brown, 1890:xxvii; Agassiz, 1833–1844, vol. 3, p. 189, pl. 22a, fig. 10, pl. 24, figs. 10, 13).

The third genus of hybodont shark from the Middle Triassic of Monte San Giorgio is *Asteracanthus*. The genus was again first introduced by Agassiz in 1837, based on tuberculated fin spines from the Upper Jurassic (Brown, 1890:xxvii; Agassiz, 1833–1844, vol. 3, p. 31). The name *Asteracanthus* derives from the Greek terms for star, and spine,

referring to the asteroid tubercles adorning the fin spines. Agassiz himself realized that this genus may be synonymous with another one he introduced in a fascicle published the following year (1838), *Strophodus*, based on distinctive teeth found from the Triassic to the Cretaceous (Brown, 1890:xxvii; Agassiz, 1833–1844, vol. 3, p. 116). However, in a fascicle published in 1839, Agassiz commented: "I was able to convince myself that the fin rays of the asteracanthids I described are the fin rays of fishes of which I described the teeth under the name *Strophodus*" (Brown, 1890:xxvii; Agassiz, 1933–1844, vol. 3, p. 171). The name *Asteracanthus* thus takes priority. The treatment of *Strophodus* as a junior synonym of *Asteracanthus* was confirmed by Arthur Smith Woodward in 1888, based on hitherto undescribed material from the Oxford Clay (Middle to Late Jurassic) near Peterborough in southeast England (Woodward, 1888:341). The nomenclatorial consequences were formalized by A. S. Woodward in his 1889 *Catalogue* and further commented upon by Bernhard Peyer in his 1946 monograph on *Asteracanthus* remains kept in Swiss museum collections, including a new find from the Ammonitico Rosso (upper Lias, Lower Jurassic) from the Breggia-gorge in the Canton Ticino (Woodward, 1889:307; Peyer, 1946). The *Asteracanthus* teeth collected in the Grenzbitumenzone at the Val Porina locality on Monte San Giorgio (fig. 2.6) had not yet been recognized for what they were (see also Kuhn-Schnyder, 1974:84ff.). The specimen from Val Porina indeed marks the first appearance of the genus in the fossil record (Capetta, 1987:34).

The fin spines of *Acrodus* and *Hybodus* are typically costate, those of *Asteracanthus* tuberculate. However, in 1855, Sir Philip Grey Egerton

2.5. Isolated tooth of *Hybodus* cf. *H. plicatilis* (Paleontological Institute and Museum, University of Zurich T2497) from "Meride" (collected in 1924) in lingual view. The tooth (root and crown) measures 11 mm in height (Photo © Heinz Lanz/Paleontological Institute and Museum, University of Zurich).

2.6. Disarticulated dentition of *Asteracanthus* cf. *A. reticulatus* (Paleontological Institute and Museum, University of Zurich T3617) from the Val Porina. The largest teeth measure 22 mm in width (Photo © Heinz Lanz/ Paleontological Institute and Museum, University of Zurich).

(1806–1881) published the illustration of a fin spine[2] from the middle Purbeck Beds (the Purbeck Formation straddles the Jurassic–Cretaceous boundary) of a species of shark he had named *Asteracanthus semiverrucosus* the year before (Egerton, 1854:434). That fin spine shows the tubercles loosely aligned in a longitudinal series which, in the upper half of the spine, coalesce or are replaced by distinct ribs (costae) (fig. 2.7). This represents an intermediate condition between the typical *Acrodus/Hybodus* and the *Asteracanthus* spines, which has been linked to ontogeny: the juvenile growth-pattern would have been costate, with increasing age the costate condition would have been replaced by a tuberculate pattern: "This transition occurred early in *Asteracanthus*, but only occurred in very old (possibly gerontic) *Hybodus* and *Acrodus*" (Maisey, 1978:658; see also pl. 72, fig. 5). But even if in view of such intermediate conditions the unequivocal identification of isolated fin spines might prove difficult, the tooth histology of *Asteracanthus* is very distinct and different from *Acrodus/Hybodus* (Peyer, 1946; Radinsky 1961).

The Val Porina specimen, tentatively referred to *A. reticulatus*—described by Agassiz in 1838 under the name *Strophodus reticulatus*—consists of a complete (two quadrants), yet disarticulated dentition, associated with patches of skin, that is, of densely packed placoid scales (Brown, 1890:xxvii; Agassiz, 1833–1844, vol. 3, p. 123, pl. 17, figs. 1–21). Heterodonty is very distinctively expressed in this species. The large, flat, lateral crushing teeth have a rectangular or slightly rhomboidal outline, with a slightly concave labial edge and an equally slightly convex lingual margin. The surface of the low crown shows vermiculate enamel striation.

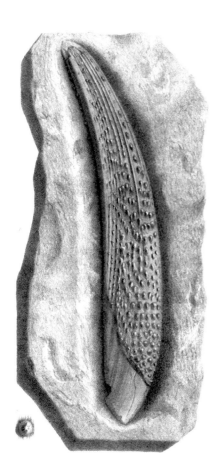

2.7. Fin spine of *Asteracanthus semiverrucosus* figured by Sir Philip Grey Egerton in 1855, showing a combined tuberculate and costate ornamentation.

The lingual margin of the crown projects into a thick and transversely plicated rim that distinctly overlaps the root and interlocks with a corresponding groove on the neighboring tooth within the same tooth family. Agassiz considered this interlocking mechanism a diagnostic character of the species *A. reticulatus*. Toward the symphysis, the teeth become progressively smaller, lozenge-shaped in outline, the anteriormost ones raising into a blunt tip. Two rows of small distal crushing plates complete the dentition (Rieppel, 1981:336, fig. 7). Tooth histology of *Asteracanthus* differs from *Acrodus* in that the vascular canals traversing the osteodentine are not branching but instead are vertically arranged, running in parallel perpendicular to the occludal surface of the tooth. Unlike other hybodonts (with the exception of *Bdellodus* from the Lower Jurassic Posidonienschiefer of southern Germany) the dentine in the teeth of *Asteracanthus* is hypermineralized, resulting in a hard tissue known as pleromin (Reif, 1973:234). Such hypermineralization of the dentine is generally considered an adaptation of a crushing dentition, although it is absent in other hybodonts that are also durophagous (e.g., *Acrodus*, *Palaeobates*).

The placoid scales associated with the *Asteracanthus* dentition from Val Porina are of the non-growing type. Non-growing placoid scales are otherwise typical of neoselachians; other Mesozoic sharks show both, growing (composite) as well as non-growing scales (Reif, 1978a). The placoid scales of *Asteracanthus* carry a posteriorly recurved, spatulate crown that tapers to a blunt posterior tip, its surface ornamented by five to nine ridges that converge toward, but do not reach the posterior tip. The wide spacing of these ridges, or striae, suggests an epibenthic, rather slow-moving durophagous predator (Reif, 1978b; see also Dick, 1978). Given the (near) anoxic conditions in the deep waters of the Grenzbitumen basin, *Asteracanthus* must have foraged in near-shore waters.

The fourth, and last hybodont recorded from the Grenzbitumenzone, all of its fossils coming from the Point 902 locality, is *Palaeobates*, a genus first erected by Hermann von Meyer in 1849 on the basis of isolated teeth from the Muschelkalk (lower Anisian) of Upper Silesia (Meyer, 1849). The taxonomy of this genus remained unresolved until in 1889 Otto Jaekel studied the histology of the teeth. These he could show to be clearly set apart from all other hybodonts on the basis of dentine histology. Jaekel conceded that on the basis of their external morphology, *Palaeobates* teeth compare closely with those of *Acrodus lateralis* (Jaekel, 1889:327). He was the first to investigate and compare the tooth histology of these two taxa, however, studies which provided astounding insights. The teeth of *Acrodus* show the typical osteodentine so characteristic of sharks, although the detailed differentiation of the osteodentine may differ between taxa, as it does between *Acrodus* and *Asteracanthus*. *Paleobates*, however, shows a strikingly different structure of dentine. Osteodentine is differentiated only in the root of the tooth; the crown in contrast is made up by orthodentine, the latter capped by single-crystallite enamel. Orthodentine shows densely packed, parallel dentine tubules that radiate from the pulp cavity to the enamel covering. Since it is a term commonly applied to tetrapods, the dentine of *Palaeobates* has also been referred to as "pallial dentine," the dentine tubules being vertically oriented in the elongate, rectangular, yet narrow lateral crushing teeth (for a more detailed discussion of "pallial dentine" see Rieppel, 1981:345ff.). But here again, there is heterodonty, the more medially positioned teeth rising to form a blunt tip. The Monte San Giorgio material shows head spines as well as fin spines associated with *Palaeobates* teeth (fig 2.8). The fin spines show the costate mantle ornamentation that is also characteristic of *Acrodus* and *Hybodus*. A row of downward curved denticles runs along the weakly concave posterior surface of the spine.

Endoskeletal elements made up of calcified cartilage are variously preserved in Monte San Giorgio hybodonts, most notably the upper (palatoquadrate) and lower (Meckel's cartilage) jaws, and the pectoral girdle (scapulocoracoid). Notably, the jaws in *Acrodus* and *Palaeobates* are closely similar and also resemble those of *Asteracanthus* (Peyer, 1946:9, fig. 1, pl. I). Meckel's cartilage is relatively deep, with an evenly convex ventral margin. The palatoquadrate is somewhat more delicately built, with strongly reduced orbital and posterior quadrate processes. The dorsal margin is nearly straight, the posterior margin slants posteroventrally and projects into the mandibular condyle that is received in an articular

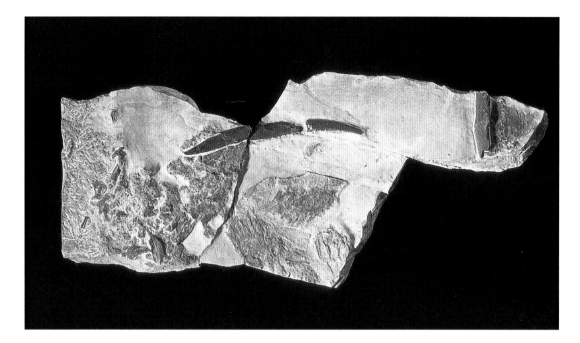

facet at the posterior end of Meckel's cartilage. The morphology of the palatoquadrate indicates a hyostylic jaw suspension in hybodonts, derived from the more primitive amphistylic jaw suspension of non-hybodont Paleozoic sharks (Maisey, 2008).

Konservatlagerstätten such as the Grenzbitumenzone of Monte San Giorgio are important not only because of the sheer number of fossils preserved but also because of the often excellent preservation. Articulated, or even only associated, skeletal material can illuminate the taxonomy of fishes (or reptiles) that had previously been known from isolated elements only. Such was the case for the most frequently found shark in the Grenzbitumenzone, *Acronemus tuberculatus* (Rieppel, 1982). In his 1886 account on the fossils collected in the bituminous layers from Besano, Bassani referred to an unpublished manuscript, the catalogue of fossil fishes kept at the Milan museum and written in 1873 by Cristoforo Bellotti (see above), who had recognized two shark species in the Besano Formation (fig. 2.9 a, b), one based on teeth and named *Acrodus bicarinatus*, the other based on fin spines and named *Nemacanthus tuberculatus* (Bassani, 1886:18). The manuscript can no longer be located, nor the type material, since all the fossils from the Besano Formation housed in the Milan Natural History Museum were destroyed during an Allied air raid in August 1943.[3] What remains in the archives of the museum are Bellotti's unpublished drawings, including those of *Nemacanthus tuberculatus*. A formal description of the latter taxon, as also of *Acrodus bicarinatus*, was later published by Bassani, who did not include illustrations of the specimens, however (Bassani, 1886:30ff.). Both taxa were later figured, in plate I of Alessandri's 1910 monograph on the Triassic fishes of Lombardy, including the type specimen of *Nemacanthus tuberculatus* (Alessandri, 1910:36, and pl. I, fig. 10, misspelt the name as *Nemacanthus tubercolatus*).

2.8. Teeth and anterior fin spine of *Palaeobates angustissimus* (Paleontological Institute and Museum, University of Zurich T3838) from Mirigioli, Point 902. As preserved, the fin spine measures 138 mm in height (Photo © Heinz Lanz/ Paleontological Institute and Museum, University of Zurich).

2.9a. Teeth of *"Acrodus bicarinatus"* (*Acronemus tuberculatus*, Paleontological Institute and Museum, University of Zurich T3821); the largest teeth measure 10 mm in width.

2.9b. Anterior fin spine of *"Nemacanthus tuberculatus"* (*Acronemus tuberculatus*, Paleontological Institute and Museum, University of Zurich T1289); the fin spine measures 118 mm in height (Photos © Heinz Lanz/Paleontological Institute and Museum, University of Zurich).

The first fossil sharks from the Grenzbitumenzone, all of which referred to the genus *Acrodus*, were described by Kuhn-Schnyder in 1945 (Kuhn-Schnyder, 1945). One of the specimens he included in this paper (his specimen *d*) was at the time not yet fully prepared. However, a radiograph of the specimen highlighted fin spines of a proportion very different from those of *Acrodus*, as well as a cleaver-shaped palatoquadrate with a well-developed posterior (quadrate) process that is absent in hybodonts. Later, Peyer (1957) mentioned an *Acrodus* from Monte San Giorgio with a tuberculate fin spine, while in 1974, Kuhn-Schnyder published drawings of two fin spines of *Nemacanthus tuberculatus* which he referred to "*Acrodus tuberculatus* (BELLOTTI)" (Kuhn-Schnyder, 1974:87, fig. 69). The confusion resulted from the simple fact that fin spines of *Nemacanthus tuberculatus* BELLOTTI emendatum BASSANI, 1886, were associated with teeth typical of *Acrodus bicarinatus*. This association is striking in a beautifully preserved, complete shark fossil, collected in 1957 at the Point 902 excavation in the Grenzbitumenzone. The specimen was designated the neotype of the species *tuberculatus*, the latter referred to a new genus named *Acronemus* (Rieppel, 1982:400).

The exceptional fossil preservation in the sediments of the Grenzbitumenzone frequently allows correction of past taxonomic errors. Isolated teeth of this shark had previously been described under the name of *Acrodus bicarinatus*; isolated fin spines had been named *Nemacanthus tuberculatus*. A complete specimen collected at Monte San Giorgio revealed that these teeth and fin spines belong to the same species, named *Acronemus tuberculatus*. This small shark reached a total length of 30 to 35 cm. As in hybodonts, prominent fin spines strengthen the leading edge of the two dorsal fins. Most recently, sophisticated modern technology such as high-resolution X-ray computed tomography has been applied to the study of this fossil, especially its braincase. However, the classification of this species remains uncertain, because the taxon is characterized by a curious mixture of hybodontiform and neoselachian characteristics.

© Beat Scheffold / PIMUZ

2.10. The neotype of *Acronemus tuberculatus* (Paleontological Institute and Museum, University of Zurich T1548) from Mirigioli, Point 902. The total length of the specimen is estimated to have been from 30 to 35 cm (Photo © Heinz Lanz/Paleontological Institute and Museum, University of Zurich).

Acronemus was a small shark just over one foot long (30–35 cm) (fig. 2.10). The neotype preserves the posterior part of the braincase on which the upper jaw (palatoquadrate) articulates. The braincase shows a prominent postorbital process on which the posterior (quadrate) process of the palatoquadrate articulated (see also discussion in Maisey, 2008). The lower jaw (Meckel's cartilage) is preserved in articulation with the palatoquadrate. Fragments of the jaw suspension apparatus (hyomandibula and ceratohyal) are preserved behind the mandibular articulation, along with fragments of labial cartilages that lie alongside the convex ventral margin of Meckel's cartilage. The dentition between the jaws is disarticulated but reveals distinct heterodonty in size and shape of the teeth. Previously known under the name *Acrodus bicarinatus*, the teeth are characterized by a centrally located, blunt cusp. The old name derived from the fact that a relatively prominent transverse crest intersects at right angles with a central crest that extends from the cusp on to the lingual and labial side of the crown. From those crests fine striations radiate toward the margins of the crown. The main cusp lingually overhangs the neighboring tooth in the same family. The largest teeth appear to have been located midway along the jaw, tooth size decreasing toward the symphysis, and even more pronouncedly so toward the back end of the jaw. The enamel is of the single-crystallite type without a superficial shiny layer (see Reif, 1973, for further discussion).

Above the pectoral fin is located the scapulocoracoid, a curved bar of calcified cartilage, broad at its base but tapering to a blunt tip at its apex. At least one proximal radial is preserved in articulation with the base of the scapulocoracoid. The contours of the specimen suggest a distinctly heterocercal caudal fin. The notochordal sheath was not calcified, however. The two dorsal fins are each supported by a robust fin spine, the anterior one larger than the posterior one. The fin spines, isolated specimens of which were previously referred to *Nemacanthus tuberculatus*, are relatively short and broad. A distinctly tuberculated "crown" is set off from a "root" with a smooth surface, the latter embedded in the epaxial musculature of the trunk. The distinct and rounded tubercles, covered with shiny enamel, align along straight vertical rows. In some fin spines, the tubercles may coalesce into short segments of ribs, or costae, that curve in a posterodorsal direction toward the posterior margin of the spine. They correspond in their orientation to the growth lines frequently seen on the enameloid mantle of euselachian fin spines (Maisey, 1977). The posterior wall of the fin spines is concave, but retrorse posterior denticles are absent. In cross section, the fin spines of *Acronemus* reveal a posteriorly displaced central cavity, which is posteriorly open along the "root" portion of the spine. There is an inner lamellar trunk layer, surrounded by a well-vascularized outer trunk layer. The tubercles are formed from the mantle layer, which is not distinctly set off from the outer trunk layer, however (for further discussion of selachian fin spine terminology see Maisey, 1979). The scales of *Acronemus* are again of the non-growing placoid type. They bear a characteristic, posteriorly recurved, lanceolate crown with an obtuse posterior end, the surface of which is ornamented by three to five widely spaced longitudinal ridges.

Acronemus tuberculatus shows a curious mixture of characters that renders its systematic placement difficult, or at least problematic: "The incertae sedis status of *Acronemus tuberculatus* within the Cohort Euselachii reflects its ambiguous combination of neoselachian and hybodontiform characteristics" (Maisey, 2011:419). Those characteristics include features revealed by a recent, detailed study of the braincase of *Acronemus* using high resolution x-ray computed tomography but are also evident in the macroscopic structures described above (Maisey, 2011). The teeth, indistinguishable from those of *Acrodus*, are clearly of hybodontiform nature, whereas the thick and shiny enamel covering of the fin spine tubercles would rather suggest neoselachian affinities. Cephalic spines, characteristic of male hybodonts, have never been found in *Acronemus*. The fin spines also lack the retrorse posterior denticles so typical of hybodont sharks. In summary: "*Acrocnemus* is one of the many problematic late Paleozoic and early Mesozoic sharks with features evocative of neoselachians and/or hybodontiforms." Its phylogenetic relationships and classification remain elusive until such time that "a much better understanding of the morphology of articulated pre-Jurassic sharks" has been obtained (Maisey, 2011:426).

The Actinopterygians from the Middle Triassic of Monte San Giorgio

Top predators among the ray-finned fishes from the Middle Triassic of Monte San Giorgio are the widely distributed genus *Birgeria*, and the genus *Saurichthys* with species of both large and small body size (Tintori, Hitij et al., 2014:400). The equally widely distributed, semi-durophagous genus *Colobodus* likewise attains large body size. Among the taxa represented in the Monte San Giorgio biota, *Birgeria* is known to have reached up to 200 cm in total length, *Saurichthys* up to 150 cm, and *Colobodus* up to 70 cm (Brinkmann, 1994). The paleoichthyologist Andrea Tintori from the Universita degli Studi di Milano colloquially and aptly captured the remaining bewildering diversity of mid-Triassic ray-finned fishes as found at Monte San Giorgio and elsewhere by the operative term "small fishes." Traditionally, actinopterygian fishes had been classified in three "grades," or levels of evolution: the Chondrostei (with the sturgeon and paddle-fish as extant representatives), the Holostei (with the bowfin and gars as extant representatives), and the Teleostei (the host of living actinopterygians: e.g., Romer, 1966). The great majority of the "small fishes" from the Middle Triassic of Monte San Giorgio would have fallen into such paraphyletic grade groups as the Holostei, or the more advanced Subholostei—both notoriously paraphyletic assemblages when these diverse fossil taxa of "small fishes" are included (although a new usage of the name Holostei for a monophyletic clade has been introduced in a recent study of living and fossil gars by Grande, 2010). It is thus best to consider the "small fishes" from Monte San Giorgio that do not qualify as stem-group neopterygians as "chondrosteans" that have been variously allocated to families and higher categories whose content and relationships ultimately remain fluid and disputed (Bürgin, 1992; see the recent discussion in Xu et al., 2015; and the classification in Gardiner, 1993).

The first actinopterygian from Monte San Giorgio that became the subject of a monographic description was *Birgeria stensioei*, based on a beautifully preserved and prepared specimen (fig. 2.11). The find, a specimen of 120 cm (approx. four feet) total length, was first announced by Emil Kuhn-Schnyder in 1946, at the twenty-fifth annual meeting of the Swiss Paleontological Society. The specimen had been collected back in 1934 at the Val Porina locality; Kuhn-Schyder estimated that the preparation of this delicate skeleton might take a single preparator up to a full year (Kuhn-Schnyder, 1946:364). At the time of its monographic description in 1970, it still remained the first and only almost complete and nearly articulated specimen of any *Birgeria* species known (Schwarz, 1970). Isolated teeth of *Birgeria* had been known since the middle of the nineteenth century, mostly from the Germanic Muschelkalk, but also from Rhaetian deposits near Bristol (UK), and from the southern Alpine Triassic. Yet these teeth were commonly referred to the genus *Saurichthys*, as had been done by Louis Agassiz in his monumental *Recherches sur les Poissons Fossiles*, in a fascicle published in 1843. However, when visiting the Senckenberg Museum in Frankfurt am Main in the fall of 1919, Stensiö found a complete maxillary bone from the Muschelkalk outcrops near Bayreuth in Bavaria, Germany (Anisian, Middle Triassic),

2.11. A complete, articulated specimen of *Birgeria stensioei* (Paleontological Institute and Museum, University of Zurich T4780), from the Val Porina. The total length of the specimen is 120 cm (Photo © Heinz Lanz/Paleontological Institute and Museum, University of Zurich).

carrying teeth of the type Agassiz had described as *Saurichthys mougeoti*, but of a morphology (short and with prominent ascending process) that could not possibly be referred to the genus *Saurichthys* (Agassiz, 1833–1844, vol. 2, pt. 2, pp. 85ff.). Stensiö consequently erected a new genus, *Birgeria*, with *mougeoti* as its genotypical species based on the specimen from the Muschelkalk of Bayreuth (Stensiö, 1919:178; named after Birger Sjöström, a friend and companion on the 1915 expedition to Spitsbergen).

The first record of *Birgeria* teeth from the southern Alpine Rhaetian (Late Triassic) was reported in the 1860s by Stoppani, who referred them to *Saurichthys* (now *Birgeria*) *acuminatus* (now *acuminata*) (Stoppani, 1860–1865). In 1886, Bassani reported jaw fragments and teeth from the Besano Formation near Basano, Lombardy, which he referred to *Belonorhynchus*, but which more probably have to be referred to *Birgeria* (Bassani, 1886; Schwarz, 1970:7; the problem, again, is that Bassani offered no illustrations of his specimens). The first report of the genus *Birgeria* in the Grenzbitumenzone of the Cava Tre Fontane locality of Monte San Giorgio is Aldinger's paper of 1931. The cranial remains at his disposal Aldinger referred to a new species, *Birgeria stensiöi*, named after the famous Stockholm paleoichthyologist Erik A. Stensiö (Aldinger, 1931:177). Considered from a more global perspective, up to 11 species of *Birgeria* have been described from the Lower Triassic of East Greenland, Spitsbergen, British Columbia, and Madagascar; from the Middle Triassic of central Europe and California; as well as from the Upper Triassic of China, southern Bolivia, and Europe (Romano and Brinkmann, 2009, and references therein). This testifies to the pelagic habits of this large predatory fish, which achieved cosmopolitan distribution during the Triassic. *Birgeria* disappears from the fossil record at the close of the Triassic.

Birgeria stensioei was a large, pelagic predatory chondrostean closely related to sturgeons and their kin; indeed, it has been classified as sister taxon to extant and fossil Aciperseriformes on the basis of a reduced opercular bone, a posteriorly elongated parasphenoid, and a severely reduced body squamation (Bemis et al., 1997:43). The almost complete reduction of the scaly covering reveals *Birgeria stensioei* to be a fast-swimming creature of the open ocean, thus pointing to an open marine connection of the marginal basin at the bottom of which the sediments forming the Grenzbitumenzone accumulated. The fusiform body terminates in a morphologically heterocercal, yet externally symmetrical caudal fin. The dorsal and anal fins are displaced posteriorly, supporting the caudal fin in creating propulsive force through lateral undulation of the posterior part of the body, the so-called suboscillatory type of lateral body undulation (Brinkmann and Mutter, 1999:123). The pectoral fin is displaced dorsally on the lateral body wall, providing efficiency in turns and stops. The jaws are furnished with a heterodont dentition, where smaller teeth fill the space between larger, fang-like teeth, all of them conical, pointed, and characteristically topped by a shiny acrodin cap. A closely similar tooth morphology is observed in the genus *Saurichthys*, species of which generally co-occur with *Birgeria*, which explains the early confusion in separating the two genera from one another on the basis of isolated teeth only (see also the discussion in Mutter et al., 2008:119ff.). Following Emanuele Gozzi, "it is impossible to distinguish *Birgeria* and *Saurichthys* teeth" (cited in Lombardo and Tintori, 2005:3).

Box 2.3. Actinopterygians from the Middle Triassic of Monte San Giorgio

Among the top predators of ray-finned fishes (actinopterygians) in the Grenzbitumenzone basin was the semi-durophagous *Colobodus bassanii*. The body of this fish, which reached a total length of 70 cm, is covered by thick scales. The caudal fin is of the abbreviated-heterocercal type, but symmetrical in outward shape. The pectoral and pelvic fins are inserted low on the ventrolateral body wall. The small fish at the bottom of the right corner represents *Ctenognathichthys bellottii* of 21 cm total length. Its elongated, densely set teeth form a grasping dentition. Among the "small fishes," the Middle Triassic of Monte San Giorgio provides evidence of early experimentation with body shape. The fish with the deep, rhomboidal body shape shown in the lower left corner represents *Bobasatrania ceresiensis,* a species that could reach a total length of 25 cm. Note the elongate dorsal and anal fins, the outwardly symmetrical (but anatomically heterocercal), deeply bifurcated caudal fin, and the insertion of the pectoral fin relatively high on the body wall. Species of this genus are known from the Lower Triassic of East Greenland, Madagascar, British Columbia, and Alberta. The Middle Triassic of Monte San Giorgio is the geologically youngest occurrence of the genus.

© Beat Scheffold / PIMUZ

Another large actinopterygian frequently found in the Grenzbitu-menzone is *Colobodus bassanii* (fig. 2.12), first described on the basis of material from the Besano location by Giulio de Alessandri in his seminal monograph of 1910. Nearly complete specimens of up to 70 cm total length of this semidurophagous fish have been collected at Monte San Giorgio (Bürgin, 1996:559). The systematics of fossil fishes that in the past have been referred to the perleidiform ("chondrostean") genus *Colobodus* is rather confused, and the original type material on the basis of which the genus was introduced by Louis Agassiz in 1844—a tooth battery on an unidentified bone from the Upper Mus-chelkalk of Lunéville (France), which Agassiz did not illustrate—was destroyed in a fire in the Department of Paleontology and Geology of the Louis Pasteur University in Strasbourg (Mutter, 2004). The fossil fishes from Besano, described by de Alessandri in 1910 and kept at the Museo Civico di Storia natural di Milano, were likewise destroyed in an Allied air raid in 1943. A petition was thus submitted to the Inter-national Commission on Zoological Nomenclature to declare *Colo-bodus bassanii* de Alessandri, 1910, the type species of its genus and to designate a well-preserved specimen from the Grenzbitumenzone of Monte San Giorgio as the neotype for that species (Mutter 2003; ICZN, 2005). The heavy scales and robust teeth of *Colobodus* are quite frequently preserved in Middle Triassic marine deposits of the western Tethyan faunal province, including the Germanic Basin (Muschelkalk deposits), but reports of *Colobodus* from other ages or areas are judged highly questionable, with the exception of an early Carnian (early Upper Triassic) occurrence in northeastern Italy (Rusconi et al., 2007). More recently, a Pelsonian (Anisian, early Middle Triassic) occurrence

2.12. A complete, articulated specimen of *Colobodus bassanii* (Paleontological Institute and Museum, University of Zurich T1804). The total length of the specimen is 45 cm (Photo © Heinz Lanz/Paleontological Institute and Museum, University of Zurich).

of *Colobodus* has been established in southwestern China as part of the so-called Panxian Fauna (upper Member of the Guanling Formation, Xinmin District, Panxian County, Guizhou Province; for further discussion of the Panxian-Luoping fauna, see chap. 11). The significance of this occurrence is that it establishes *Colobodus baii* not only as the geologically earliest occurrence of its genus but also the only undisputed occurrence outside the western Tethyan faunal province (Sun, Tintori et al., 2008).

Colobodus was a fish of fusiform body shape, with a relatively high and rounded snout, thick and ridged ganoid scales, low-set pectoral fins, and an abbreviated heterocercal, externally symmetrical caudal fin. Lingual to a palisade of peg-like, pointed teeth on maxillaries and dentaries are located enlarged, molariform teeth with a distinct acrodin knob on their apex and a vertically striated flank, testifying to durophagous habits (Guttormsen, 1937). In his unpublished PhD thesis defended in 2002, Raoul J. Mutter distinguished several species of *Colobodus* that follow one another in close succession through the middle part of the Grenzbitumenzone. The variation detected and analyzed by Mutter concerns mostly the ganoin ornamentation on the body scales (Mutter, 2004, fig. 4). The taxonomic conclusions based on this pattern of variation have not been universally accepted, however, but have instead been interpreted as an expression of a high degree of individual variability observed in the genus *Colobodus*, as well as in the genus *Crenilepis*, considered a close relative of *Colobodus* (Sun, Tintori et al., 2008:366). Indeed, the specimens recently described from the Anisian of Panxian, Guizhou Province (southwestern China), show remarkable variation of the ganoin ornamentation on scales covering different areas of the body (Sun, Tintori et al., 2008, fig. 10). This has in the past led to a situation where "*Colobodus* has been used as a wastebasket-genus: almost all Triassic striated molariform teeth and ornament-rich ganoin scales were determined as *Colobodus* or *Colobodus*-like remains" (Sun, Tintori et al., 2008:372). There are currently four valid species recognized in the genus, one from the Grenzbitumenzone/Besano Formation of Monte San Giorgio/Besano (*Colobodus bassanii*), two species from the Upper Muschelkalk (Ladinian) of Germany, and one species from the Pelsonian (Anisian) of southwestern China (Mutter, 2004; Sun, Tintori et al., 2008).

Another signature actinopterygian from the Grenzbitumenzone of Monte San Giorgio is the highly species-rich genus *Saurichthys*. The genus *Saurichthys* was introduced by Agassiz in the second volume (second part) of his *Recherches sur les Poissons Fossiles*, in a fascicle published in 1834, designating *S. apicalis* as the genotypical species, based on jaw fragments from the Muschelkalk (Anisian) of Bayreuth (Agassiz, 1833–1844, vol. 2, pt. 2, p. 84 (*Saurichthys*) published in 1834; p. 85 (*S. apicalis*) published in 1835). Species of the genus *Saurichthys* are a prominent faunal element in Triassic biotas worldwide, found in both marine and freshwater deposits (Mutter et al., 2008; Romano et al., 2012). Species of *Saurichthys* vary drastically in size. Among the smaller species rank *S. striolatus* (Raibl shales near Villach, Carinthia, Austria; Carnian, Late Triassic) with a total length of 10 to 18 cm, and *S. (Sinosaurichthys) minuta* from the Pelsonian (Anisian) of Luoping County, Yunnan Province (China), with a total length of 10 to 21 cm (Griffith, 1959:589; F. Wu et al., 2011). The largest specimens of *Saurichthys deperditus* known, reaching a total length of 160 to 180 cm, come from the Norian (Late Triassic) *Calcare di Zorzino* from Lombardy, such as the quarried outcrops at Cene in the Bergamo Prealps (Tintori, 2013:295; "species A" in Tintori, 1990; see also Romano et al., 2012:553). Several genera, subgenera, or species groups have been recognized among Triassic saurichthyids, with a general consensus emerging that all species should be referred to a single genus, *Saurichthys* (Romano et al., 2012; see also Tintori, 2013; Maxwell, Romano et al., 2015). Other than the latter genus, entirely restricted to the Triassic, the family Saurichthyidae also includes the monotypic genus *Eosaurichthys* from Late Permian marine deposits near Zhejiang (southern China), and two species in the genus *Saurorhynchus* from early Jurassic marine deposits of Europe and North America (for more details and references see Romano et al., 2012:553). In a broad sense, the Saurichthyidae is related to *Birgeria*, and the sturgeon and its kin: in the most recent phylogenetic analysis, Saurichthyidae came out in an unresolved trichotomy with *Birgeria* and *Acipenser* (Maxwell, Romano et al., 2015; see also Rieppel, 1992).

Species of the genus *Saurichthys* are a common faunal element in both marine and freshwater deposits of Triassic age. The genus is very species rich and achieved a cosmopolitan distribution. A total of six species of *Saurichthys* have been described from the Middle Triassic of Monte San Giorgio. The body is slender and elongate, the jaws are drawn out into a prominent pointed rostrum, the scale covering is reduced to a variable degree, the dorsal and anal fins are located toward the posterior end of the trunk, and together with the symmetrical caudal fin form a propulsive mechanism indicating an ambush predator. The spectacular preservation of fossils in the Middle Triassic of Monte San Giorgio allows unique insights into the life history of *Saurichthys*, such as evidence of internal fertilization and viviparity. Species of *Saurichthys* that coexisted in time and space often show habitat partitioning as indicated by different overall body size and dentition. A good example is the sympatric occurrence of *Saurichthys curionii* (relatively small, with a gracile dentition) and *Saurichthys macrocephalus* (relatively large, with a robust dentition) in the lower Meride Limestone of Ladinian age. Rieppel (1985a) documented that a large individual of *Saurichthys macrocephalus* preyed on a smaller individual of *Saurichthys curionii*.

© Beat Scheffold

Saurichthys is characterized by a slender and elongate body shape (fig. 2.13). The skull is elongated as well, with the upper and lower jaws (premaxillaries and dentaries) drawn out to form a long and pointed rostrum. The whole skull, including the opercular apparatus, can make up 25% to 30% of the total body length. Premaxilla and dentary are characteristically ornamented by vertical striations and carry teeth of two or three different size classes. The prominent dorsal and anal fins are placed symmetrically in the posterior part of the body, close to the morphologically and externally symmetrical (homocercal) caudal fin; an abbreviated heterocercal caudal fin is still differentiated in basal (early) species of the

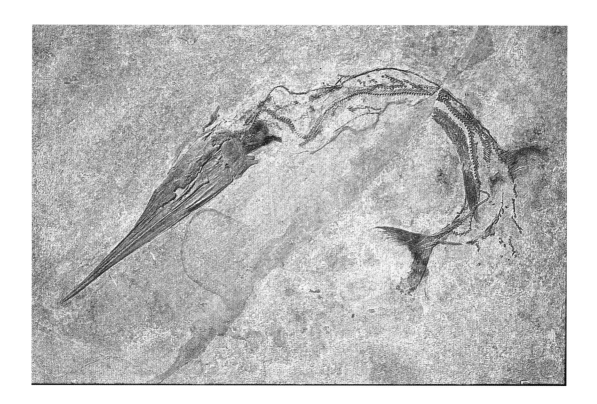

genus. Basal (early) members of the group also sport distinct fulcral scales stiffening the leading edge of the unpaired fins, which disappear in more advanced species. The fan-shaped paired pectoral and pelvic fins are inserted low along the ventrolateral body ridge. There is, however, the report of a dorsally inserted pectoral fin, high up on the lateral body wall in a position above the dorsal tip of the cleithrum, in some *Saurichthys* (*Sinosaurichthys*) species from the Anisian of southwestern China (F. Wu et al., 2011). Given developmental and functional constraints, such interpretation of these specimens remains controversial, however. In early and basal species such as *Saurichthys madagascariensis* from the Lower Triassic of Madagascar the entire body is covered by large ganoid scales, but the body squamation is reduced to a variable degree in more derived species (Piveteau, 1944–45; Rieppel, 1980a; Kogan and Romano, 2016). It is thus interesting to note that the reduction of the body squamation occurred convergently in *Saurichthys*, *Birgeria*, and *Acipenser*. Vertebral centra did not ossify in *Saurichthys*, the notochord being flanked by neural arches dorsally, and hemal arches in the caudal region ventrally. The overall organization of species in the genus *Saurichthys* resulted in their classification as "pike-like predators," performing what has also been called "fast-start or ambush predation" (Kogan et al., 2015). The density of *Saurichthys* fossils found in some localities might indicate that some species, especially the smaller ones, lived in schools.

Six different species of *Saurichthys* have been described from the Middle Triassic of Monte San Giorgio (Rieppel, 1985a, 1992; Maxwell,

2.13. The neotype of *Saurichthys curionii* (Paleontological Institute and Museum, University of Zurich T3913) from the Alla Cascina beds. The total length of the specimen is 35 cm (Photo © Heinz Lanz/Paleontological Institute and Museum, University of Zurich).

Romano et al., 2015). The first *Saurichthys* to be described from the *schisti bituminosi di Besano* was *Ichthyorhynchus Curioni* by Bellotti in 1857; the currently valid species name is *Saurichthys curionii*, also found in significant numbers in the Alla Cascina beds of the lower Meride Limestone of Ladinian age(see fig. 2.13). It is a gracile fish of 30 to 45 cm total length, characterized by a slender rostrum carrying a delicate dentition, a broad and rounded opercular bone, and a total of six scale rows along the body: a mid-dorsal and mid-ventral row, a lateral scale row on either side of the body transmitting the lateral line canal, and a ventrolateral scale row again on either side of the body. The occurrence of *S. curionii* in the Grenzbitumenzone proper remains equivocal, which is surprising in view of the abundance of the species in the Alla Cascina beds, while the original type material was collected in the Besano Formation near Besano (see discussion in Rieppel, 1985a, 1992).

The second species of *Saurichthys* described from the Middle Triassic of Monte San Giorgio is *Saurichthys macrocephalus* (fig. 2.14). The species was first described under the name *Belonorhynchus macrocephalus* by Deecke in 1889, based on material from the *Calcare di Perledo* of Ladinian age (possibly lowermost Carnian in its uppermost part) (Tintori, Muscio et al., 1985:199; Deecke, 1889:127). In his unpublished catalogue of the fossil fishes in the Museo Civico di Storia Naturale di Milano of 1873, Bellotti listed *Belonorhynchus robustus* as a faunal element in the black shales of Perledo (Bassani, 1886:33). Bassani's account of 1886 does not offer a diagnostic description of the taxon he listed with reference to Bellotti's manuscript as *Belenorhynchus* cfr. *Robustus*, a species he considered synonymous with *Ichthyorhynchus Curionii*, the latter described by Bellotti in 1857. In 1910, Alessandri listed *Belonorhynchus macrocephalus* as a junior synonym of *Belonorhynchus robustus*. Today the original material dealt with by Bellotti, Bassani, and Alessandri is either lost or destroyed. Given the unpublished status of Bellotti's manuscript, and the lack of illustrations in Bassani's paper of 1886, Deecke's original description of *Belonorhynchus macrocephalus* is the first treatment of this species from the black shales of Perledo that satisfies the requirements of the International Code of Zoological Nomenclature (ICZN). Today, the species is referred to the genus *Saurichthys*, *Belonorhynchus* being a junior synonym of the latter. At Monte San Giorgio, *Saurichthys macrocephalus* is known from the Alla Cascina beds and was consequently sympatric with *Saurichthys curionii* in its spatial and temporal distribution.

The sympatric occurrence of these two species in the lower Meride Limestones of Ladinian age is interesting, as their respective morphology suggests trophic habitat partitioning. In contrast to *S. curionii*, *S. macrocephalus* reaches a total length of minimally 55 to 57 cm. The rostrum of *S. macrocephalus* is relatively shorter, more robustly built, and furnished with a distinctly more robust dentition than that of *S. curionii*. The whole animal is of a sturdier appearance, but as in *S. curionii* there are six longitudinal rows of scales along the body. Trophic habitat partitioning

between these two species may have been driven by the fact that embedded as they are in the fish fauna of the Cascina beds, they both belong to the dominant predators in their biota (Stockar, 2010:109).

A third species from Monte San Giorgio, this one from three horizons within the Grenzbitumenzone, is *Saurichthys costasquamosus* (Maxwell, Romano et al., 2015, fig. 1; see also Rieppel, 1985a). This is a large species known to have reached a total length of 80 cm. The rostrum is relatively long but not as slender as in *S. curionii*, and the dentition is more robustly differentiated. As in the two previously mentioned species, there is a total of six scale rows along the body, whereby the elements in the mid-lateral scale rows are of a very peculiar, highly diagnostic shape. Their expanded middle part is pierced by the canal that carries the lateral line organ. From it extend slender dorsal and ventral projections that extend over most of the height of the body and could thus be misidentified on the basis of cursory observation as ossified ribs.

There exists, however, in the upper part of the Grenzbitumenzone a second species with such rib-like mid-lateral scales, which is of much smaller size than *S. costasquamosus*, reaching a total length of 21 to 26 cm. The rostrum is very long and delicate, and the dentition diminutive by comparison to *S. costasquamosus*. This second species, known as *Saurichthys paucitrichus* (fig. 2.15), is further characterized by a lower number of fin rays (lepidotrichia) in the unpaired fins, and by a lesser degree of segmentation of these compared to the other species of Monte San Giorgio. It is interesting to note that again as a consequence of their sympatric occurrence in the Grenzbitumenzone, the morphological differences between *S. costasquamosus* and *S. paucitrichus* suggest trophic habitat partitioning (Rieppel, 1992).

More recently, two additional species of *Saurichthys* have been described from the Grenzbitumenzone of Monte San Giorgio (Maxwell, Romano et al., 2015). *S. breviabdominalis* from the upper part of the Grenzbitumenzone is characterized by a relatively short abdominal

2.14. Skull and trunk region of a specimen of *Saurichthys macrocephalus* (Paleontological Institute and Museum, University of Zurich T4106) from the Alla Cascina beds. Gut contents in the anterior trunk region represent a small specimen of *Saurichthys curionii*. The length of the fossil-bearing slab is 59.5 cm (Photo © Heinz Lanz/Paleontological Institute and Museum, University of Zurich).

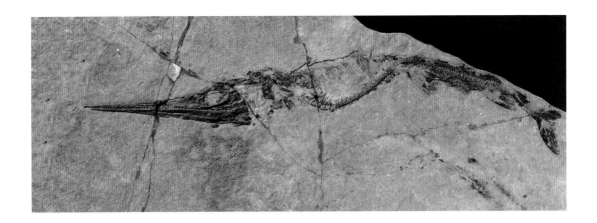

2.15. The holotype of *Saurichthys paucitrichus* (Paleontological Institute and Museum, University of Zurich T322) from Mirigioli, Point 902. The total length of the specimen is 26 cm (Photo © Heinz Lanz/Paleontological Institute and Museum, University of Zurich).

region, which is the region between the pectoral girdle and fins and the anterior edge of the dorsal/anal fins. It shares with *S. costasquamosus* and *S. paucitrichus* the rib-like mid-lateral scales, but with a total adult length of 35 cm is intermediate in size between the latter two species. The three species of *Saurichthys* from the Grenzbitumenzone with the rib-like mid-lateral scales form a monophyletic clade within the genus, the so-called *Costasaurichthys* Group (Maxwell, Romano et al., 2015, fig. 8; subgenus *Costasaurichthys* in Tintori, 2013:291). Similar rib-like scales also occur in *Saurichthys yunnanensis* from the Anisian (Middle Triassic) of Luoping County, Yunnan Province, southwestern China (Zhang, Zhou et al., 2010).

The last *Saurichthys* species that has been described from Monte San Giorgio, again from the Grenzbitumenzone, is *S. rieppeli* (fig. 2.16), with an adult length of around 60 cm (Maxwell, Romano et al., 2015). It sports a delicate dentition set in a slender rostrum. Small fringing fulcral scales line the leading edge of the unpaired fins. The hemal arches in the anterior caudal region form rectangular plates that span the width of two neural arches. Only a single, mid-dorsal scale row is present in the abdominal region in front of the pelvic plate. The six scale rows otherwise typical for the *Saurichthys* from Monte San Giorgio appear only in the caudal region, that is, behind the dorsal and anal fins. *S. rieppeli* documents an advanced stage in the reduction of the squamation, which otherwise characterizes *Saurichthys* species that were recovered from sediments that formed in deeper basins (e.g., *S. grignae* from the early Ladinian of the northern Grigna mountain near Lecco, Lombardy, with a total length of 130 cm and only two longitudinal scale rows, the mid-dorsal and mid-ventral one: Tintori, 2013). Since the Grenzbitumenzone formed in a restricted intraplatform basin, *S. rieppeli* was interpreted as "a transient visitor to the platform rather than a resident" (Maxwell et al., 2015:906; see also discussion in Rieppel, 1992).

Given the wealth of the material from Monte San Giorgio, some unique insights into the mode of life of *Saurichthys* could be gained. One is the potential for interspecific predation. A large specimen of *Saurichthys macrocephalus* is preserved with a small specimen of *Saurichthys*

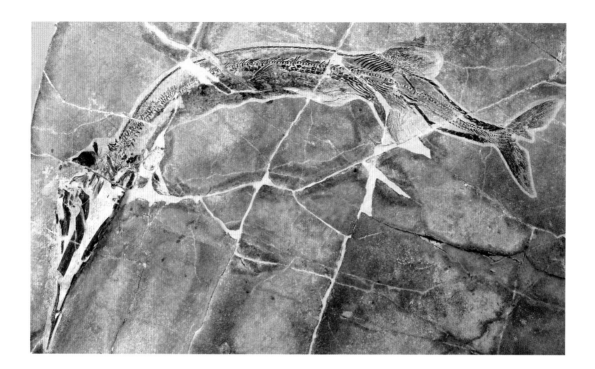

2.16. The holotype of *Saurichthys rieppeli* (Paleontological Institute and Museum, University of Zurich T61) from Mirigioli, Point 902. The total length of the specimen is 60 cm (Photo © Heinz Lanz/Paleontological Institute and Museum, University of Zurich).

curionii as its stomach content (Rieppel, 1985a:53f; see fig. 2.14). More recent excavations in the Cascina beds by the Museo Cantonale di Storia Naturale in Lugano have yielded a specimen of *Saurichthys curionii* with a stomach content comprising a small *Saurichthys*, ingested head first, and remains of four other, small, and unidentified halecomorph fishes. It is impossible to identify the ingested *Saurichthys* to the species level, but if it is a small individual of *S. curionii* (one of the only two species known from the Cascina beds), the new find would document a case of intraspecific cannibalism (Renesto and Stockar, 2015:103; see also Stockar, 2010). *Saurichthys* was described as an ambush predator capable to achieve high levels of activity if only of limited duration. Such at least can be inferred from the spiral intestine with a high spiral valve turn count (Argyriou et al., 2016).

Another interesting life history trait is evidence of internal fertilization and viviparity in *Saurichthys*. In two (possibly three) specimens of *S. curionii*, as well as in one specimen of *S. macrocephalus*, an unpaired, cylindrical, gonopodium facilitating internal fertilization is observed located in a mid-ventral position between the pelvic and anal fins (fig. 2.17a). Given its position, the gonopodium is believed to be derived from modified mid-ventral scales (Bürgin, 1990:387; see also Rieppel, 1985a). The presence of a gonopodium in males combines with the presence of embryos in females (fig. 2.17b). A gravid female of *Saurichthys curionii* contains at least seven embryonic skulls, whereas a gravid female of the larger *S. macrocephalus* contains some 16 embryos (Bürgin, 1990:348; see also Rieppel, 1985a). New excavations by the Museo Cantonale di Lugano in the Alla Cascina beds

2.17a. The gonopodium in *Saurichthys curionii* (Paleontological Institute and Museum, University of Zurich T3914; Alla Cascina), located in a mid-ventral position between the pelvic and anal fins. The gonopodium measures 13 mm in length.

2.17b. Skulls of two embryos in the abdominal cavity of *Saurichthys curionii* (Paleontological Institute and Museum, University of Zurich T3917; Alla Cascina). The embryonic skulls measure 13 mm in length (Photos © Heinz Lanz/Paleontological Institute and Museum, University of Zurich).

yielded additional gravid females of *S. curionii*, one specimen of approx. 34 cm total length containing "sixteen very small skulls" (Renesto and Stockar, 2009:327; see also Stockar, 2010). A recent comprehensive survey revealed small specimens of *Saurichthys* in the abdominal cavity of males, the latter characterized by the presence of a gonopodium. These evidently cannot be embryos and must have been the subject of predation. Special criteria are therefore required to distinguish cannibalized juveniles from true embryos. The latter are characterized by a phosphatized notochord, a position parallel to the axial skeleton and to each other, dorsal to the gastrointestinal tract, in the posterior two-thirds of the abdominal region, their skull pointing anteriorly (Maxwell, Argyriou et al., 2017). Arguments have been put forward that *Birgeria* was a viviparous fish as well, but the evidence adduced in support of this hypothesis remains unconvincing (Beltan, 1977, 1980). This leaves *Saurichthys* as the only Mesozoic actinopterygian for which undisputed direct evidence exists of a live-bearing mode of reproduction (see discussion in Bürgin, 1990:388).

Given the high level of species diversity on the genus *Saurichthys*, and its frequent occurrence throughout the Triassic, a number of authors have identified evolutionary trends within the genus that are manifest progressing from the Lower through the Middle and into the Upper Triassic (Stensiö, 1925; Griffith, 1957; Rieppel, 1985a; for a recent discussion see Romano et al., 2012). These trends concern the body elongation; the reduction of the number of discrete elements in the skull (dermatocranium); the reduction of the body squamation from a complete covering to six rows, four rows, and eventually a mere two (mid-dorsal and mid-ventral) scale rows; the disappearance of fulcral scales on the leading edge of the unpaired fins; and the reduction of lepidotrichia, both in numbers and in segmentation, in the unpaired fins. A novel developmental mechanism, namely, the doubling of the number of neural arch elements per myomeric segment, was proposed to explain body elongation in *Saurichthys* (Maxwell, Furrer et al., 2013). Drawing on experimental work in developmental genetics in modern actinopterygians, simple mutations in gene regulatory networks were invoked to explain morphological diversity in the genus *Saurichthys*, expressed in such characters as the reduction of body squamation and the reduced segmentation of lepidotrichia in the unpaired fins (Schmid and Sánchez-Villagra, 2010). Whereas these developmental explanations certainly fit the observed pattern of morphological diversification in the genus *Saurichthys*, the idea that such morphological diversification shows a trend-like pattern through the Triassic had to be abandoned in the light of the discovery of new species throughout the Triassic (Tintori, 2013). What once was interpreted as an evolutionary trend now is considered a result of convergent evolution (Maxwell, Romano et al., 2015).

Among the "small fishes" from the Middle Triassic of Monte San Giorgio (the genera are listed in table 2.1) a remarkable diversity of body shapes, locomotor adaptations and feeding mechanisms becomes apparent, evidence of a distinct radiation of actinopterygians in the western Tethyan faunal province (Bürgin, 1992; Tintori, Hitij et al., 2014:401). The

Table 2.1. "Small Fishes"—List of genera known from the five fossiliferous beds of Monte San Giorgio

'Chondrostei'	Aetheodontus
	Besania
	Bobasatrania
	Caelatichthys
	Ctenognathichthys
	Daninia
	Gracilignathichthys
	Gyrolepis
	Meridensia
	Pholidopleurus
	Platysiagum
	Ptycholepis
Neopterygii	Allolepidotus
	Altisolepis
	Archaeosemionotus
	Broughia
	Cephaloxenus
	Dipteronotus
	Ducanichthys
	Eoeugnathus
	Eosemionotus
	Furo
	Habroichthys
	Luganoia
	Legnonotus
	Ophiopsis
	Peltoperleidus
	Peltopleurus
	Peripeltopleurus
	Perleidus
	Placopleurus
	Prohalecites
	Ticinolepis
Teleostei	Gen. et sp. nov.

Source: Based on Furrer, 2003:63; Bürgin, 1999a; and Mutter and Herzog, 2004. Bürgin, 1999b:492–494, lists the fishes as distributed across the five fossiliferous levels)

taxonomy remains still rather fluid, as many new species and even genera continue to be described. A robust phylogenetic analysis of their interrelationships remains elusive to the present day (Gardiner and Schaeffer, 1989; Xu et al., 2015).

One interesting feature observed in many species of Middle Triassic "small fishes" of fusiform shape is the occurrence of deepened flank scales (fig. 2.18). Although a comprehensive and well-supported phylogenetic analysis of the interrelationships of these fishes remains unavailable, it seems that deepened flank scales evolved convergently more than once for reasons that still remain obscure. A single row of deepened flank scales, extending all the way to the caudal fin, is found in species only that grow to no more than 6 cm standard length. This may indicate some correlation of the development and distribution of deep flank scales with body size (Mutter and Herzog, 2004).

As among the species of *Saurichthys*, a tendency toward trophic habitat partitioning can be observed, for example, among the perleidid fishes (Perleididae) from Monte San Giorgio (Bürgin, 1996). The genera known from there that have been assigned to that family include *Aetheodontus*, *Colobodus*, *Ctenognathichthys*, *Meridensia*, *Peltoperleidus*, and *Perleidus*. As was detailed by Toni Bürgin, expert on Triassic "small fishes," species in these genera can be grouped into three size classes: large with a total length of 30 cm and more (*Colobodus*), medium sized with a total length of 15–30 cm, and small with a total length of less than 15 cm. And again among these species, three different types of dentition can be distinguished: a seizing type of dentition (*Peltopleurus*), a grasping type of dentition (*Peltopleurus*, *Ctenognathichthys*), and a crushing type of dentition (*Aetheodontus*, *Colobodus*, *Meridensia*) (fig. 2.19). This suggests trophic habitat partitioning among fishes that "were living in the vicinity of a richly structured habitat, such as the nearby algal-reef of Monte San Salvatore" north of the Monte San Giorgio basin (Bürgin, 1996:562; on the Middle Triassic Salvatore Dolomite and its relation to the Triassic of Monte San Giorgio, see Zorn, 1971). Interestingly, this diverse radiation of perleidid fishes disappears from Late Triassic marine deposits, by which time they seem to have been replaced by advanced neopterygians such as semionotids. Again according to Bürgin, this pattern suggests—in contrast to habitat partitioning—"a competitive model of faunal replacement by ecological displacement" (Bürgin, 1996:562).

The list given above of actinopterygian genera represented among the "small fishes" from Monte San Giorgio remains incomplete today,

2.18. A specimen of *Habroichthys minimus* (Paleontological Institute and Museum, University of Zurich T2917) from Acqua Ferruginosa, showing deep flank scales. The specimen measures 26 cm in length (Photo © Toni Bürgin/ Paleontological Institute and Museum, University of Zurich).

and new discoveries continue to be made. For example, a new genus
from the uppermost Grenzbitumenzone, *Ticinolepis*, was only recently
described, comprising two coexisting species, *T. longaeva* and *T. crassidens*
(López-Arbarello et al., 2016). The two conspecific species again provide
evidence for habitat partitioning, as one of them, *Ticinolepis crassidens*
shows a marked tritorial dentition indicating a durophagous lifestyle (fig.
2.20). Trophic habitat partitioning quite generally reduces interspecific
competition among fishes with respect to food resources, and in the case
of the two sibling species in the genus *Ticinolepis* has been interpreted as
evidence for sympatric speciation (López-Arbarello, 2016).

Body shape, ranging from elongate to fusiform to deep-bodied forms,
as well as the structure and position of fins, vary extensively among the
"small fishes" from Monte San Giorgio, suggesting a broad range of
variation. It is, perhaps, noteworthy that the description of *Bobasatrania
ceresiensis* from the Grenzbitumenzone constituted the first record of a
deep-bodied fish in Middle Triassic marine deposits (Bürgin, 1992:38ff;
see also Tintori, Hitij et al., 2014:401). Particularly mesmerizing are modifications in various ways of the anal fin in several genera, such as *Aetheodontus*, *Peltoperleidus*, *Habroichthys*, *Peltopleurus*, *Peripeltopleurus*,
and *Cephalonexus*. Except for *Aetheodontus* and *Peripeltopleurus* that
show only minor change, the modified anal fin is always characterized by
hooklet-like structures on the leading edge of the fin (Bürgin, 1992:158).
The anal fin is most extensively modified in *Peltopleurus lissocephalus*
(fig. 2.21), where it shows a tripartite structure that also involves a gonopodium of conical shape (Bürgin, 1990). Whether this is evidence for viviparity must remain conjectural in the absence of actual embryos being

2.20. The holotype of *Ticinolepis crasssidens* (Paleontological Institute and Museum, University of Zurich T273) from Mirigioli, Point 902. Note the durophagous dentition. The specimen measures 10 cm in length (Photo © Heinz Lanz/Paleontological Institute and Museum, University of Zurich).

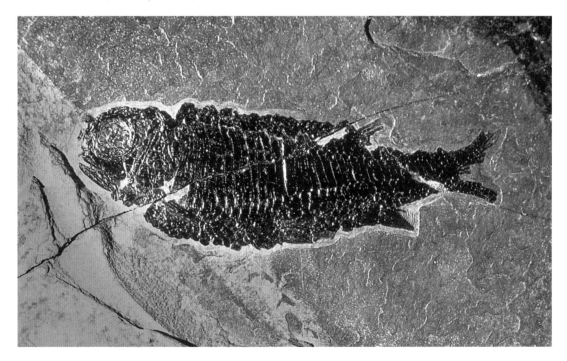

preserved. Modifications of the anal fin that involve highly specialized gonopodia are known in males of extant fishes with females that are oviparous (Bürgin, 1990:387). Still, these modified anal fins in the "small fishes" from Monte San Giorgio may have functioned in courtship and mating behavior (Bürgin, 1992:158).

2.21. A specimen of *Peltopleurus lissocephalus* (Paleontological Institute and Museum, University of Zurich T4332) from Mirigioli, Point 902. Note the modified anal fin. The specimen measures 39 mm in length (Photo © Toni Bürgin/Paleontological Institute and Museum, University of Zurich).

2.22. The holotype of *Ticinepomis peyeri* (Paleontological Institute and Museum, University of Zurich T3925a) from Mirigioli, Point 902. The specimen measures 18 cm in length (Photo © Heinz Lanz/Paleontological Institute and Museum, University of Zurich).

The Actinistians from the Middle Triassic of Monte San Giorgio

Three species of coelacanths are known from the Middle Triassic of Monte San Giorgio, of which one remains undescribed (Furrer, 2003:64). The best-known one is *Ticinepomis peyeri*, known from a single complete specimen from the upper part of the Grenzbitumenzone, collected at Point 902 (Rieppel, 1980b). It is a fish of up to 18 cm length with a relatively shallow head (fig. 2.22). The jaw articulation is placed far forward, under the eye. The scales are heavily ornamented with ganoin ridges or tubercles. The anterior fin rays in the anterior dorsal and caudal fin carry small denticles. The species looks very similar to the specimen of *Undina picnea* from the Norian of Salerno (Italy) described by Bassani in 1895 (1896), and might even be synonymous with the latter (Bassani, 1895, 1896:179; see also Rieppel, 1980b). The Italian specimen is poorly preserved, however, rendering a detailed comparison difficult. Indeed, the validity of the taxon *Undina picnea* has been questioned (Forey, 1998:361). Remains of a larger coelacanth found in the Middle Triassic (Ladinian) of the Prosanto Formation of the Ducan and Landwasser area near Davos, Switzerland, was referred to *Ticinepomis* cf. *peyeri* (Cavin et al., 2013). A recent phylogenetic analysis places *Ticinepomis* rather close with the extant coelacanth in the family Latimeriidae (Cavin et al., 2013:174).

The second actinistian described from the Middle Triassic Monte San Giorgio is based on two much less well-preserved, fragmentary specimens, again collected at Point 902. There are scattered skull remains, a somewhat poorly preserved but highly characteristic caudal fin (fig. 2.23), and scales

bearing blunt tubercles on their posterior part. On the basis of a closely similar ornamentation of the scales, the Monte San Giorgio taxon was tentatively referred to cf. *Undina picnea*, re-described by Bassani on the basis of material from the *Dolomia Principale di Giffoni* (Norian of Salerno) (Rieppel, 1985b; Bassani, 1895 [1896]). The problem here, again, is that the type material of *Undina picnea* from the Norian of Salerno is too poorly preserved to be diagnostic, a situation that is not improved by the fragmentary material recovered from the Grenzbitumenzone (Forey, 1998:361).

2.23. Part and counterpart of the caudal fin of cf. *Undina picnea* (Paleontological Institute and Museum, University of Zurich T14589) from Mirigioli, Point 902. The maximal width of the fossil-bearing slab is 17 cm (Photo © Heinz Lanz/Paleontological Institute and Museum, University of Zurich).

Notes

1. The same specimen is also illustrated in Woodward, 1889, p. 293, fig. 10, again under the name of *Acrodus anningiae*, but Woodward recognized the synonymy of *Acrodus undulatus* with *A. anningiae* implied by Agassiz (Woodward, 1889:289). Woodward apparently did not consider the different publication dates of separate fascicules and plates of Agassiz (1833–44).

2. Memoirs of the Geological Survey of the United Kingdom. Figures and Descriptions Illustrative of British Organic Remains, Decade VIII. London: Brown, Green, and Longmans. 1855. Decade VIII, Plate III, followed by a two-page description. The figure is reproduced in Woodward, 1916, p. 17, fig. 8.

3. Bassani (1886:31) is the first published occurrence of the species name *Acrodus bicarinatus*. The specific epithet is spelled *bicarenatus* in Alessandri (1910:34), Deecke (1926, p. 26), and Kuhn-Schnyder (1945:664).

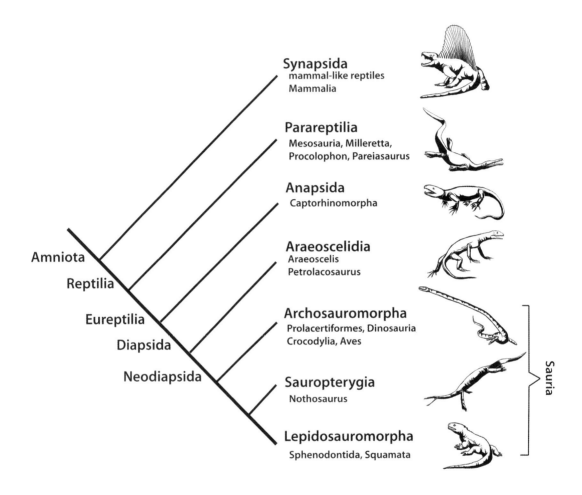

Synapsida
mammal-like reptiles
Mammalia

Parareptilia
Mesosauria, Milleretta,
Procolophon, Pareiasaurus

Anapsida
Captorhinomorpha

Araeoscelidia
Araeoscelis
Petrolacosaurus

Archosauromorpha
Prolacertiformes, Dinosauria
Crocodylia, Aves

Sauropterygia
Nothosaurus

Lepidosauromorpha
Sphenodontida, Squamata

Amniota

Reptilia

Eureptilia

Diapsida

Neodiapsida

Sauria

3.1. The current
understanding of amniote
relationships (as discussed in
the text; artwork by Marlene
Donnelly, the Field Museum,
Chicago).

A Sketch of Reptile Evolution

3

With their descendants, the birds, included, the reptiles (Reptilia) form the most species-rich monophyletic clade among tetrapods. This is true even if some groups of reptiles, such as the rhynchocephalians (to which belongs the extant tuatara, *Sphenodon punctatus*, from New Zealand), turtles, and crocodiles, are represented in modern biota by a relatively small number of species given their rich evolutionary past. Others like the squamates (the "lizards" and "snakes") presumably are at or near their evolutionary peak today, with more than 10,000 extant species described, distributed across the globe except in cold climate zones (Uetz and J. Hosek (eds.), The Reptile Database, http://www.reptile-database.org, accessed April 7, 2016). But overall, and not counting birds, the reptiles reached the greatest degree of taxonomic diversification during the Mesozoic, which is also known as the "Age of Reptiles." The rich fossil record collected in the Middle Triassic marine deposits of Monte San Giorgio, augmented today by the incredible riches of late Lower, Middle, and early Upper Triassic marine reptiles collected in southern and southwestern China, offer unique insights into the rapid diversification of marine reptiles, the initial invaders of the Triassic seas, and therewith into the reconstitution of the marine biota after the end-Permian mass extinction (Tintori and Felber, 2015). But to better understand reptilian diversification during the Mesozoic, it is helpful to first sketch the phylogeny of the reptile clade, and the characteristics of its major subgroups.

When talking about phylogeny and clades, the concept of monophyly moves center stage, since only monophyletic groups, or clades, are considered natural. A monophyletic taxon includes its most recent common ancestor, and all, and only, its descendants. When birds are found to have descended from theropod dinosaurs, the birds must therefore be included in the Reptilia in order to render that clade monophyletic. Understood in this way, Reptilia is one of the two major clades of amniotes. The other major amniote clade, sister group of the Reptilia, is the Synapsida, which includes mammals and their fossil relatives (fig. 3.1).

The Paleozoic stem-reptiles had previously been classified as Anapsida (Williston, 1917). In that context, the term "anapsid" acquired two different meanings. One characterizes a reptile as one that belongs to the Anapsida. The other points to the anatomical structure of the skull that characterizes the Anapsida. The predicate "anapsid" is used today only as an anatomical term, as it applies to a reptile skull in which the temporal or cheek region of the skull (dermatocranium), located behind the eye socket, is not fenestrated. Taxonomically, the Anapsida have

An Outline of
Reptile Phylogeny

been recognized as a paraphyletic assemblage. Anapsid Paleozoic stem-reptiles are today classified in two separate clades, the Parareptilia and the Captorhinomopha. The Parareptilia includes a number of lineages, comprising the saltwater-inhabiting mesosaurs, the herbivorous pareiasaurs, the lizard-like insectivorous millerettids, and the procolophonids, the latter once claimed to be related to turtles (Laurin and Reisz, 1995). Alternatively, the pareiasaurs have been considered to be close to turtle ancestry as well (Lee, 1993, 1996). Of all the parareptiles, only the terrestrial procolophonids extend into the Triassic. The parareptiles are consequently of no importance for the marine Middle Triassic biota of Monte San Giorgio/Besano.

The Paleozoic (Pennsylvanian and Permian) Captorhinomorpha are an anapsid stem-lineage of Eureptilia, most of them relatively small, terrestrial, and insectivorous reptiles that form the sister group of the Diapsida. They too have at one time been related to turtle origins (Gaffney and McKenna, 1979). Clearly, the vast majority of reptiles—extant and extinct—belong to the Diapsida, a term first introduced by Osborn in 1903 (Osborn, 1903). Like Anapsida originally, the terms "Diapsida" and "diapsid" have a dual meaning, namely, as the name of a monophyletic taxon, and as an anatomical term referring to the structure of the skull. In diapsids, the temporal region of the skull (dermatocranium) sports two openings, or temporal fenestrae, located behind the eye socket, an upper and a lower one. The upper temporal fenestra is separated from the lower temporal fenestra by the upper temporal bar, composed of the postorbital anteriorly, and the squamosal posteriorly. The lower temporal fenestra is closed ventrally by the lower temporal bar, composed of the jugal anteriorly, and the quadratojugal posteriorly. The most basal diapsids are the Araeoscelidia, relatively small, lizard-like animals with slender limbs and a long tail from the Late Carboniferous and Early Permian. *Araeoscelis* is interesting as it has an upper temporal fenestra only, whereas the lower cheek region is closed. It remains unknown whether this represents a primitive diapsid condition or a secondary closure of the lower temporal fenestra. The functional reasons for the development of temporal fenestration in diapsids remain rather speculative to the present day.

The Araeoscelidia is the sister group of the Neodiapsida, that is, all remaining diapsids. There are a number of fossil forms, from the Permian and Permo-Triassic, that are neodiapsids but fall outside the large archosauromorph-lepidosauromoph dichotomy that is embedded within the Neodiapsida. The term Neodiapsida refers to this greater assemblage, whereas use of the term Sauria is restricted to the clade comprising archosauromophs and lepidosauromorphs. There is some evidence that the lower temporal arch was lost early in neodiapsid evolution, such that the occurrence of a complete lower temporal arch in archosauromorphs and lepidosauromorphs is a reversal, that is, a secondary condition, not the primitive one (Müller, 2003). And again, it is within the Sauria so conceived that we find the greatest diversification of reptiles, living and fossil. Indeed, with turtles now accepted as diapsids, there are no extant reptiles falling outside the Neodiapsida, or even the Sauria (Schoch and Sues, 2015; Bever et al., 2015; for a recent review see Burke, 2015).

The Sauria divide into two large clades, the Archosauromorpha and the Lepidosauromorpha. The latter clade includes the very species-rich scaly reptiles, the Squamata (the lizards, amphisbaenians, and snakes), the tuatara (*Sphenodon punctatus*) and its fossil relatives (the Rhynchocephalia), and an enigmatic group of "gliding," lizard-like creatures from the Late Triassic (Norian) of southwest England and eastern North America, the Kuehneosauridae. These forms were characterized by elongated ribs that supported a patagium; they were not active flyers but rather passive gliders, much like the extant agamid lizards in the genus *Draco*. The fossil record of rhynchocephalians stretches back to the Middle Triassic; for squamates, there is no confirmed occurrence in the Triassic, their fossil record reaching back to the Lower Jurassic (Evans, 2003; Bonaparte and Sues, 2006; Hutchinson et al., 2012; Jones et al., 2013).

Taxonomically, the archosauromorph lineage is far more diverse than the lepidosauromorph lineages, as it includes readily recognized forms of Archosauria such as crocodiles (fossil and extant), dinosaurs, and birds. Among basal archosauromorphs, that is, placed outside the Archosauria, nest the Protorosauria. The protorosaurs (sometimes also referred to as prolacertiforms) constitute an important radiation of terrestrial and marine Permo-Triassic reptiles with a worldwide geographic distribution. The relationships of protorosaurs within advanced diapsid reptiles (Sauria) are still the subject of debate, but the current consensus is that protorosaurs nest at the base of the archosauromorph clade. The monophyly of the Protorosauria has recently been called into question, as the Permian genus *Prolacerta* from South Africa and Antarctica may share a more recent common ancestor with basal archosauriforms such as *Euparkeria* and *Proterosuchus*, rather than with other protorosaurs (Dilkes, 1998; Ezcurra, 2016). At the same time, protorosaur monophyly seems to require the inclusion of drepanosaurs, which in turn include such strange forms from the Late Triassic of northern Italy as *Drepanosaurus unguicaudatus*, characterized by an enormous claw on the second digit and a spine at the tip of the tail, and *Megalancosaurus preonensis*, which is characterized by a delicate head with a pointed snout, and opposable digits in hands and feet that indicate arboreal habits (Pinna, 1980, 1984, 1993; Benton and Allen, 1997; Dilkes, 1998; Renesto, 2000; Renesto, Spielmann et al., 2010; but see Pritchard and Nesbitt, 2017, for an alternative view). Clearly, the monophyly of protorosaurs, as also their position at the base of Archosauromorpha, requires further testing (Dilkes, 1998; Rieppel, Fraser et al., 2003; Ezcurra, 2016; Pritchard and Nesbitt, 2017).

Other basal archosauromorphs include the Choristodera, the Rhynchosauria, and the Trilophosauria (Brochu, 2001). The lineage of the semiaquatic Choristodera, also sometimes called champsosaurs, extends from the Late Triassic to the early Miocene (Gao et al., 2000; Evans and Klembara, 2005). They were convergent on crocodiles, with slender and much elongated jaws, and broadly expanded temporal arches. Their distribution extends across North America, Europe and Asia into Japan. Interestingly, "choristoderes were the only major group of Mesozoic reptiles that survived the K-T transition event [the Cretaceous-Tertiary extinction event] and then became extinct" (Gao et al., 2000:417).

The herbivorous, heavily built rhynchosaurs were globally distributed during the Triassic. The skull is wedge-shaped, broad posteriorly, pointed anteriorly, with the premaxillaries projecting downward to form a distinct beak. Multiple densely set rows of blunt teeth in the upper and lower jaw were used in grinding up plant material. Large claws on the digits of the hind feet might have been used in digging up roots (Dilkes, 1998; Benton, 2005). Trilophosaurs are another group of Triassic herbivores, which during the Late Triassic seem to have occupied the ecological niche in North America that rhynchosaurs occupied in other parts of the world (Heckert et al., 2006).

More derived archosauromorphs than these stem-group taxa are included in the monophyletic Archosauriformes (Nesbitt, 2011). The proterosuchids nest at the base of the archosauriforms, Late Permian and Early Triassic forms characterized by a downturned premaxilla and a crocodiloid body. The erythrosuchids were large, heavy-set predators that occur in the Lower Triassic. The Proterochampsidae from the Late Triassic of South America are of uncertain relationships within archosauriforms (Carroll, 1988). They were "quadrupedal carnivores with a powerful bite" (Brochu, 2001:1187). The genus *Euparkeria* from the Early Triassic of South Africa has in the past played a pivotal role in paleontologists' understanding of early archosaur phylogeny (Gower and Weber, 1998; Sobral et al., 2016). Once considered the sister taxon of the crown-group Archosauria (Brochu, 2001, fig. 1), a more recent phylogenetic analysis found it to represent a more basal archosauriform (Nesbitt, 2011). With its hind limbs distinctly longer than its fore limbs, it might have been facultative bipedal (Carroll, 1988:271).

The Archosauria proper divide into two major clades, the Pseudosuchia or "crocodylomorph" lineage comprising the crocodiles and their fossil relatives, and the Avemetatarsalia comprising the birds and their fossil relatives (Benton, 1999; Nesbitt, 2011; Nesbitt, Butler et al., 2017). The "rauisuchids" from the Middle Triassic (Anisian) are a group of particular importance for the Monte San Giorgio fauna, where it might be represented by *Ticinosuchus*; the putative monophyly of Rauisuchia remains controversial, however (Brochu, 2001:1187). They were considered the earliest representatives of crown-group Archosauria that join other basal archosaurs such as the Phytosauria and Aetosauria on the crocodylomorph stem (Brochu, 2001, fig. 2). A more recent phylogenetic analysis found the Rauisuchia to be non-monophyletic, however, with *Ticinosuchus* and *Rauisuchus* nesting at different places on the crocodylomorph stem (Nesbitt, 2011). The phytosaurs (Upper Triassic) are rather closely convergent on crocodiles in body form and lifestyle, but unlike crocodiles, the external nares are retracted to a posterior position, located closely in front of the eyes in phytosaurs. The aetosaurs (Upper Triassic) were quadrupedal herbivores with a delicate dentition comprising "small, leaf-shaped teeth"; their body was covered with an extensive osteoderm armor (Carroll, 1988:273). "Protosuchians," long-limbed terrestrial forms from the Late Triassic and Early Jurassic, approach crocodiles in their skull structure, except for a still rather anterior placement of the internal nares in the dermal palate. The "mesosuchians," with terrestrial and aquatic forms ranging from the Lower Jurassic into the Upper Cretaceous, have loosely been interpreted as "intermediate" between "protosuchians" and the Eusuchia, the true ("modern-style") crocodiles, as their internal nares are more posteriorly positioned, but not quite as far back yet as in modern crocodiles (Carroll, 1988:282). Neither the "protosuchians" nor the "mesosuchians" constitute monophyletic lineages, however; they rather represent successive grades in the evolution of crocodylomorphs.

Pterosaurs, which first appear in the Upper Triassic (Norian) of northern Italy, represent the earliest actively flying vertebrates. They radiated extensively throughout the Jurassic and Cretaceous, and included the largest flying animals ever to be known. They are considered as a basal clade on the avemetatarsalian lineage (Nesbitt, Butler et al., 2017), although this hypothesis has recently been challenged by controversially relating pterosaurs to protorosaurs (prolacertiforms) (Peters, 2000; see the review and discussion in Brochu, 2001:1192).

The Late Triassic (Carnian) genus *Scleromochlus* from Scotland is generally placed as the basal-most dinosauromorph (Benton, 1999). Dinosaur (Dinosauria) evolution and diversification in general proceeded along two major lineages, the Ornithischia and the Saurischia. Some early forms such as *Eoraptor* (Carnian, Late Triassic of Argentina), and *Herrerasaurus* (again Carnian, Late Triassic of Argentina) have not been conclusively classified as either saurischian or ornithischian and may well lie outside these two clades (see the review in Brochu, 2001). This traditional classification has recently been challenged, however, in that theropod dinosaurs were found most closely related to the Ornithischia, while *Herrerasaurus* grouped with the Sauropodomorpha (Baron et al., 2017). As is amply corroborated by feathered dinosaurs from the Lower Cretaceous Jehol biota of Liaoning Province, northeast of Beijing, birds originated from advanced theropods, the Maniraptora, a crown-clade in the saurischian lineage.

The above sketch of reptile evolution so far eschewed most Mesozoic marine reptiles. There are, of course, marine representatives of rhynchocephalians (the Lower Jurassic pleurosaurs), of squamates (the Cretaceous aigialosaurs, dolichosaurs, and mosasaurs), of Triassic protorosaurs (such as those from Monte San Giorgio, to be discussed later), and of Mesozoic crocodiles, the Jurassic teleosaurs and metriorhynchids (Steel, 1973; Carroll, 1985a; DeBraga and Carroll, 1993; M. T. Young et al., 2010). But in addition, there is a diverse radiation of Triassic marine reptiles that group into four major clades, the ichthyosauromorphs, the sauropterygians, the thalattosaurs, and the saurosphargids; only the first two of these clades survived beyond the Triassic, with representatives in the Jurassic and Cretaceous. Especially the post-Triassic ichthyosaurs and sauropterygians are easily recognized and widely known clades, whereas the Thalattosauria and Saurosphargidae, as well as the basal ichthyosauromorphs from the late Early Triassic, are more enigmatic groups, which remain generally unknown to the broader public. The phylogenetic relationships of these four clades of marine reptiles, whether archosauromorph or lepidosauromorph, remain to some extent debatable (Motani, Jiang, Chen et al., 2015; Schoch and Sues, 2015).

Recent discoveries and work on Early Triassic stem-ichthyosauromorphs from southern China led to the recognition of an expanded clade named Ichthyosauromorpha, of which the Ichthyopterygia and Ichthyosauria are successively nested subgroups; the oldest, short-snouted ichthyosauromorphs are from the Spathian of Chaohu in Anhui Province (Motani, Jiang, Chen et al., 2015; Jiang, Motani, Huang et al., 2016). The Ichthyosauria proper diversified in the Middle Triassic (middle Anisian) at a time when the stem-ichthyosauromorphs from the Lower Triassic had already become extinct. Ichthyosaurs in turn became extinct at the end of the Cenomanian (mid-Cretaceous), that is, before the end-Cretaceous mass extinction (V. Fischer et al., 2016). The standard picture of an ichthyosaur is that of a marine reptile convergent on dolphins in body shape. Especially in advanced representatives of the group from the Lower Jurassic onward, the caudal vertebral column shows a distinct downward turning tailbend indicating the presence of a vertically positioned caudal fin. The same is confirmed by specimens with skin preservation, known from the lower Toarcian (Lower Jurassic) deposits of Holzmaden in southern Germany; the same specimens also confirm the presence of a dorsal fin devoid of bony support structures (Hauff and Hauff, 1981). The limbs of ichthyosaurs were transformed into flippers and the jaws drawn out into an elongated rostrum. The skull in ichthyosaurs is diapsid, but the lower temporal arch is absent. The cheek region below the upper temporal fossa is nevertheless more or less covered by bone, due to a shortening of the postorbital skull and a broadening of the postorbital skull elements. Ichthyosaurs are known to have been viviparous, which accords well with their marine lifestyle. The tail-first birth posture of ichthyosaurs, convergent on whales and sea cows, is believed to be a secondary feature. A gravid female of one of the oldest marine reptiles, the ichthyosauromorph *Chaohusaurus* from the Lower Triassic (Spathian) of Chaohu, Anhui Province (China), shows an articulated embryo in a head-first birth position (Motani, Jiang, Tintori et al., 2014). Ichthyosaurs were certainly pelagic animals, roaming the open sea, except for the smaller forms in the Triassic, which may have preferred habitats closer to the coastline. Ichthyosaur diversification progressed rapidly in the Triassic, with large-bodied forms in existence by the end of the Olenekian (late Early Triassic) already (Scheyer, Romano et al, 2014; T. Watson, 2017). The largest ichthyosaurs, with an estimated total length of 20 meters, appeared in the Norian (Late Triassic) of British Columbia (Nicholls and Manabe, 2004). A wide range of dentition types, including durophagous specialization, as well as differences in overall body size suggest a wide range of trophic adaptations among ichthyosaurs that suggests habitat partitioning among the species (Ji, Jiang, Motani et al., 2016, and references therein).

The earliest undisputed occurrence of a sauropterygian is again in the Early Triassic (Spathian) of Chaohu, Anhui Province, China (Jiang, Motani, Tintori et al., 2014). The sauropterygians underwent a wide diversification during the middle Triassic, most forms becoming extinct in the Late Triassic (Rieppel, 2000a). The major clades among the Triassic sauropterygians were the placodonts, the pachypleurosaurs, the nothosauroids, and the pistosauroids. A multitude of new taxa recovered from the Middle and Late Triassic of southern and southwestern China complicates this picture, however, such that a robust new phylogenetic analysis of all the known taxa still remains to be completed. Only one lineage survived into the Jurassic, the Plesiosauria, which underwent a second radiation during the Jurassic and Cretaceous (O'Keefe, 2001). Together with the dinosaurs, the Plesiosauria went extinct during the end-Cretaceous mass extinction. Whereas the Triassic sauropterygians were inhabitants of coastal waters, intraplatform basins, or shallow epicontinental seas, the plesiosaurs were pelagic animals, their limbs transformed to form flippers. Along with some basal lineages including *Plesiosaurus*, the Plesiosauria comprises four major derived clades—the Rhomaleosauridae and Pliosauridae (within the Pliosauroidea), as well as the Cryptocleidoidea and Elasmosauridae (within the Plesiosauroidea) (O'Keefe, 2001:30; but see also Benson and Druckenmiller, 2014).

The placodonts (Placodontia) are a clade of medium- to large-size reptiles characterized by a highly derived dentition comprising large crushing tooth plates on the upper (maxilla) and lower jaws (dentary), and on the dermal palate (palatine). Procumbent anterior teeth (if present) probably served to remove hard-shelled invertebrate prey from the substrate. Among the placodonts there were an unarmored (Placodontoidea) and an armored (Cyamodontoidea) lineage, the latter distinctly more species-rich than the former. The body-armor of a cyamodont placodont consists of a carapace-like dorsal shield, composed of osteoderms, to which may be added a separate tail shield covering the pelvic and proximal tail regions, and a plastron-like ventral shield. A small skull only 20.5 mm long from the Middle Triassic (Anisian) of Winterswijk (the Netherlands) was interpreted as representing a juvenile stem-placodont (Neenan, Klein et al., 2013). This species, named *Palatodonta bleekeri*, had not yet acquired the blunt crushing dentition so characteristic of placodonts. Instead, it had conical pointed teeth, set in two parallel rows on the maxilla and on the enlarged palatine, in between which fit the single tooth row on the dentary at jaw closure. The occurrence of this stem-placodont in marine Triassic (Anisian) deposits in the Netherlands led to the suggestion that placodonts originated in the western Tethyan faunal province, from where they dispersed eastward into the eastern Tethys, where today they are found in Middle and Late Triassic (Anisian and Carnian, respectively) deposits in southwestern China (Zhao, Li et al., 2008). Placodonts are absent, or at least have so far never been found, in the eastern Pacific faunal province (western North America).

The small- to medium-sized pachypleurosaurs were lizard-like creatures foraging in shallow water. Some species, such as *Keichousaurus hui* from the early Late Triassic of southwestern China, or *Neusticosaurus* ssp. from the Monte San Giorgio, are known from hundreds of specimens, comprising embryos as well as gravid females (e.g., Sander, 1989a; Y.-N. Cheng, Wu and Ji, 2004). Many species of pachypleurosaurs, when adequately known, show distinct sexual dimorphism expressed in humerus proportions and body size, the males being somewhat larger and more robustly built than the females (Y.-N. Cheng, Holmes et al., 2009). In contrast to the placodonts, it has been suggested that the pachypleurosaurs originated in the eastern Tethys, from where they dispersed westward. They would have entered the Anglo-Germanic basin through the Carpathian Gate in the wake of the late Olenekian–Anisian marine transgression, which itself progressed from east to west within the Anglo-Germanic basin. Through the Burgundian Gate they would have entered the intraplatform basins to the south, such as the Monte San Giorgio (Grenzbitumen) basin (Rieppel, 1999a:10; see also Rieppel and Hagdorn, 1997). New material from the late Lower, Middle, and early Upper Triassic of southern China blurs this neat picture, however, and when included in phylogenetic analyses, even called the monophyly of Pachypleurosauridae as previously understood into question (see Holmes et al., 2008; J. Liu, Rieppel, Jiang et al., 2011; Jiang, Motani, Tintori et al., 2014; Ma et al., 2015). Many Chinese marine reptiles show character combinations that disrupt tree topologies that had been based on European (western Tethyan) material only.

Tree topologies of Triassic sauropterygians quite generally remain rather unstable in phylogenetic analyses that include the wealth of the new Chinese taxa. Such is to be expected if you consider Chinese fossils like *Wangosaurus brevirostris* from the Middle Triassic (Ladinian) from Wusha, which essentially sports a nothosaur skull on a pistosaur neck, or nothosaurs that combine features otherwise known from two genera, *Nothosaurus* and *Lariosaurus*, which in most cases are easily recognized as distinct and different in the western Tethyan faunal province (J.-L. Li, Liu et al., 2002; Rieppel, J.-L. Li et al., 2003; J.-L. Li and Rieppel, 2004; Ma et al., 2015; for a more detailed account see chap. 11). Generally within the nothosauroid clade, there is an evolutionary trend from a relatively high skull with a short snout and a moderately elongated temporal region (as in *Simosaurus* from the Germanic Muschelkalk, Ladinian) toward a very flat skull with an elongated rostrum and a very distinct elongation of the temporal region (as in advanced members of the genus *Nothosaurus*). These changes in skull proportions required a reorganization of the jaw adductor musculature to maintain functional efficiency (Rieppel, 2002a; but see Araújo and Polcyn, 2013, for a dissenting view). At the same time, anteriorly positioned teeth became elongated and strongly procumbent, functioning as a fish trap. Given the low height of the skull and the elongated rostrum, fish most likely were caught with a rapid sideways strike. Among the western Tethyan nothosauroids, this evolutionary trend is exemplified by the genera *Simosaurus* (Germanic Muschelkalk, Ladinian), *Germanosaurus* (Germanic Muschelkalk, early Anisian), *Nothosaurus* (western Tethyan faunal province, Middle to early Upper Triassic), and *Lariosaurus* (western Tethyan faunal province, late Anisian and Ladinian). The postcranial skeleton is more conservative in this clade. The front limbs tend to be longer and more robustly built than the hind limbs; a mild degree of hyperphalangy may have developed among the most advanced nothosauroids. There is, again, evidence of habitat partitioning among nothosaurs, particularly well expressed in the Makhtesh Ramon (Negev, Israel) deposits of late Anisian or early Ladinian age. There, two species of the genus *Nothosaurus* coexisted that show marked differences in both size and dentition (Rieppel, Mazin et al., 1997).

The pistosauroid clade is the only Triassic sauropterygian lineage that is represented in the New World, where only two species have been recorded of Triassic age. The earliest known pistosauroid is *Corosaurus alcovensis* from the Alcova Limestone (late Spathian, Lower Triassic, or earliest Anisian, Middle Triassic) of Casper, Wyoming (Storrs, 1991a; Rieppel, 1998a; Lovelace and Doebbert, 2015). *Corosaurus* is related to the genus *Cymatosaurus* from the western Tethyian faunal province, the latter marking the early occurrence (lower Anisian) of pistosauroids in the Germanic Triassic (Muschelkalk). The second Triassic New World sauropterygian known is the pistosauroid *Augustasaurus hagdorni* from the Middle Triassic (late Anisian) of Augusta Mountains, northwestern Nevada (Sander, Rieppel et al., 1997; Rieppel, Sander et al., 2002). *Augustasaurus* is related to the genus *Pistosaurus* from the western Tethyan faunal province (Germanic Muschelkalk, upper Anisian), both distinctly larger than the earlier occurring pistosauroids *Corosaurus* and *Cymatosaurus*. Complete skeletons of large pistosaurs have recently also been reported from China: *Yunguisaurus* from the late Middle Triassic of Chajiang, Guizhou Province, and *Wangosaurus* from the late Middle Triassic of Xingyi, Guizhou Province (N.-Y. Cheng, Sato et al., 2006; Sato et al., 2014a; Ma et al., 2015). With only three known individual occurrences in the Germanic Muschelkalk, *Pistosaurus* is a rare faunal element in the western Tethyan faunal province. A detailed analysis of these occurrences shows that *Pistosaurus* sought out deeper waters than other Muschelkalk sauropterygians (Dr. Hans Hagdorn, Ingelfingen [Germany]: personal communication). This is interesting, since the pelagic plesiosaurs, with a first occurrence near the Triassic–Jurassic boundary, but with an origin possibly going as far back as the early Carnian (Fabbri et al., 2014), are thought to nest within the pistosauroids, possibly close to *Pistosaurus*.

The thalattosaurs (Thalattosauriformes, Thalattosauria) form a rather enigmatic clade of Triassic marine reptiles (Nicholls, 1999). Thalattosaurs are known from the western Tethys, from western North America (eastern Pacific faunal province), and from the eastern Tethys (southwestern China), where they are particularly speciose and abundant. The earliest occurrence of thalattosaurs is in the Lower to Middle Triassic of British Columbia, western North America; their latest remains are found in Carnian to Norian deposits at several localities along the western coast of North America, from Nevada and California to British Columbia and Alaska (Bardet et al., 2014). The most basal thalattosaurs, the askeptosauroideans known from the western Tethys and southwestern China, superficially resemble slender, medium-sized crocodiles in body shape, the jaws drawn out into an elongate and slender rostrum furnished with a palisade of pointed, slightly recurved teeth ideal for capturing free-swimming prey such as cephalopods and fish. More derived thalattosaurs such as claraziids vary in size but generally remain smaller, growing to not much more than one meter in length. The skull is drawn out into a relatively shorter rostrum that converges to a narrow tip anteriorly. Their durophagous dentition is adapted to crushing hard-shelled invertebrate prey. Some claraziids sport remarkable specializations in their feeding mechanism, which will be discussed in greater detail below, with reference to a thalattosaur from Monte San Giorgio (Rieppel, Müller et al., 2005). Within the thalattosaur clade represented by the genus *Xinpusaurus* from the Ladinian and Carnian (Middle to Late Triassic) of southwestern China there is again evidence of habitat partitioning, different species within the same genus (or in two very closely related genera) differing predominantly in size and dentition.

The saurosphargids (Saurosphargidae) are an even more enigmatic clade of Triassic marine reptiles. The genus and species *Saurosphargis voltzi* was based on a fragmentary trunk skeleton from the lowermost Germanic Muschelkalk (lower Gogolin Formation, lower Anisian of Upper Silesia, now Poland; see Rieppel, 2000a, for more discussion). The specimen was accessioned in the Museum of Breslau (now Wroclaw), but has been lost since World War II. In 1936, Friedrich von Huene published excellent photographs of the specimen together with a description, and credited Fritz Drevermann as author of the name, without indication of a published source, however (Huene, 1936). Although very incomplete, the fossil still shows highly characteristic features, which are much-elongated transverse processes of the dorsal vertebrae to which articulate ribs with a distinct uncinate process, the latter associated with osteoderms. It is for that reason that *Saurosphargis* was believed by Bernhard Peyer to possibly represent a placodont (Huene, 1936; Peyer and Kuhn-Schnyder, 1955a). Today, complete skeletons with skulls are known from the Anisian of Xinmin District, Panxian County, Guizhou Province, southwestern China. These specimens very closely resemble *Saurosphargis*, so much so that their classification as a separate genus remains debatable. With their skull and dentition, however, they prove that *Saurosphargis* is not a placodont, indeed not even a sauropterygian (C. Li, Jiang, Cheng et al., 2014, and references therein). A small saurosphargid (*Eusaurosphargis dalsassoi*) is known from the Besano Formation, a taxon that has so far not been reported from the Monte San Giorgio (Nosotti and Rieppel, 2003). A complete specimen of *Eusaurosphargis dalsassoi* has recently been reported from the slightly younger (Ladinian) Prosanto Formation of the eastern Swiss Alps (Scheyer, Neenan et al., 2017). Saurosphargid remains (aff. *Eusaurosphargis*) have also been reported from the lower Muschelkalk (early Anisian) of Winterswijk, the Netherlands (Klein, 2009:666; Sander, Klein et al., 2014:57; Klein and Sichelschmidt, 2014). The most intriguing representative of the clade no doubt is *Sinosaurosphargis* from the Middle Triassic (Anisian) of Luoping, Yunnan Province, China. It is an animal at least 70 to 80 cm long (without the missing tail) that in a curious way mimics turtles (C. Li, Rieppel, Wu et al., 2011; Hirasawa et al., 2013; see chap. 11 for more discussion).

Other than the more pelagic and hence more widely distributed ichthyosaurs, the Triassic marine reptiles mostly occur in three faunal provinces (Rieppel, 1999a). The Monte San Giorgio biota is part of the western Tethyan faunal province, with fossiliferous localities known in central and southern Europe, and circum-Mediterranean territory. Among those many localities, the most important ones are the Middle Triassic Germanic Muschelkalk, Monte San Giorgio and the various Middle and Late Triassic localities in Lombardy (northern Italy), followed perhaps by the Middle Triassic Makhtesh Ramon locality in the Negev (Israel). The record of Triassic marine reptiles from Turkey, Spain, and Tunisia remains rather scanty; the specimens recovered are generally of poor quality (Rieppel, 1999a, 2000a). More potentially very important Middle Triassic fossiliferous outcrops remain unexplored in the Middle East (e.g., Saudi Arabia, Jordan) (Thomas H. Rich, Eitan Chernov, personal communications; see also Vickers-Rich et al., 1999). The marine deposits in the western Tethyan realm yielding reptiles span the period from the uppermost Lower Triassic (uppermost Olenekian, or Spathian) to the Late Triassic (Rhaetian) (Rieppel, 2000a).

The eastern Tethyan faunal province is largely represented by deposits that formed on the Yangtze Plate, also known as the South China Block, which at that time (Anisian) started to slowly dock on to the main Asian Plate. Important localities at which excavations proceed today are located in southwestern China, in Guizhou (Guanling Biota; Xingyi and Panxian Faunas), Yunnan (Luoping Fauna), and Anhui (Chaohu Fauna) Provinces (X.-F. Wang et al., 2008; Jiang, Motani, Hao et al., 2009; Stone, 2010; Benton et al., 2013). Add to this record a smattering of other localities in Hubei Province, Guangxi Zhuang Autonomous Region, and Xizang (Tibet) Autonomous Region, most of those only yielding very rare reptile fossils of generally poor quality (J.-L. Li, 2006). The exception is the occurrence of well-preserved representatives of the Hupehsuchia in the late Lower Triassic (Spathian) of Yuan'an and Nanzhang Counties, Hubei Province (Chen, Motani et al., 2014a, 2014b). Collectively, these localities cover the time period from the late Lower Triassic (Olenekian, Spathian) to the Norian.

Triassic marine reptiles of the eastern Pacific faunal province are found in several localities in Wyoming, Nevada, northern California, and British Columbia. One of only two sauropterygians known from the New

World is *Corosaurus alcovensis* from the Alcova Limestone near Casper, Wyoming. The Alcova Limestone has been dated as uppermost Lower Triassic (upper Spathian), or lowermost Anisian (Storrs, 1991a; Lovelace and Doebberd, 2015). The second sauropterygian known from the New World is *Augustasaurus hagdorni* from the late Anisian of northwestern Nevada, deposits that otherwise have yielded a variety of ichthyosaurs (Sander, Rieppel et al., 1997; Fröbisch et al., 2013). Earlier, late Spathian (late Lower Triassic) ichthyosaurs with affinities to contemporaneous faunal elements in Japan and Spitsbergen come from the Prida Formation at Fossil Hill, northwestern Nevada (Kelly et al., 2016). Late Carnian deposits near Berlin (Nevada), yielded large ichthyosaurs on display at the Berlin-Ichthyosaur State Park, some specimens reaching an overall length of close to 50 feet (15 meters). Ichthyosaurs and thalattosaurs have been collected in the Late Triassic (Carnian, Norian) Hosselkus Limestone of northern California (Hilton, 2003). Farther north, outcrops of the Lower Triassic (Griesbachian to Spathian) Vega-Proso Siltstone Member of the Sulphur Mountain Formation near Wapiti Lake in British Columbia (Canada) have yielded ichthyosaurs and thalattosaurs (e.g., Nicholls and Brinkman, 1993a, 1993b). Thalattosaur remains from the Upper Triassic (Carnian) Natchez Pass Formation of the Humboldt Range in northwestern Nevada currently remain undescribed (Storrs, 1991b). The youngest occurrence of thalattosaurs is in the Norian (Upper Triassic) Pardonet Formation limestone of Pink Mountain, British Columbia (Storrs, 1991b).

Other Triassic marine reptile localities not covered in the above account mostly concern ichthyosaurs that have been found in Tibet, Japan, and Spitsbergen (Svalbard). Among these localities, Spitsbergen (Svalbard) is historically certainly the most important one (Harland, 1997).

Adaptations to a Life in the Sea

"The mode of life of the marine iguana [*Amblyrhynchus cristatus*] has entailed surprisingly little departure from capacities for activity and locomotion characterizing terrestrial iguanids" (Dawson et al., 1977:894). Indeed, a comprehensive review of physiological attributes of secondarily aquatic (also marine) reptiles concluded: "Reinvasion of aquatic habitats by members of previously terrestrial lines is not surprising because the basic reptilian design incorporates many physiological properties that would well serve aquatic behaviour" (Seymour, 1982:1). The Mesozoic saw a remarkable diversification of (secondarily) marine reptiles; the earliest representatives of which start to show up in the fossil record in the Spathian (late Lower Triassic). Invading the sea required physiological adaptations with respect to three major factors: buoyancy, sodium and water balance, and gas exchange.

To achieve negative buoyancy, Triassic marine reptiles possibly pursued two strategies, the acquisition of gastroliths and/or the addition of bone ballast. Gastroliths have been claimed to be used to regulate buoyancy in the Nile crocodile, but a more recent review of the occurrence and possible function of gastroliths in fossil marine reptiles rejected their function in buoyancy control (Seymour, 1982:9; Taylor, 1993:168). Crocodiles are known to actively ingest pebbles during excursions on land, but their role in buoyancy control has been experimentally rejected. They also seem not to contribute to any significant degree to stability of posture in or under water. But it seems conceivable that gastroliths increase diving time in crocodiles, as their weight compensates a greater lung volume (Grigg and Kirshner, 2015).

Adding bone ballast is achieved through pachyostosis, which, where it occurs, mostly affects ribs and vertebrae, thus adding weight to the skeleton. A marine reptile typically faces a trade-off situation: achieving negative buoyancy allows staying submerged, even walking on the sea floor, but requires more energy to raise to the surface for respiration. Conversely, increasing the lung volume may prolong a dive as there is more room for oxygen storage, but it requires more energy to stay submerged (Seymour, 1982:9; see also Heatwole, 1978:601).

Reptiles entering, or living in, saltwater must excrete excess salt yet retain water that is prone to be lost by osmosis (Grigg and Kirshner, 2015). The osmotic sodium and water balance in marine reptiles is regulated "by salt-secreting glands in the head that compensate for the inability of the kidneys to secrete high concentrations of electrolytes" (Heatwole, 1978:598). These glands evolved convergently in different groups of reptiles: marine turtles have lachrymal salt glands (associated with the adnexae of the eye), lizards have nasal salt glands (associated with the external naris), sea snakes have sublingual salt glands (associated with the tongue sheath), the eustarine crocodile (*Crocodylus porosus*) has lingual salt glands (associated with the tongue) (Dunson, 1976; Cramp et al., 2008). Even small, juvenile specimens of *Crocodylus porosus* with a large surface to volume ratio manage to survive in hypersaline conditions (Grigg and Kirshner, 2015:396). Given the frequent convergent evolution of salt glands in extant reptiles, one can easily imagine that salt glands evolved repeatedly in Triassic marine reptiles as well.

Gas exchange is another challenge marine reptiles must face. Quite generally, the low metabolic rates in reptiles equip them well for prolonged breath-holding, and while routine activity is generally aerobic, strenuous situations engage them in anaerobic metabolism (Seymour, 1982:13, 15ff.). But lung breathing is only one way for marine (aquatic) reptiles to exchange gas; another modus of gas exchange is with water. Indeed, "the ability to exchange gases with water is of such significant adaptive value that it appears in aquatic representatives of practically every major animal group (except the endotherms)" (Seymour, 1982:35). It appears that some turtles and snakes can satisfy most if not all their needs in gas exchange through non-pulmonary pathways. These include primarily the skin in snakes, but even more efficiently the walls of the pharynx and the cloaca in turtles (Seymour, 1982:38).

Other challenges faced by Triassic reptile invaders of the sea concern locomotion and feeding in a dense medium. In terms of locomotion, a vast spectrum of adaptations and modifications can be identified. Among the Triassic marine reptiles, ichthyosaurs with their fusiform body shape, caudal fin (without a tailbend in the caudal vertebral column in many, but not all Triassic taxa), and limbs transformed into flippers are certainly the most highly adapted to a marine environment. Indeed, ichthyosaurs (as also plesiosaurs) appear to have been endothermic, allowing them to invade—at least periodically—cold climate seas (Bernard et al., 2010). Generally long-bodied, the Triassic ichthyosaurs are thought to have used an anguilliform swimming mechanism (Gowan and Motani, 2003:3). Their superior loco-motion abilities also account for their wide geographic distribution.

Other than ichthyosaurs, Triassic marine reptiles—with a few exceptions—retained a generalized reptilian, that is, crocodile or lizard-like body form. In some clades, most notably sauropterygians, there is a tendency to increase the length and sturdiness of the fore limbs relative to the hind limbs, suggesting an increased importance of the fore limbs in locomotion providing paraxial thrust (Storrs, 1993a:78; see also discussion in Carroll, 1985b). Indeed, a trend away from lateral undulation and toward increasing paraxial propulsion seems to have prevailed in Triassic marine reptiles, especially in sauropterygians. This could have been cor-related with the development of a rigid gastral rib basket, which not only provided bone ballast but also constrained lateral undulation of the trunk (Storrs, 1993a:82ff.). The tail, however, may be elongated and the neural spines as well as the chevrons prominent, to form a laterally compressed tail providing improved propulsion. Slight hyperphalangy may develop in the manus and pes of some taxa, but among Triassic marine reptiles, only ichthyosaurs developed true flippers. Experimentation with body form occurred only in select taxa, most notably the armored placodonts (Cyamo-dontoidea) and a saurosphargid. Cyamodontoids tend to develop a circular, discoidally shaped body, which is most extremely developed in *Henodus chelyops* from the lower Carnian (early Upper Triassic) of southwestern Germany (Tübingen: Huene, 1936). Equally mesmerizing is the circular and discoidal body shape of the thalattosaur *Sinosaurosphargis yunguiensis* from the Pelsonian (Anisian, Middle Triassic) of Luoping County, Yunnan Province, southwestern China (C. Li, Rieppel, Wu et al., 2011). Here, the dorsal ribs are transversely broadened so as to establish contact with one another along their length, thus forming a rib cage that is completely cov-ered by osteoderms. An intriguing question is whether there are in these taxa similar developmental mechanisms at work as are known to shape the turtle shell (Hirasawa et al., 2013). A precondition for such experimentation with body shape is, of course, an antecedent shift to paraxial locomotion.

There is a broad range of trophic specializations among Triassic marine reptiles (Rieppel, 2002a). Feeding in a dense medium like water creates its special problems, to which a variety of solutions have been real-ized. Approaching a prey item with an open mouth, then trying to engage the teeth in the prey by rapidly closing the jaws creates a jet stream of water expelled from the buccal cavity that carries the prey away from the predator. One way to avoid this effect is through suction feeding, which has been described for elasmobranchs, actinopterygians, lungfishes, the coelacanth, larval and adult salamanders, pipid frogs, caecilians, and aquatic chelonians (Lauder, 1985). The mechanism is, broadly speaking, the same in all these groups: the rapid opening of the mouth, coupled with a depression of the gular/pharyngeal region, creates a suction effect that carries the prey/food item into the buccal cavity. Across lower vertebrates, two distinct patterns of suction feeding have been identified. The unidirectional pattern occurs in fishes, where water is expelled from the buccal cavity posteriorly through the gill slits or gill chamber, and the bidirectional pattern is seen in lower tetrapods, except for larval salamanders that still retain gills (Lauder, 1985; Lauder and Schaffer, 1986; Lauder and Pendergast, 1992; Damme and Aerts, 1997).

The alternative mode of underwater feeding is the so-called raptorial feeding practiced by predators who develop elongate "pincers jaws" in a dorsoventrally flattened skull (Taylor, 1987; Werth, 2000). Catching prey with a "sideways strike of the head" is a "typical prey capture technique" in crocodiles (Grigg and Kishner, 2015:220). A paradigmatic example is the gharial, with slender and elongate pincers jaws furnished with densely set needle-shaped teeth. Fish are captured by a quick sideways sweeping of the head with the jaws open, a movement pattern that allows water to freely pass between the open jaws. Among Triassic marine reptiles, some thalattosaurs (i.e., askeptosauroids) as well as nothosauroid sauropterygians are typical examples of this feeding behavior. In nothosauroids in particular, a progressive elongation and narrowing of the rostrum can be observed, which anteriorly carries a battery of procumbent fang-like teeth. Progressing from *Simosaurus*, to *Germanosaurus*, and on to different species of *Nothosaurus*, as well as pistosauroids, a transformation can be observed from suction feeding (in *Simosaurus* with a broad and rounded snout) to a "fish trap" dentition in pincers jaws (*Nothosaurus*, *Cymatosaurus*), to a puncturing dentition on pincers jaws (in *Pistosaurus*) (Rieppel, 2002a:58, table 2). This does, however, by no means exhaust the feeding specializations of Triassic marine reptiles.

Many placodonts have procumbent premaxillary and anterior dentary teeth that they used to pluck hard-shelled invertebrate prey from the substrate, which was then crushed between the more posteriorly positioned tooth plates. Advanced cyamodontoid (armored) placodonts such as *Psephoderma* with their delicate, edentulous rostrum have been interpreted as predators feeding on endobenthic shelled invertebrates (Pinna and Nosotti, 1989; Stefani et al., 1992; Mazin and Pinna, 1993). This interpretation was criticized on the grounds that the *Calcare di Zorzino* (Norian, Upper Triassic) in Lombardy, which yielded several specimens of *Psephoderma*, did not also yield a "shelled" endofauna (Renesto and Tintori, 1995). Some thalattosaurs such as *Hescheleria* from Monte San Giorgio (of which more later) or *Nectosaurus* from the Hosselkus Limestone (Carnian, Upper Triassic) of northern California, show a most remarkable trophic specialization. The premaxillaries form an elongated rostrum that points vertically downward, overhanging the anterior tip of the lower jaw (Rieppel, Müller et al., 2005). This rostrum must have been used to scrounge up endobenthic invertebrates from the bottom sediments.

During the Middle and Upper Triassic, respectively, lived the only two Mesozoic marine reptiles that were interpreted as herbivores (on the recovery of primary producers in the Triassic marine biota following the end-Permian mass extinction, see Payne and van de Schootbrugge, 2007). The existence of non-skeletal algae in the Lower Triassic is inferred from the occurrence of "Spongiostromata," "concentric laminae [that] originate from fine grains of sediment adhering to the mucilaginous surface of cyanobacteria and algae" (Flügel, 2004:122; see also Flügel, 1991, fig. 3). Calcareous algae are absent in the Lower Triassic but undergo an explosive radiation beginning in the early middle Anisian (Flügel, 1991:481). The earliest herbivorous marine reptile known is *Atopodentatus unicus* from the Anisian (Middle Triassic) of Luoping County, Yunnan Province, China (C. Li, Rieppel, Cheng et al., 2016). Its skull and dentition are uniquely specialized, the transverse expansion of the jaws giving the skull a hammerhead shape. Spatulate teeth on the anterior margin of the premaxillaries must have been used to scrape algae off the substrate. A sudden opening of the mouth would have created suction, bringing the loose plant material into the buccal cavity. The algae were then sieved from the water as it was expelled through a palisade of tall, densely set, needle-shaped teeth, according to the bidirectional pattern of suction feeding. The second herbivorous marine reptile was the cyamodontoid placodont *Henodus chelyops* from the Gipskeuper (lower Carnian, Upper Triassic) of Lustnau, a suburb of Tübingen, Germany. Although without a hammerhead, *Henodus* is strikingly convergent on *Atopodentatus*, with a broad, wide skull. The snout is delimited anteriorly by transversely oriented premaxillaries carrying a row of densely set denticles on their anterior ventral margins (Reif and Stein, 1999). Of the typical placodont dentition, only four diminutive tooth plates persist, one on each side on the posterior end of the palatine and dentary. Instead, the maxilla and dentary form deep grooves along the margins of the jaw, inside of which baleen-like structures have been observed (Huene, 1936; Reif and Stein, 1999). The functional interpretation is again for the premaxillary teeth to have scraped plant (algal) material off the substrate, and then to have filtered it through the baleen-like structures lining the upper and lower jaws. *Henodus* seems to have had quite some tolerance for salinity fluctuations, as the sediments in which it has been collected indicate variations from brackish to hypersaline conditions (Reiff, 1942; W. Fischer, 1959).

Based on the currently available evidence, reproduction in Triassic marine reptiles has been described as viviparous. Extant marine turtles bury their clutches in the sand of beaches, that is, on land. Such a mode of reproduction, which cannot necessarily be ruled out for all Triassic marine reptiles, would not be expected to leave traces in the marine fossil record. In extant reptiles, two modes of viviparous reproduction have been recognized: lecithotrophic viviparity (formerly known as ovoviviparity) with nutrients provided by the yolk, and matrotrophic viviparity with nutrients provided by the gravid female (e.g., Blackburn, 1992). Distinction of these two modes of viviparity in the fossil record is again impossible (Sander, 1988; Renesto, Lombardo et al., 2003). Nevertheless, the distinction of lecithotrophic versus matrotrophic viviparity is interesting with regard to ichthyosaurs, with respect to which viviparous reproduction has been discussed since 1842 (Böttcher, 1990:2; see also Jäger, 1852). In his survey of 48 gravid females of the genus *Stenopterygius* from the lower Toarcian (Early Jurassic) of Holzmaden (Germany), Ronald Böttcher found the late embryos to lie outstretched parallel to the vertebral column in the female body (1990). However, there is the rare isolated ichthyosaur embryo curled up as if contained in a non-calcified eggshell (Fraas, 1891, pl. 6, fig. 3). Böttcher interprets this latter specimen as regurgitated prey (Böttcher, 1990:18). Whether or not this interpretation is correct, it seems possible that at least the post-Triassic ichthyosaurs were matrotrophic viviparous.

Whereas viviparity in ichthyosaurs is well documented, this is not the case for other Triassic marine reptiles (Deeming et al., 1993; Motani, Jiang, Tintori et al., 2014; Maxwell and Caldwell, 2016). An isolated, "curled-up" embryo of the pachypleurosaur *Neusticosaurus* was reported from Monte San Giorgio, as well as gravid females of *Keichousaurus* from the Triassic (Ladinian) of the Xingyi area, Guizhou Province, China (Sander, 1988; Y.-N. Cheng, Wu and Ji, 2004). Curled-up embryos of *Keichousaurus* have also been collected, suggesting lecithotrophic viviparity for pachypleurosaurs (Li Chun, personal communication; see also Lin and Rieppel, 1998). The same conclusion seems to apply to nothosaurids, based on embryos of *Lariosaurus* collected in the late Ladinian Kalkschieferzone at the Ca' del Frate locality in Lombardy, northern Italy (Renesto, Lombardo et al., 2003). Another case of a gravid female suggesting viviparity has been reported for the protorosaur *Dinocephalosaurus orientalis* (J. Liu, Organ, Benton et al., 2017). Viviparity has furthermore been documented for post-Triassic plesiosaurs (O'Keefe and Chiappe, 2011). For obligatory marine reptiles, the question arises whether they inherited viviparity from their terrestrial ancestors or whether this life history trait evolved only after they had invaded the sea. A pregnant female of the basal ichthyopterygian *Chaohusaurus* from the Early Triassic (Spathian) of Chaohu, Anhui Province, China, shows embryos with the skull pointing backward, that is, emerging head-first during parturition, in contrast to the Liassic (Early Jurassic) ichthyosaur *Stenopterygius*, where the skulls of the embryos point craniad relative to the pregnant female. This was taken as evidence indicating that viviparity in the ichthyopterygian lineage evolved in a terrestrial environment, such that a reorientation of the embryos inside the body of pregnant females occurred only after the invasion of the sea (Motani, Jiang, Tintori et al., 2014). A pregnant female of the Triassic ichthyosaur *Mixosaurus* from the Grenzbitumenzone of Monte San Giorgio shows embryos in the presumed primitive position, that is, their skull pointing rearward relative to the female body, such that head-first parturition would have ensued (Brinkmann, 1996).

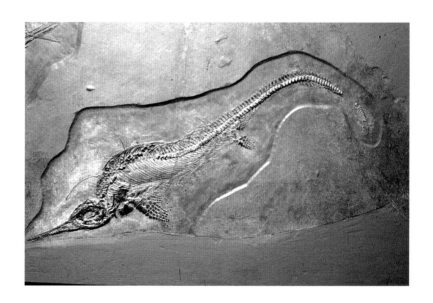

4.1. *Mixosaurus cornalianus* (Paleontological Institute and Museum, University of Zurich T4858) from the Grenzbitumenzone of Monte San Giorgio. The specimen measures 100 cm in length (Photo © Heinz Lanz/ Paleontological Institute and Museum, University of Zurich).

Ichthyosaurs

4

Remains of the small ichthyosaur *Mixosaurus* constitute the most frequently encountered fossil reptiles in the Grenzbitumenzone. After all, it was the finding of a well-preserved *Mixosaurus* fore fin on the premises of the local mining company in 1919 that led Bernhard Peyer to recognize the potential of these deposits for vertebrate paleontology. The ichthyosaur lineage appeared more or less simultaneously in the late Lower Triassic (Spathian) in Asia (southern China, Japan), in western North America (Nevada, British Columbia), and on the arctic (Norwegian) island of Svalbard (Spitsbergen) (Kelley et al., 2016). Whereas some of the stem forms did not survive into the Middle Triassic, the ichthyosaur lineage underwent rapid diversification during the Anisian; ichthyosaur diversity peaked during the Middle Triassic. As indicated by an isolated humerus from Bear Lake County, Idaho, large ichthyosaurs of up to 11 meters in length existed already by the end of the Lower Triassic (Spathian), suggesting the rapid recovery of a marine biota characterized by a diversity of trophic levels including higher ones after the end-Permian mass extinction (Wiman, 1910; Scheyer, Romano et al., 2014:11). Trophic diversity among ichthyosaurs is also indicated by the evolution of multiple types of dentition. Most ichthyosaurs have conical tooth crowns with pointed to blunt tips. This contrasts with *Thalattoarchon*, a large (> 8.6 meters total length) macropredatory ichthyosaur from the middle Anisian (Middle Triassic) of northwestern Nevada, with large, labiolingually flattened teeth that have serrated anterior and posterior cutting edges (Fröbisch et al., 2013). Durophagous adaptations evolved at least three times in the early diversification of the ichthyosauriform lineage, that is, in the stem-ichthyopterygians *Chaohusaurus* and *Grippia* from the late Lower Triassic, and in the Middle Triassic ichthyosaurian *Mixosaurus* (of which more later). They all show crushing teeth with a rounded crown in the posterior part of the tooth row only, whereas anterior teeth remain conical and pointed (Motani, 2005a).

The ichthyosaur lineage acquired pelagic adaptations early in its evolutionary history, which resulted in a widespread geographical distribution of species, rendering the reconstruction of the paleobiogeographical pattern of their diversification difficult. A review of ichthyosaur distribution and phylogeny lists four major localities for the late Lower Triassic (Spathian), five major localities in the Middle Triassic, and three localities for the Upper Triassic (Maisch, 2010:152). The Lower Triassic (Spathian) localities are Spitsbergen (Svalbard; Sticky Keep Formation), British Columbia (Sulphur Mountain Formation), Japan (Osawa Formation), and Chaohu, Anhui Province, China (Majiashan Formation). A new late Lower Triassic

characterized *Mixosaurus cornalianus* as isodontous and hence different from a second, much less frequent mixosaurid taxon represented in the Monte San Giorgio material at their disposal. They tentatively identified this second species as *Mixosaurus* cf. *nordenskioeldii*. The species *Mixosaurus nordenskioeldii* was first described on the basis of material from the Middle Triassic of Spitsbergen (Svalbard) (Hulke, 1873). Parallel to the study of mixosaur material, Maisch and Matzke described a new ichthyosaur from the Middle Triassic of Monte San Giorgio, which they named *Mikadocephalus gracilirostris* (1997b). Larger than *Mixosaurus*, the species is characterized by an elongate and slender rostrum. The following year, Maisch and Matzke described yet another new genus and species of ichthyosaur from Monte San Giorgio, named *Wimanius odontopalatus* (1998a). Of similar size as *Mixosaurus*, *Wimanius* was distinguished from the latter on the basis of details of its dentition. The following year, Maisch and Matzke published a reconstruction of the skull of *Mikadocephalus* and *Wimanius*, maintaining both genera as valid in spite of notable similarities shared by *Mikadocephalus* and *Besanosaurus leptorhynchus* (1999). The latter is a shastasaurid from the black shales of Besano (Dal Sasso and Pinna, 1996; Ji et al., 2016).

In 1997, Winand Brinkmann from the Paleontological Institute and Museum of the University of Zurich published his first review paper on mixosaurs from Monte San Giorgio (Brinkmann, 1997). In the year before, he had already described a gravid female of *Mixosaurus cornalianus* (Brinkmann, 1996). In his 1997 review, Brinkmann identified a specimen that was collected in 1962 at Point 902 as a representative of the genus *Phalarodon*, which he found to differ from *Mixosaurus cornalianus* in details of the dentition, that is, the shape and spacing of the posterior crushing teeth. He further commented on habitat partitioning of durophagous mixosaurs (*Mixosaurus*, *Phalarodon*) hunting free-swimming prey (fishes, cephalopods), and the equally durophagous placodonts subsisting on benthic prey. The following year, Brinkmann described yet another new genus and species of mixosaur from Monte San Giorgio, *Sangiorgiosaurus kuhnschnyderi*, with a more pronounced durophagous dentition than is typical of *Mixosaurus cornalianus* (1998a). At the same time, he recognized the similarities the new genus and species shares with the specimen he had previously identified as *Phalarodon*, for which reason he now identified that latter specimen as aff. *Sangiorgiosaurus*. In a later monographic treatment of some mixosaur material from Monte San Giorgio, Brinkmann seemingly reverted to his earlier opinion, however, and confirmed the identification of the one specimen from Point 902 as a juvenile representative of the genus *Phalarodon* (2004). However, the manuscript that was published as the 2004 monograph had been completed as early as 1998; its publication hence had been severely delayed (Brinkmann, 2004:84). A synonymy of Brinkmann's *Phalarodon* sp. with *Mixosaurus kuhnschnyderi* was formalized by McGowan and Motani (2003:69).

In another review paper, Brinkmann declared *Sangiorgiosaurus* a subjective junior synonym of *Mixosaurus* but retained the species *kuhnschnyderi* as distinct from *M. cornalianus* (1998b:170; independently recognized by Motani, 1999a). The synonymy was motivated by a new specimen that had been donated to the Museo Cantonale di Storia Naturale in Lugano, Switzerland. Upon preparation, this specimen revealed the presence of a sagittal crest, a feature characteristic of *Mixosaurus* but that, given its preservation, could not be ascertained in the holotype of *M. kuhnschnyderi*. Beyond *Mixosaurus* cf. *nordenskioeldii* and *M. kuhnschnyderi*, Brinkmann recognized two distinct morphotypes within *Mixosaurus cornalianus* (1998b; also recognized by Maisch and Matzke, 1998b). He did not draw any taxonomic consequences, however, as he had not had the opportunity yet to study the "neoholotype" of *Mixosaurus cornalianus* designated by Pinna and kept in the Milan museum, nor had he himself yet clarified to which of the two morphotypes the specific epithet *cornalianus* would apply. This would be resolved only after he had designated a valid neotype for *M. cornalianus* based on a well-preserved specimen from Monte San Giorgio (Brinkmann, 1999).

Given the many new taxa described for the Grenzbitumenzone of Monte San Giorgio, mostly based on differences in the dentition, it comes as no surprise that various synonymies have been proposed by different authors. Motani noted the close similarities shared by *Wimanius* and *Mikadocephalus* and called for a phylogenetic approach to species delimitation of Monte San Giorgio ichthyosaurs (1999b). Sander confirmed *Sangiorgiosaurus* to be a junior synonym of *Mixosaurus* and proposed a synonymy of *Mikadocephalus* with *Besanosaurus*, thus establishing the presence of the latter genus at Monte San Giorgio (2000). He furthermore doubted that *Wimanius* was validly diagnosed by Maisch and Matzke (Sander, 1998b). In contrast, McGowan and Motani tentatively considered *Mikadocephalus*, and possibly also *Wimanius*, as subjective junior synonyms of *Pessosaurus*, the latter a genus first erected by Wiman on the basis of material from the Middle Triassic of Spitsbergen (Svalbard: Wiman, 1910; see discussion in McGowan and Motani, 2003:127f). Clearly, and as emphasized by Ryosuke Motani and P. Martin Sander, clarification of the taxonomy of mixosaurid ichthyosaurs at Monte San Giorgio still awaits a comprehensive review of the extensive collections, coupled with a phylogenetic approach to species delimitation (Motani, 1999a, 1999b; Sander, 2000).

The large ichthyosaur *Cymbospondylus buchseri* is here shown among a group of the smaller ichthyosaur *Mixosaurus cornalianus*. *Mixosaurus* is one of the most frequently found fossils in the Grenzbitumenzone. The genus achieved a cosmopolitan distribution during the Middle Triassic. The semi-durophagous *Mixosaurus cornalianus* typically reached a total length of around 1.5 meters. The skull of *Mixosaurus* shows a mixture of primitive and advanced traits, such as the relatively large maxilla and the small upper temporal fossa. From each upper temporal fossa, a distinct depression of the skull roof extends anteriorly on either side of a sagittal crest. The fins in this Triassic ichthyosaur are again primitive, as they retain elongated epipodials (humerus, radius, ulna in the fore

Box 4.1. Ichthyosaurs from the Middle Triassic of Monte San Giorgio

fin). A tailbend, so characteristic of Jurassic ichthyosaurs, is absent, but the presence of a caudal fin in *Mixosaurus* can still be inferred from the structure of the caudal vertebral column. A gravid female collected in the late Anisian part of the Grenzbitumenzone documents viviparity for the genus *Mixosaurus*, a life history trait shared by all ichthyosaurs.

© Beat Scheffold / PIMUZ

© Beat Scheffold

Two species of mixosaurs are currently recognized in the Besano Formation (Grenzbitumenzone) (McGowan and Motani, 2003): *Mixosaurus cornalianus* from Besano–Monte San Giorgio, and *Mixosaurus kuhnschnyderi* from Monte San Giorgio. The monophyly of mixosaurs is well supported through both cranial and postcranial synapomorphies, such as the presence of a sagittal crest extending anteriorly up to the nasals, a large anterior terrace or depression that extends from the anterior margin of the upper temporal fenestra all the way to the posterior end of the nasal, and an increased height of the mid-caudal vertebral centra (Motani, 1999a:930). The genus is plesiomorphic, however, in that a tailbend is absent in the caudal vertebral column. *Mixosaurus cornalianus* reached an adult total length of up to 1.5 m. The skull represents about 25% of the total length. Compared to more advanced ichthyosaurs, the rostrum is relatively short. The external naris opens laterally, the orbit is enlarged as in all ichthyosaurs, but the upper temporal opening is small. In his study of maxillary teeth, Besmer noted the absence of plicidentine in *Mixosaurus*, another plesiomorphic feature (Besmer, 1947). Callaway estimated the vertebral column to comprise 110 to 122 vertebrae, of which between 45 and 50 are presacrals (1997:54). The neural spines in *Mixosaurus cornalianus* are tall and vertically oriented throughout the presacral and proximal caudal regions. A tailbend is absent, but the vertebral centra are heightened in the mid-caudal region, carrying more pronouncedly elongated neural spines that slant forward. Such a morphology has been identified as a "caudal peak" by Motani (1999b: 493, character 96; 2005b:400). Together with the elongated chevrons, the caudal skeleton supported a laterally compressed caudal fin (Kuhn-Schnyder, 1974:78). Scapula and coracoid are both fan shaped and often difficult to distinguish if disarticulated (Callaway, 1997:52). The interclavicle is characteristically Y-shaped. In the pelvic girdle, both pubis and ischium retain the plesiomorphic broad and plate-like dimensions. The ilium is short and expands somewhat distally. The comparatively plesiomorphic structure of the fore fin in *Mixosaurus cornalianus* was already recognized by Baur (1887a). The humerus is short and plate-like, its anterior margin expanded into a "broad, semicircular flange," although some humeri show an excavated anterior margin, which "may reflect some degree of sexual dimorphism, [although] this seems unlikely" (Callaway, 1997:52). Describing specimens of *Mixosaurus cornalianus* kept in the Institute of Geology and Paleontology of the University of Tübingen, Maisch and Matzke found all available humeri to have a well-developed anterior lamina, but they did find two distinct morphologies based on humerus proportions (length versus width) (1998b). The same was confirmed by Brinkmann, who for *Mixosaurus cornalianus* distinguished two morphotypes, A and B (1998). The radius and ulna are both longer than broad. The radius is more robustly built than the ulna, with a biconcave shaft; the ulna is of a curved appearance. The two elements remain separated by a spatium interosseum. The proximal carpal elements—radiale, intermedium, ulnare—remain identifiable as do the five digits (see also Motani, 1999c, fig. 5D; Kuhn-Schnyder, 1974, fig. 57b). In the hind limb, Callaway was able to identify the astragalus and calcaneum, and five distal tarsals, but noted the absence of a centrale (Callaway, 1997:53; see also Brinkmann, 1996, fig. 5).

Stomach contents of mixosaurs from Monte San Giorgio comprise fishes and cephalopod hooklets. A gravid female (fig. 4.2) collected in 1957 at Point 902 contains at least three embryos in the body cavity (Brinkmann, 1996). As discussed in chapter 3, the embryos are oriented in a way that indicates head-first parturition (Brinkmann, 1996:138). This differs from Liassic and younger ichthyosaurs, where tail-first parturition has been documented (Böttcher, 1990). Clearly, head-first parturition is the more primitive condition, also known from the late Lower Triassic ichthyopterygian *Chaohusaurus* from southern China (Motani, Jiang, Tintori et al., 2014).

The Cymbospondylidae

The type species for the genus is *Cymbospondylus piscosus* Leidy, 1868, from the Prida Formation (Anisian, Middle Triassic) of the West Humboldt Range and New Pass Range, northern Nevada. Two additional species are currently recognized in the genus *Cymbospondylus*: *C. buchseri* from the Middle Triassic (Anisian-Ladinian boundary) of Monte San Giorgio, and *C. nichollsi* from the Middle Triassic (Anisian) of Augusta Mountains, northwestern Nevada (Leidy, 1868; Sander, 1989b; McGowan and Motani, 2003; Fröbisch et al., 2006; Ji et al., 2016). One of the early large ichthyosaurs, *Cymbospondylus* quickly reached a cosmopolitan distribution (Sander, 1992).

Box 4.2. *Cymbospondylus buchseri* from the Grenzbitumenzone of Monte San Giorgio

One of the early large bodied ichthyosaurs, *Cymbospondylus buchseri*, is known from a single incomplete specimen collected in 1927 at Monte San Giorgio (Grenzbitumenzone). The posterior part of the skeleton had fallen victim to earlier mining operations at the Cava Tre Fontane locality. The total length of the animal is estimated to have been 5.5 meters. This renders *C. buchseri* a relatively small species in its genus, although it could also be an immature individual, as indicated by the unfinished ossification of the humerus. *Cymbospondylus pisco-sus* from the Anisian of northern Nevada is estimated to have reached 9.1 meters in length. The third species in the genus, *Cymbospondylus nichollsi*, is known from the Anisian of the Augusta Mountains, northwestern Nevada. As is known from well-preserved specimens of *Cymbospondylus piscosus*, a tailbend is again absent in this Triassic ichthyosaur of generally slender contours. The occurrence of the genus at Monte San Giorgio testifies to a wide geographic distribution of this early large ichthyosaur. Fragmentary material from the southern Alpine Triassic, as well as from Middle Triassic Muschelkalk deposits of central Europe, have been assigned to the genus *Cymbospondylus*, sometimes on rather tenuous grounds. *Cymbospondylus* surprisingly appears to be absent in the marine Middle and Upper Triassic deposits of southwestern China.

© Beat Scheffold

Like almost all other ichthyosaur remains from Monte San Giorgio, the single specimen of *Cymbospondylus buchseri* (fig. 4.3) comes from the Grenzbitumenzone (rare remains of *Mixosaurus* sp. have been found in the Meride Limestone, Ladinian: Kuhn-Schnyder, 1964:402). The specimen was collected in 1927 at the Cava Tre Fontane locality. It consists of the anterior half of a skeleton, the posterior half having been destroyed through mining activity (Sander, 1989b:163). The species is named after Fitz Buchser, the first preparator at the Paleontological Institute and Museum of the University of Zürich, and longtime field-worker at Monte San Giorgio (fig. 4.4). A photograph of the specimen was published by Kuhn-Schnyder, who suggested affinities with *Cymbo-spondylus* (1964, pl. 23; see also 1974, fig. 58). The preserved part of the skeleton of *Cymbospondylus buchseri* is 2.3 m long; Sander estimated a total length of 5.5 m (1989b:164). This is considerably smaller than a specimen of *Cymbospondylus piscosus* that was at least 9.1 m long (Merriam, 1908:105). Despite the large size of the latter specimen, the proximal head of its humerus was still rather poorly ossified, which would seem to indicate some degree of immaturity. The humerus of *Cymbospondylus buchseri* has a proximal head that is very poorly ossified, which again points to an immature individual (McGowan

4.5. The skull of *Besanosaurus leptorhynchus* (Museo Civico di Storia Naturale di Milano BES SC 999) from the Sasso Caldo quarry, Besano. The skull measures approximately 52 cm in length (Photo © Giorgio Teruzzi/Museo Civico di Storia Naturale di Milano).

Province, southwestern China (Cao and Luo in Yin et al., 2000). '*Callawayia*' *wolonggangensis* is from the Carnian (Upper Triassic) of Guizhou Province, southwestern China (Chen et al., 2007). *Shastasaurus* is from the Carnian (Upper Triassic) of California and from the Norian (Upper Triassic) of British Columbia; the validity of its occurrence in southern China remains controversial (Merriam, 1895; Yin et al., 2000; McGowan and Motani, 2003; Chen, Cheng et al., 2007). *Guanlingsaurus* is again from the Carnian (Upper Triassic) of Guizhou Province, southwestern China (Yin in Yin et al., 2000). And finally, *Shonisaurus* is known from the Carnian (Upper Triassic) of Nevada, and from the Norian (Upper Triassic) of British Columbia (Camp, 1976; Nicholls and Manabe, 2004). This early radiation of large pelagic ichthyosaurs was the first to quickly achieve a cosmopolitan distribution by Late Triassic times (Bardet et al., 2014:881).

Besanosaurus leptorhynchus was collected in the Sasso Caldo quarry near Besano in 1993 (Dal Sasso and Pinna, 1996). The species is represented by a single specimen, a gravid female 5.5 m in length (fig. 4.5). The skull is relatively small, only about 9.5% of the body length. The rostrum is long and slender, 67% of the skull length, and furnished with small teeth. There are 60 presacral vertebrae, of which approximately 11 are counted as cervicals, two sacrals, and 139 caudal vertebrae. The cervical ribs are dichocephalous, the dorsal ribs are holocephalous. The dorsal neural spines are tall, anterior caudal neural spines are posterodorsally inclined, and terminal caudal neural spines are anterodorsally inclined, suggesting the presence of a caudal peak. The humerus is rounded and short, with a distinct anterior notch and a length/width ratio of 0.95. The radius is notched anteriorly and posteriorly; the ulna is of rounded contours, not longer than wide. More distal ossifications of the fore fin are disarticulated and widely scattered. The femur is more delicately built than the humerus, and proximally somewhat expanded; its length/width ratio is 1.22. The tibia is somewhat longer than the fibula, with a biconcave shaft. The fibula is distally expanded. The pelvic fin is reconstructed as tetradactylous, and with a hyperphalangy of up to 20 phalanges. *Besanosaurus* is characterized as different from *Shonisaurus* "in many basic characters" (Dal Sasso and Pinna, 1996:12), and also from *Shastasaurus*,

4.6. An undescribed ichthyosaur referred to Shastasauridae indet. (Paleontological Institute and Museum, University of Zurich T4376) from the Grenzbitumenzone of Monte San Giorigo. The specimen measures 216 cm in length. Located above is a juvenile specimen of *Mixosaurus cornalianus* (Paleontological Institute and Museum, University of Zurich T4923) of 80 cm length (Photo © Heinz Lanz/Paleontological Institute and Museum, University of Zurich).

in particular, in the structure of the tetradactylous pelvic fin. It does, however, share both similarities as well as differences with *Cymbospondylus*, the similarities in this case explained as instances of convergence by Dal Sasso and Pinna (1996). In contrast, Sander and Faber concluded: "It would not be surprising if, upon further study of the Besano specimen, it should turn out to belong to *Shastasaurus*" (1998, p. 159).

Radiographs of *Besanosaurus* revealed "three or four clusters each comprising a tiny ichthyosaurian vertebral chain" (Dal Sasso and Pinna, 1996:18). These vertebrae are interpreted as representing embryos, the specimen therefore representing a gravid female. Cannibalism that had earlier been suggested to occur in ichthyosaurs is dismissed by Dal Sasso and Pinna in the case of *Besanosaurus* on the grounds that these embryonic vertebrae correspond in size to those of an adult *Mixosaurus*—a highly unlikely prey for a "teuthophagous ichthyosaur, equipped with such a thin rostrum and so small teeth" (Böttcher, 1990; Dal Sasso and Pinna, 1996:189).

Interestingly, Kuhn-Schnyder mentioned a second large ichthyosaur skeleton from the Grenzbitumenzone of Monte San Giorgio of 2.16 m length (1964:399). That specimen (fig. 4.6), according to Brinkmann, is a shastasaurid currently under study, but a description has not yet been published (1997:70). This specimen, on exhibit at the Paleontological Museum of the University of Zurich, was figured by Brinkmann in the guide to the exhibition (1994:46, fig. 28). On the basis of that photograph, and taking the possibility of different ontogenetic stages into account, Dal Sasso and Pinna dismiss the identification of that specimen as *Besanosaurus*, from which it differs in body proportions and in the configuration of the pectoral girdle (1996:14). However, the presence of *Besanosaurus* in the Grenzbitumenzone of Monte San Giorgio could still be established if the synonymy of *Mikadocephalus* with *Besanosaurus* is accepted (Sander, 2000. See discussion above; *Besanosaurus* would take priority).

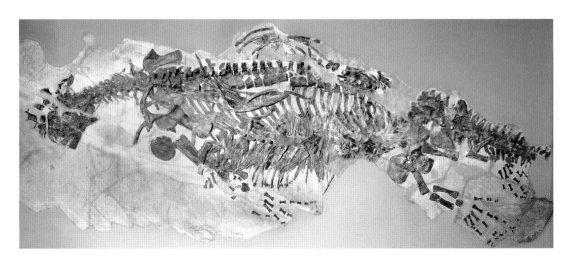

5.1. The holotype of
Helveticosaurus zollingeri
(Paleontological Institute and
Museum, University of Zurich
T4352) from the Cava Tre Fontane.
As preserved, the specimen
measures 250 cm in length; total
length is estimated at 360 cm
(Photo © Heinz Lanz/Paleontological
Institute and Museum, University of
Zurich).

Helveticosaurus, Eusaurosphargis, and the Placodonts

5

The Reverend William Daniel Conybeare (1787–1857), together with Henry De la Beche (1796–1855), were the first to scientifically name in 1821 a sauropterygian, *Plesiosaurus*, based on incomplete and disarticulated material without a skull from the Lower Lias (Lower Jurassic) of Street, Somerset, England (De la Beche and Conybeare, 1821; Taylor, 1997). Interestingly, they titled their report "*Notice of the discovery of a new Fossil Animal, forming a link between the Ichthyosaurus and Crocodile . . .*" (De la Beche and Conybeare, 1821:559). The choice of the name *Plesiosaurus* they justified "as expressing [the animal's] near approach to the order Lacerta," the latter at the time also sometimes called Sauria (De la Beche and Conybeare, 1821: 560). The first nearly complete and articulated *Plesiosaurus* skeleton was collected in 1823 by Mary Anning (1799–1847), a scion of a famous commercial fossil-collecting enterprise run by the Anning family of Lyme Regis, England, and purchased by the Duke of Buckingham for 150 guineas (Buffetaut, 1987: 73). The Duke of Buckingham placed the specimen at the disposal of theologian, geologist, and paleontologist William Buckland (1784–1856) for scientific description, who in turn alerted Conybeare of that find. In his report on the specimen, Conybeare noted with satisfaction that "the magnificent specimen recently discovered at Lyme has confirmed the justice of my former conclusions in every essential point" (1842:381).

Meanwhile, the frequent occurrence of conspicuous black and shiny placodont tooth plates signaled the occurrence of vertebrate fossils in the Muschelkalk (Middle Triassic) quarries of the Lainecker Höhenzug and Bindlacher Berg east of Bayreuth in southern Germany (Freyberg, 1972). Although the earliest descriptions and figures of such teeth date back as far as the late eighteenth century, the systematic collecting of this fauna began in 1809 (Weiss, 1983), and, for sauropterygians, culminated in a magnificently illustrated folio compendium published by H. v. Meyer during the years 1847 through 1855. In Italy, sauropterygians again rank among the earliest Mesozoic vertebrate fossils to be described. The first sauropterygians were found in black shales (of Ladinian age) near Perledo above Varenna at Lake Como in northern Italy and were referred to as members of the "famiglia dei Paleosauri" by Balsamo-Crivelli (1839). But whereas sauropterygians played an important role in the early development of vertebrate paleontology as an autonomous science, the concept of a monophyletic taxon Sauropterygia, and insight into their relationships among reptiles in general, was slow in coming.

Sauropterygia and Placodontia—a History of Changing Relationships

In 1860, Richard Owen first recognized the relationships that prevailed among some of the Muschelkalk reptiles described by H. v. Meyer, such as *Nothosaurus*, *Simosaurus*, *Pistosaurus*, with *Placodus*, and also with *Plesiosaurus*, and consequently included these taxa together with some other enigmatic fossils such as the (prolacertilian) genus *Tanystropheus* also described by H. v. Meyer, within a separate order, which he named Sauropterygia (Owen, 1860: 209). Subsequently, Cope considered the Sauropterygia to be closely related to chelonians and rhynchocephalians, and he formalized this opinion by the erection of the Synaptosauria, to include the aforementioned taxa (1885:246). However, Cope's classification was criticized by Baur, who grouped the Rhynchocephalia with squamates (lizards, mosasaurs, and snakes), but retained the Synaptosauria to include the chelonians and Sauropterygia only (1887). In his catalogue of the fossil reptiles at the British Museum (Natural History), Lydekker retained the "synaptosaurian branch" as one that in his view again included the Sauropterygia and Chelonia (1889). In the Sauropterygia, Lydekker included the nothosaurs, lariosaurs, simosaurs, and plesiosaurs, but not the placodonts, which he classified as an order of uncertain position (Lydekker, 1889; 1890:2).

Placodonts did, in fact, pose special problems. First referred to pycnodont fishes by Agassiz, the reptilian nature of the placodonts was first recognized by Owen, who initially classified them with the Sauropterygia (Agassiz, 1833–1844; Owen, 1858, 1860). Later, however, Owen (1879) raised the possibility that placodonts are to be referred to anomodonts (a group of "mammal-like reptiles"), a hypothesis that was formally adopted by Seeley (1889).

In the meantime, the concept of classifying reptiles on the basis of the structure of the temporal region of the skull was gaining momentum. Putting an end to some extended controversy, Osborn introduced the influential proposition that the class Reptilia should be subdivided into two main branches, the Synapsida (with none or only a single temporal fenestra) which eventually would give rise to mammals, and the Diapsida (with two temporal fenestrae) which would give rise to birds (1903). The publication of a full account of the osteology of the genus *Araeoscelis* from the Lower Permian of Texas, published by Williston, emphasized the presence of a single (upper) temporal fenestra in this lizard-like form, and prompted a comparison with all other fossil and extant reptiles characterized by a single temporal opening, the "Ichthyosauria, Sauropterygia, 'Pelycosauria,' Placodontia, Therapsida, Squamata and . . . the Proganosauria" (Williston, 1914a: 391). It was not until 1917 that Williston formalized his view of reptile phylogeny and classification: "One thing is evident: even earlier than the origin of the lower vacuity in the Diapsida a simple upper vacuity, as in lizards, had developed, but with the temporal region imperforate below" (Williston, 1917:418). This conclusion he thought to be supported by *Araeoscelis* from the Lower Permian of Texas. In his 1917 classification, Williston included the Theromorpha, Therapsida, Sauropterygia, and Placodontia in Osborn's Synapsida, whereas he introduced the new term Parapsida for a division including the Ichthyosauria, Squamata, and Protorosauria (the latter comprising the Araeoscelidia and Acrosauria). In 1925, finally, Williston synthesized the prevailing thoughts of the time in a synoptic reptile classification that recognized five major subdivisions, namely, the Anapsida, Synapsida, Synaptosauria, Parapsida, and Diapsida. Williston resurrected Baur's Synaptosauria to include Sauropterygia (nothosaurs and plesiosaurs) and Placodontia along with synapsids (Williston, 1925; Baur, 1887b:93). This to Romer indicated "a rather wistful hope that some relationship [of sauropterygians and placodonts] to mammal-like forms might be proved" (Romer, 1956:652). The Parapsida were a thoroughly polyphyletic assemblage, including forms as diverse as *Mesosaurus*, ichthyosaurs, *Araeoscelis*, *Protorosaurus*, *Saphaeosaurus*, and squamates.

This classification of reptiles proved unsatisfactory, particularly after the description of *Prolacerta* by Parrington, a diapsid reptile from the early Triassic of South Africa, with an incomplete lower temporal arcade (1935). The fossil was initially interpreted as a lizard precursor in the process of reducing the lower temporal arch, and thus refuted Williston's assumption that the lizard skull never possessed a lower temporal fenestra but instead reduced the dermal covering of the cheek region by ventral emargination (1914a). Romer reminisced that it was generally acknowledged that Williston in his 1925 classification had correctly grouped the Sauropterygia and Placodontia together on the basis of a single upper temporal opening, and how E. H. Colbert, "after discussion with me and others, reasonably suggested Euryapsida as a substitute" for the "awkward" term Synaptosauria (Romer, 1968:113; Colbert, 1955). *Araeoscelis* was included within the Euryapsida as their earliest representative on the basis of the upper temporal fenestra, a position that was supported by Vaughn (1955). On that basis, Romer developed the concept of an early radiation of euryapsids, poorly represented in the Paleozoic fossil record as opposed to the Mesozoic record, and including the closely related sauropterygians and placodonts (1968).

The concept of the Euryapsida as supported by Romer stands in sharp contrast to the hypothesis of diapsid affinities of sauropterygians. A diapsid relationship of nothosaurs was first proposed by Jaekel, based on the configuration of the upper temporal opening and its surrounding bones in the pachypleurosaur *Anarosaurus*, which he found to resemble the upper temporal fossa of diapsids (1910). Jaekel furthermore noted a free ending posterior projection of the jugal and maxilla in the nothosaurian genus *Simosaurus*, which he compared to the posterior projection of the jugal in a lizard with a reduced lower temporal arch (1910). Diapsid relations of nothosaurs were further supported by Kuhn-Schnyder, who based his arguments on a reinvestigation of the skull of *Simosaurus* (1967, 1980). He believed the shape of the jugal bone in this taxon to be best explained by the loss of the lower temporal arcade in a diapsid reptile, rather than by ventral emargination of the cheek region, a view that was rejected by Romer (1968, 1971). However, Kuhn-Schnyder forcefully rejected sauropterygian relationships of placodonts (1980, 1989a, 1989b). He found himself unable to place placodonts in any of the categories of Williston's or Romer's classifications because he accepted Pinna's reconstruction of the temporal region of the placodont skull, in which the postorbital would meet the quadratojugal along the lateral margin of the upper temporal fossa (Pinna, 1976, 1989). He therefore maintained an isolated status of placodonts among reptiles and even postulated a separate origin of placodonts from amphibians (Kuhn-Schnyder, 1963a). He postulated the same for sauropterygians, ichthyosaurs, and turtles; indeed he believed that the Reptilia comprised seven lineages which all had crossed the boundary between amphibians and reptiles independently of one another: "The class of Reptiles is a 'grade' and polyphyletic" (Kuhn-Schnyder, 1963:81). Similarly, he thought that the transition from reptiles to mammals occurred several times independently, rendering the latter a polyphyletic assemblage as well.

Kuhn-Schnyder's views on sauropterygian and placodont relationships reveal a deep bias toward polyphyly that dominated German, indeed Continental European, paleontology from the late 1950s well into the 1970s. It was mainly the American paleomammalogist George G. Simpson, with his 1953 book *The Major Features of Evolution*, and the British evolutionary biologist Julian S. Huxley, who motivated this bias (Forey, 2004:167). Simpson portrayed the origin of major groups, such as of tetrapods from fishes, or of birds from reptiles, as transitions between *adaptive zones*. Evolution was driven by natural selection and adaptation, and major groups originated when organisms invaded a new ecosystem, a new adaptive zone, a step that required major evolutionary transformation. Huxley in turn introduced the distinction of *clades* versus *grades* (J. S. Huxley, 1958). Groups of monophyletic origin Huxley called clades. Groups that occupy a certain adaptive zone, and hence share certain evolutionary adaptations, he called grades. For example, it is generally accepted today that birds evolved from theropod dinosaurs. As a consequence, birds are more closely related to crocodiles than to lizards among living reptiles. A group that includes crocodiles and birds, but not lizards, is therefore a clade. In contrast, a group that contains crocodiles and lizards but not birds represents a grade, the "reptilian" grade of evolution. Grades need not be clades; the invasion of a certain adaptive zone could in principle have happened multiple times at different places. This is exactly how Kuhn-Schnyder envisaged the polyphyletic origin of reptiles and mammals (1962b, 1963a, 1965). Such a perspective, of course, creates a stark contrast to the search for monophyletic groups in the wake of the cladistic revolution (Hull, 1988).

Authors adhering to the principle of monophyly continued to recognize placodonts as diapsids, however, classifying them as Diapsida *incertae sedis* that are not closely related to sauropterygians, the latter including pachypleurosaurs, nothosaurs, and plesiosaurs (e.g., Sues, 1987a, 1987b). In their study of the pachypleurosaur *Neusticosaurus* ("*Pachypleurosaurus*") *edwardsii*, Carroll and Gaskill (1985) noted an important character supporting the monophyly of the Sauropterygia so conceived (i.e., comprising pachypleurosaurs, nothosaurs, and plesiosaurs), which is the inverse relationship of the clavicle relative to the scapula. Whereas the clavicle is applied to the anterior margin or lateral surface of the scapula in the plesiomorphic condition, it is applied against the anteromedial or medial surface of the scapula in sauropterygians. And while some authors invoked a close relationship of placodonts with ichthyosaurs, Rieppel noted that Drevermann had described the same inverse relationship of the scapula and clavicle in *Placodus* (Drevermann, 1933; Mazin, 1982; Pinna, 1989; Rieppel, 1989a). Zanon corroborated a sister-group relationship of placodonts with sauropterygians within diapsids, but correctly observed that, honoring the priority established by Owen, the name Sauropterygia should be used to encompass both the placodonts and all other sauropterygians (Owen, 1860; Zanon, 1989). Although the close relationships of placodonts and other sauropterygians are no longer in doubt at the present time, there continued some debate as to the proper placements of the placodonts, either nested within the other sauropterygians, or as sister-group of the Eosauropterygia (Storrs, 1991a, 1993b; Rieppel, 1998a, 2000a). A recent comprehensive phylogenetic analysis of Triassic marine reptiles firmly established placodonts as sister group of all other sauropterygians, the latter referred to as Eosauropterygia (Neenan, Li et al., 2015).

The Placodontia

The *locus classicus* to study placodonts is the Middle Triassic Germanic Muschelkalk, where isolated placodont teeth have always been a common find, first described by Count Georg Ludwig Friedrich Wilhelm zu Münster of Bayreuth in 1830, who believed they represent durophagous fish (Münster, 1830). He did so following the suggestion of Louis Agassiz, who in his *Recherches sur les Poissons Fossiles*, on the basis of these teeth named the genus *Placodus*, which he referred to pycnodont fishes (1833–44). It was left to Richard Owen to recognize the reptilian nature of the genus *Placodus* (1858). Important historic placodont material from the Muschelkalk comprises the armored placodonts from the upper Muschelkalk (Anisian) of Bayreuth, such as *Cyamodus rostratus* described by Münster in 1839, or the nearly complete skeleton of the unarmored *Placodus gigas* from the upper Muschelkalk (Anisian) of Steinsfurt, located between Heidelberg and Heilbronn, Germany, that was first discovered in 1915 and described by Drevermann in an incomplete monograph published posthumously in 1933 (see also Drevermann, 1915).

The discovery of placodont material at Monte San Giorgio dates back to the first field campaign under Peyer's direction, targeting the Grenzbitumenzone at the Val Porina location in 1924. The excavations produced the skeleton of an armored placodont, in part rather poorly preserved, which Peyer described in 1931—after difficult and time-consuming preparation—under the name of *Cyamodus hildegardis*, the specific epithet honoring his wife (Peyer, 1931d). Continued excavations in the Grenzbitumenzone at the Val Porina locality in 1929 yielded remains of a second placodont from a dolomitic layer, sporting teeth that differed drastically from the dentition known in any other placodont. Peyer published a preliminary description of the tooth-bearing elements in 1931, which he referred to a new genus and species named *Paraplacodus broilii*, thereby honoring his colleague, friend, and mentor Ferdinand Broili (1874–1946), the famous paleontologist from Munich who lent early support to Peyer's work at Monte San Giorgio (Peyer, 1931e). The monographic treatment of *Paraplacodus* followed in 1935, along with the description of newly found *Cyamodus* remains (Peyer, 1935). These descriptions of placodont material from Monte San Giorgio by Peyer made reference to two additional taxa, which he controversially related to placodonts: the enigmatic *Saurosphargis* from the early Anisian of Gogolin, Upper Silesia (now Poland), and *Helveticosaurus* from the Grenzbitumenzone of Monte San Giorgio (Huene, 1936; Peyer, 1955; see also Peyer and Kuhn-Schnyder, 1955a, and chap. 3).

Helveticosaurus zollingeri PEYER

Helveticosaurus zollingeri known from three specimens, is a highly enigmatic taxon, the relationships of which with other reptiles have not yet been satisfactorily clarified (Peyer, 1955). The holotype (fig. 5.1) was collected in 1935 in the Galeria Arnaldo of Cava Tre Fontane. A second, completely disarticulated specimen had previously been collected at the same locality in 1933. A snout fragment associated with 14 anterior teeth was collected in 1937. The skull of the holotype is badly crushed, and most of the tail is missing. The preserved skeleton is 2.5 m long; Kuhn-Schnyder estimated the total length of the specimen to have been 3.6 m (1974:73).

 Helveticosaurus is remarkable for its tall premaxilla carrying five long, dagger-shaped teeth of subthecodont implantation. The anterior part of the maxilla carries a prominent canine. Given its state of preservation, the skull offers little morphological detail; the slender upper temporal arch is formed by the postorbital and squamosal. The mandible is deep and massively built, and—like the premaxilla and maxilla—furnished with long, dagger-shaped teeth. The identification of sutures is problematic. The coronoid process is rather low; the retroarticuar process is distinct. There are 13 (perhaps 14) cervical vertebrae associated with very characteristic, robust cervical ribs. The latter are dichocephalous and carry a distinct free-ending anterior process. There are 27 dorsal vertebrae preserved in the holotype, an incomplete number, as some posterior dorsal vertebrae are missing. The neural spines in the dorsal region are upright, tall, and slender, the transverse processes characteristically elongated. The dorsal ribs are holocephalous and carry no uncinate process. The gastral rib basket is well developed, each gastral rib comprising five elements. Due to the poor preservation of the sacral region, the number of sacral vertebrae remains unknown. A series of 20 vertebral centra represents the proximal part of the tail, with perhaps one sacral element at its anterior end. In the tail, the neural spines become progressively shorter and inclined posterodorsally; caudal ribs are present at least in the base of the tail.

The interpretation of the pectoral girdle of *Helveticosaurus* remains a matter of debate. Peyer modeled his interpretation on Drevermann's account of the pectoral girdle in *Placodus gigas*, with the result that the clavicles would overlap the interclavicle, articulating in facets on its dorsal surface, as is also the case in placodonts and sauropterygians (Peyer, 1955; Drevermann, 1933). Indeed, the pectoral girdle became an important part of Peyer's argument that *Helveticosaurus* is related to placodonts. An alternative interpretation is possible, however, which results in a plesiomorphic structure of the pectoral girdle in *Helveticosaurus* (Rieppel, 1989a). The scapula is developed into a tall dorsal blade. The humerus is longer, distinctly expanded distally, and overall distinctly more robust than the femur. Radius and ulna are both rather slender, the radius a little longer than the ulna and angulated in its proximal part. In the pelvic girdle, the ilium is noteworthy for its small size, especially the small dorsal blade that tapers off to a blunt tip posteriorly, again reminiscent of the placodont and sauropterygian condition. There is no thyroid fenestra between pubis and ischium, but given the rounded contours of the latter elements, neither do they meet in a closed suture—again as in *Placodus*. The single tarsal ossification is the astragalus, located distal to the spatium interosseum between tibia and fibula. The fifth metatarsal is straight. The phalangeal formula for manus and pes remains unknown.

Peyer stipulated placodont affinities of *Helveticosaurus* (1955). Support for this hypothesis he derived mainly from the structure of the dorsal vertebrae. Their amphicoelous centrum is slender and tall, the neural canal is narrow, and the transverse processes are characteristically elongated. Similarities in the pectoral and pelvic girdles have already been mentioned. Given those shared characters, Peyer placed *Helveticosaurus* in the Placodontia, because "amongst the reptiles of the Triassic, I know of no other group that would even only passably accommodate *Helveticosaurus*" (1955:45). Kuhn-Schnyder rejected placodont affinities of *Helveticosaurus* (Kuhn-Schnyder, 1974; also Rieppel, 1989a). More recent phylogenetic analyses that included *Helveticosaurus* found it to be the sister taxon to *Eusaurosphargis dalsassoi* (Li, Jiang, Cheng et al., 2014; see also Nosotti and Rieppel, 2003). A recent phylogenetic analysis targeting placodont interrelationships found *Helveticosaurus* to fall into an unresolved trichotomy with the Ichthyopterygia and a saurosphargid-sauropterygian clade (Neenan, Li et al., 2015).

Eusaurosphargis dalsassoi NOSOTTI and RIEPPEL

This genus and species was originally based on a single specimen (fig. 5.2), a fully disarticulated and incomplete skeleton from the Besano Formation, collected in 1993 in the Sasso Caldo quarry near Besano (Nosotti and Rieppel, 2003).

The same taxon has so far not been reported from Monte San Giorgio. The animal is of small size, which is difficult to quantify, however,

5.2. The holotype of *Eusaurosphargis dalsassoi* (Museo Civico di Storia Naturale di Milano BES SC 390) from the Sasso Caldo quarry, Besano. The skeleton of this small reptile is completely disarticulated. The humerus seen in the right lower corner of the photograph measures 30 mm in length (Photo © Giorgio Teruzzi/Museo Civico di Storia Naturale di Milano).

given that the skeleton is completely disarticulated. The species is named after Cristiano Dal Sasso, curator at the Museo Civico di Storia Naturale di Milano, who recognized the importance of the find and did an incredible job preparing the delicate specimen. An isolated dentary measures 19.5 mm in length; the width of a dorsal neural arch measured across the tips of the transverse processes is 30 mm. The shape of the premaxilla is very tall, reminiscent of the premaxilla of the much larger *Helveticosaurus*. The teeth of *Eusaurosphargis*, known from the premaxilla and both dentaries, are very distinctive. The dentition is homodont, the tooth crowns leaf shaped, with a distinct lingual heel. The enamel surface is lightly striated. Equally distinctive are the neural arches of the dorsal vertebrae, with very elongate transverse processes, a feature *Eusaurosphargis* shares with *Saurosphargis*, *Helveticosaurus*, and placodonts (Huene, 1936, pl. 13). What is more, the dorsal ribs carry a distinct uncinate process, which in one case is preserved in association with an osteoderm. This is a feature *Eusaurosphargis* shares with *Saurosphargis*, and with an undescribed saurosphargid from the Anisian of Winterswijk, the Netherlands; an uncinate process on the dorsal ribs is also present in *Paraplacodus* (Huene, 1936). A thyroid fenestra is absent in the pelvic girdle, although their rounded contours do not allow the pubis and ischium to meet in a closed suture—again a feature shared with *Helveticosaurus* and *Placodus*. A preliminary phylogenetic analysis recovered *Eusaurosphargis* as sister taxon to *Saurosphargis*, and *Helveticosaurus* as sister taxon of those two. This clade comprising three taxa nests outside a thalattosaur-sauropterygian group, and hence is not related to placodonts,

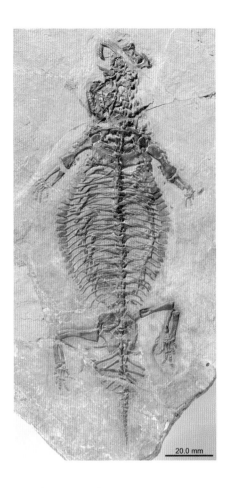

5.3. *Eusaurosphargis dalsassoi* (Paleontological Institute and Museum, University of Zurich A_III_4380) from the Middle Triassic (early Ladinian) Prosanto Formation of the eastern Swiss Alps. As preserved, the specimen measures appoximately 20 cm in length (Photo © T. Scheyer/ Paleontological Institute and Museum, University of Zurich).

20.0 mm

which nest inside sauropterygians (Nosotti and Rieppel, 2003:13). A more recent phylogenetic analysis found *Eusaurosphargis* to constitute the sister taxon of a clade that comprises *Helveticosaurus*, the Ichthyopterygia, *Sinosaurosphargis*, and the Sauropterygia (Neenan, Li et al., 2015). More recently, a complete and beautifully preserved juvenile specimen of *Eusaurosphargis dalsassoi* of approximately 20 cm total length was collected from the early Ladinian Prosanto Formation in the eastern Swiss Alps (fig. 5.3). The phylogenetic analysis based on this new specimen placed *Eusaurosphargis* as sister taxon of the Sauropterygia (Scheyer, Neenan et al., 2017).

The saurosphargids are an enigmatic clade of Triassic marine reptiles of uncertain phylogenetic relationships. The genus *Saurosphargis* was based on a partial trunk skeleton from the early Middle Triassic of Upper Silesia, which was lost during World War II. Fossils of complete skeletons of the genus, or at least a very closely related genus, are today known from the Middle Triassic of southwestern China. *Eusaurosphargis* is a very small representative of the clade that was collected

Box 5.1. *Eusaurosphargis dalsassoi* from the Besano Formation, Besano

from the Besano Formation. Due to the complete disarticulation and only partial preservation of the skeleton, the anatomy of *Eusaurosphargis* remained rather poorly known. This situation changed with the discovery of a perfectly preserved representative of the genus in the Middle Triassic Prosanto Formation in the eastern Swiss Alps. The reconstruction shown here is based on this specimen.

© Beat Scheffold

It is interesting to note that both *Saurosphargis* and *Helveticosaurus* have in the past been classified as placodonts (Peyer, 1955; Peyer and Kuhn-Schnyder, 1955). New saurosphargid material (referred to the genus *Largocephalosaurus*, but probably congeneric with *Saurosphargis*), including complete and articulated specimens from the Anisian of Xinmin District, Panxian County, Guizhou Province, southwestern China, proves that saurosphargids are not related to placodonts, nor are they themselves placodonts (Li, Jiang, Cheng et al., 2014). Whereas the postcranial skeleton does resemble that of *Placodus* to some extent, the skull shows that saurosphargids are not to be included in the Sauropterygia:

there is a distinct interpterygoid vacuity exposing the basicranium, and an open palatobasal articulation. The dentition is homodont, and the teeth of the same characteristic morphology as that described for *Eusaurosphargis*.

The initial finds of this species comprised an isolated premaxilla and two skull fragments showing parts of the dentition in ventral view. They were described and figured by Peyer before the much more complete skeleton collected in 1929 was prepared and ready for description (1931d, pl. 17). It became clear that the fragments described in Peyer (1931d) represented the same taxon as the more complete and better-preserved specimen, which Peyer designated as holotype (1931e).

The holotype of *Paraplacodus broilii* consists of scattered skeletal remains, most notably tooth-bearing elements that immediately signaled its distinctiveness from other placodonts (Peyer, 1931e; 1935, pl. 42). In addition, there are vertebrae, the distinctive dorsal ribs carrying a large uncinate process, gastralia, and girdle and limb elements. The same holds true for a second specimen described by Peyer, his specimen B, although in this case the elements in the trunk region, especially the gastral rib basket, are more closely associated (1935). The latter was collected in a bituminous layer at the Val Porina locality in 1932. An articulated skeleton of *Paraplacodus* of approximately 1.4 m length was eventually retrieved from the Galeria Arnaldo Superiore at the Cava Tre Fontane locality in 1936, and described by Kuhn-Schnyder in 1942 (fig. 5.4). Whereas this material is crucial for the analysis of the dentition and postcranial skeleton in *Paraplacodus*, the best skull is a specimen Peyer must have given to Broili.

The latter specimen, an isolated skull preserved in lateral view, is now accessioned in the Bavarian State Collections of Paleontology in Munich and was crucial in determining the diapsid origin of placodonts, prior to the description of *Palatodonta bleekeri* from the Middle Triassic (Anisian) of Winterswijk, the Netherlands (Neenan, Klein et al., 2013; see chap.3). Pinna found the ventrally deeply embayed cheek region of *Paraplacodus* to closely resemble that of early ichthyopterygians such as *Grippia* and followed Mazin in stipulating a close relationship of placodonts with ichthyopterygians (Mazin, 1982; Pinna, 1989). Since Mazin derived ichthyopterygians from diapsid reptiles, Pinna considered a diapsid origin of placodonts *"molto probabili"* (very likely) (1989:1587; the same was previously concluded by Sues, 1987a). Zanon focused explicitly on the morphology of the temporal skull elements such as the boomerang-shaped jugal and squamosal, concluding that placodonts had a diapsid origin with the lower temporal arch absent or lost (1989). The upper temporal arch is much more slender in *Paraplacodus* as compared to *Placodus*, indicating that the dermal cover of the cheek region in the latter genus is the result of a secondary expansion of the postorbital, jugal, and squamosal (Rieppel, 2000b).

5.4. An articulated specimen of *Paraplacodus broilii* (Paleontological Institute and Museum, University of Zurich T4775) from the Cava Tre Fontane. The specimen measures 140 cm in length (Photo © Heinz Lanz/ Paleontological Institute and Museum, University of Zurich).

Box 5.2. *Paraplacodus broilii* from the Grenzbitumenzone of Monte San Giorgio

Placodonts are Triassic marine reptiles characterized by large crushing tooth plates that gave the group its name. *Paraplacodus broilii* is a primitive representative of the placodont lineage, which Bernhard Peyer named after his friend, the famous paleontologist Ferdinand Broili from Munich. Broili lent early support for Peyer's fieldwork at Monte San Giorgio. *Paraplacodus* reached a total length of approximately 1.5 meters. As in *Placodus*, the anterior teeth of *Paraplacodus* are procumbent, used to pluck hard-shelled invertebrate prey from the substrate. But more posterior teeth in the upper and lower jaws, while massive and blunt, hence suitable to crush invertebrate shells, have not yet transformed to the large, flat tooth plates, which gave the lineage its name. With respect to other placodonts, *Paraplacoodus* retains a more generalized dentition. Since the deep waters in the Grenzbitumenzone basin were oxygen depleted, *Paraplacodus* must have foraged in the marginal zones of the basin. Osteoderms, ossifications embedded in the deep layer of the skin, are absent in *Paraplacodus*. A well-preserved skull of *Paraplacodus*, which Peyer donated to the Munich fossil vertebrate collection curated by Broili, played a crucial role in identifying placodonts as diapsid reptiles.

© Beat Scheffold / PIMUZ

The characters that had earlier been invoked to claim a "primitive" status of *Paraplacodus* among placodonts relate to the dentition. According to Kuhn-Schnyder, the characteristic "dentition of placodonts is revealed in *statu nascendi* in *Paraplacodus*" (1974:71). Each premaxilla carries three elongate, sigmoidally curved and pointed, strongly procumbent teeth. Two similar teeth, again procumbent, are located at the anterior end of each dentary. The anterior part of the dentition, congruent with *Placodus* in tooth count, is clearly adapted to procure hard-shelled invertebrate prey from the muddy substrate in marginal zones of the Monte San Giorgio basin. These gripping teeth are separated from the more posteriorly positioned crushing teeth by a distinct diastema.

Paraplacodus had not yet evolved the large, flat crushing tooth plates that are located on the dentary, maxilla and palatine in *Placodus* and armored placodonts, which gave the clade its name. The crushing teeth of *Paraplacodus* have a comparatively high crown; they are somewhat broader than long, and they terminate apically in a blunt tip (maxilla, dentary) or flat surface (palatine). There are seven crushing teeth located on the maxilla, parallel to which four somewhat larger teeth are aligned on the palatine. The dentary carries seven crushing teeth again of the same size as the maxillary teeth. In occlusion, the dentary tooth row fits in between the parallel tooth rows on the maxilla and palatine. Kuhn-Schnyder compared the action of this dentition to someone breaking a tree branch across one's knee (1974:71). The dentary tooth row functions like the knee across which the branch is broken, and the palatine and maxillary tooth rows function like the two hands used to break the branch.

In the postcranial skeleton, Kuhn-Schnyder counted 6+ cervical, 21 dorsal, 3 sacral, and 54 caudal vertebrae (1942:175). The tail is thus quite long, with neural spines and chevrons that undergo a curious metamorphosis in shape in an antero-posterior sequence (Rieppel, 2000b, fig. 6). In addition to pre- and postzygapophyses, the intervertebral articulation is enforced by a hyposphene-hypantrum articulation, a feature also known from *Placodus*. The transverse processes on the trunk vertebrae are characteristically elongated, again as in *Placodus*, and they articulate with dorsal ribs that carry a distinct uncinate process. The gastral rib cage is again well developed and sturdy. The vertebral neural spines are capped by osteoderms in *Placodus*, in contrast to *Paraplacodus*, in which osteoderms are absent. The scapula differs from that of other placodonts in that it carries a distinct dorsal blade of rectangular contours. The ilium of *Paraplacodus* again is unique among placodonts in that it essentially forms a vertical strut with concave margins; the ventral acetabular portion is more distinctly expanded than the dorsal head. The humerus is more sturdily built than the femur in *Paraplacodus*, and the radius is more robustly built than the ulna. In general, however, the appendicular skeleton remains rather incompletely known, especially with regard to the ossifications in manus and pes (Rieppel, 2000b).

The elongate transverse processes on the dorsal vertebrae, and the dorsal ribs carrying a distinct uncinate process, have invited a comparison of *Paraplacodus* with *Saurosphargis voltzi*, represented by a section of the trunk skeleton described and figured by Huene (1936). Pinna indeed included *Saurosphargis* in his Paraplacodontinae (1999). However, as discussed in chapter 3 and above, new and complete material from southwestern China shows that *Saurosphargis* is not a placodont, nor even a sauropterygian, but a representative of a separate clade of Triassic marine reptiles.

Isolated teeth, or dorsal ribs, from the Middle Triassic of the southern Alps, Spain, southwestern Germany, Transylvania, and Israel have been referred to *Paraplacodus*, but the material is not diagnostic, not even at the genus level (Rieppel, 2000b, and references therein).

Cyamodus hildegardis PEYER

The holotype of *Cyamodus hildegardis* was collected from a partially weathered dolomite layer at the Val Porina locality in 1924 (Peyer, 1931d). The entire skeleton is approximately 1.3 meters long. Its preservation was, in part, precarious. The skull and the trunk are exposed in dorsal view, the tail in right lateral view. The appendicular skeleton is not preserved or exposed, except for a couple of incompletely preserved or exposed elements that are tentatively referred to the fore limb. The skull and mandible are very badly dorsoventrally crushed and offer little morphological detail beyond the general contours and the dentition. The dermal armor (carapace) is incompletely preserved but covers the pelvic and proximal tail region posteriorly.

Box 5.3. *Cyamodus hildegrdis* from the Grenzbitumenzone of Monte San Giorgio

Cyamodus hildegardis is a representative of the armored placodonts. The armored placodonts, the Cyamodontoidea, are much more diverse than the unarmored placodonts, and reach up into the very late Triassic (Rhaetian); they were the last of the placodont lineage to go extinct. Cyamodontoid placodonts developed a body armor superficially resembling the turtle shell, and have at some times even been considered (erroneously) to be turtle ancestors. In cyamodontoid placodonts, the body armor is composed of osteoderms, ossifications that form in the deep layer of the skin. Such osteoderms also adorn the temporal region of the skull of many cyamodontoids. In the case of *Cyamodus hildegardis,* the dorsal body armor comprised two components, a dorsal shield covering the trunk of the animal, and a caudal shield covering the pelvic and proximal tail regions. Although present in other cyamodontoid placodonts, a ventral body armor was not developed in *Cyamodus hildegardis*. The skull of cyamodontoid placodonts is distinctly expanded posteriorly, allowing the accommodation of massive jaw adductor muscles operating the jaws furnished with large crushing tooth plates. These placodonts must have searched for their prey in the marginal zones of the Grenzbitumenzone basin, picking hard-shelled invertebrate prey off the substrate, as the deep waters were oxygen depleted.

© Beat Scheffold

A skull and mandible of *Cyamodus hildegardis*, associated with a few other skeletal elements—mostly osteoderms and incomplete limb elements—were collected at the Val Porina locality in 1931, and a tiny skull of a very immature individual was collected at the same locality in 1933 (Peyer, 1935, pl. 46). The skull and mandible are again severely dorsoventrally compressed, offering little morphological detail beyond the general contours and the dentition. The radiograph of the juvenile skull revealed an incomplete preservation of bone, such that preparation concentrated on the dentition only. The skull and mandible of a subadult individual of *Cyamodus hildegardis* were collected in 1957 at Point 902. The skulls and mandibles of juvenile, subadult, and adult individuals (fig. 5.5) allowed Kuhn-Schnyder the reconstruction of ontogenetic changes in the dentition of *Cyamodus hildegardis* (1959b). Yet another skull and mandible of a subadult individual were collected at the Besano locality in 1975, and finally the particularly well-preserved skeleton of a juvenile individual came to light in a cave near Pogliana above Besano (Pinna, 1992:3).

Except for the dentition, the cranial morphology of *Cyamodus hildegardis* remains poorly known. The skull does show the typical cyamodontoid proportions, though. The preorbital region is relatively short and tapers to a relatively narrow rostrum with a rounded anterior margin. The temporal region is broadly expanded, the upper temporal fenestra is large, and the posterolaterally expanding upper temporal arch is encrusted with osteoderms in its posterior part (Rieppel, 2001a). The dentition shows interesting ontogenetic changes in that the number of teeth on the maxillary and palatine is reduced in the adult, whereas the palatal tooth plates increase in size (Kuhn-Schnyder, 1959b; 1974, fig. 48). The juvenile specimen from Val Porina, 1933, shows two teeth on the premaxilla, three teeth on the maxilla, and three tooth plates on the palatine, of which the anteriormost one is very small. The subadult specimen from Point 902,

5.5. The skull of *Cyamodus hildegardis* (Paleontological Institute and Museum, University of Zurich T4771) from Mirigioli, Point 902. The skull is preserved in ventral view, displaying the cyamodont dentition. The skull measures 10.5 cm in length (Photo © Heinz Lanz/ Paleontological Institute and Museum, University of Zurich).

1957, shows two teeth on the premaxilla, four teeth on the maxilla, and two tooth plates on the palatine, of which the posterior one is distinctly enlarged. The adult dentition is reconstructed based on the Val Porina specimens 1924 and 1931, and it shows two teeth on the premaxilla, three teeth on the maxilla, and two tooth plates on the palatine, the posterior one again distinctly enlarged. In the dentary, two anterior procumbent teeth are separated from posterior crushing teeth by a distinct diastema. In the juvenile and subadult condition, there are four posterior crushing teeth, of which the anteriormost one is lost in the adult condition.

Much of the postcranial skeleton is obscured by the dermal armor, except in the juvenile specimen from Pogliana, in which the carapace is not yet fully formed. As far as is known, there are at least 12 dorsal, 3 sacral, and 15 caudal vertebrae (Pinna, 1992:12). The transverse processes of the dorsal vertebrae are elongated as is typical for placodonts. There is no ventral dermal armor (plastron), but the gastral rib basket is well developed. In the juvenile specimen from Pogliana, irregularly shaped osteoderms that "thin out" toward their margin have not yet coalesced to form a closed carapace (fig. 5.6). Their coalescence appears to occur in an antero-posterior gradient, as osteoderms are more densely packed in the anterior trunk region. The adult dorsal armor comprises two separate shields, formed of hexagonal osteoderms that meet in weakly interdigitating sutures. The dorsal shield (carapace) covers the trunk, and the tail

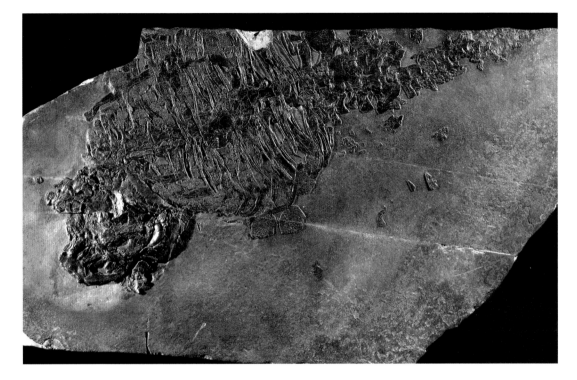

shield covers the pelvic and proximal caudal region. Distinctly enlarged pyramidal osteoderms line the margins of both the dorsal and the tail shield (Rieppel, 2002b, fig. 11).

The dermal armor offered protection, but also added "bone ballast" to counter positive buoyancy. The Nile crocodile has been claimed to ingest gastroliths to adjust buoyancy, although the evidence is considered equivocal by some (Seymour, 1982:9; Taylor, 1993:168). It has been suggested that the addition of "bone ballast" through pachyostosis as in the sauropterygian pachypleurosaurs, or the development of dermal armor as in cyamodontoid placodonts, would allow an increase in the volume of the lungs without impairing diving activity (Taylor, 2000). There is a trade-off, though, in that more energy is required to rise to the surface for respiration. The bipartition of the dorsal dermal armor in *Cyamodus hildegardis* may have added mobility, in particular dorsiflexion in diving. The precumbent premaxillary and anterior dentary teeth would have been used to pick hard-shelled invertebrate prey from the substrate. Given that much of the dermal armor appears to have developed during postembryonic growth, as is indicated by the juvenile specimen from Pogliana, *Cyamodus hildegardis* would have been vulnerable to predation during early stages of its postembryonic life cycle. Such is, indeed, indicated by the stomach contents of the sauropterygian *Lariosaurus buzzi* (Tschanz, 1989). They comprise four fragments of tooth-bearing bones of *Cyamodus hildegardis*, which correspond in size to the tiny juvenile skull collected at the Val Porina locality in 1933. Of the latter specimen Peyer noted that it represents "by far the smallest placodont skull that has ever become known" (1935:23).

5.6. A juvenile specimen of *Cyamodus hildegardis* (Museo Civico di Storia Naturale di Milano V 458) from the Rossaga quarry, Pogliana (Varese). The specimen is preserved in ventral view; the skull points to the lower left. Note the incompletely formed carapace. As preserved, the specimen measures 36 cm in length (Photo © Giorgio Teruzzi/Museo Civico di Storia Naturale di Milano).

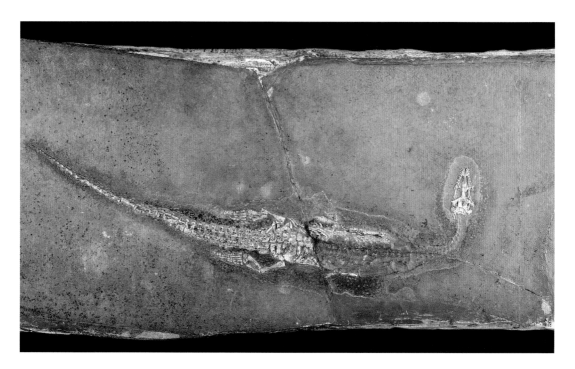

6.1. The holotype of
Pachypleurosaurus edwardsii
(Museo Civico di Storia
Naturale di Milano V 3549)
from Besano. The specimen
measures 30 cm in length
(Photo © Giorgio Teruzzi/
Museo Civico di Storia
Naturale di Milano).

Pachypleurosaurs

In 1854, Emilio Cornalia described a new species of a small, lizard-like reptile on the basis of four specimens collected near Besano and Viggiù (the Ca' del Frate locality north of Viggiù), which he named *Pachypleura Edwardsii* (Cornalia, 1854:5). Significantly, the Besano Formation is geologically older than the fossiliferous outcrops at Ca' del Frate. Cornalia published at a time when two other fossil reptiles had already been described from the Triassic of Lombardy, although from a different location and different time horizon (1854). These outcrops were located near Perledo on the western slope of the Grigna mountains east of Lake Como. The deposits are geologically younger than the Besano Formation, the historical fossils probably coming from the upper part (Perledo Member) of the Ladinian Perledo-Varenna Formation (Furrer, 1995:833; Tintori and Lombardo, 1999:495). The first fossil reptile ever to have been described from the Triassic of Lombardy came from the Perledo locality, informally ascribed to the "famiglia dei Paleosauri" by Guiseppe Balsamo-Crivelli (1800–1874) in 1839 (Balsamo-Crivelli, 1839:425). That specimen was later named *Lariosaurus balsami* by Curioni, and thus became the holotype of its species; *L. balsami* then is the type species of the genus *Lariosaurus* (1847:166). In the same paper, Curioni described a second saurian from Perledo under the name *Macromirosaurus Plinj* (1847:161; *Macromerosaurus* in Cornalia, 1854, and Curioni, 1863). When in 1854, Cornalia described the four small saurian specimens from Besano and Viggiù (Ca' del Frate) under the name *Pachypleura Edwardsii*, he did not designate a holotype. The third specimen in his sample, his "esemplare minore" said to come from Viggiù (Ca' del Frate) and originally belonging to the collection of Count Vitaliano Borromeo, was already mentioned in passing by Curioni, and it was selected as holotype for the species by Boulenger (Curioni, 1847; Boulenger, 1898:7, n.1; see also Broili, 1927:218; Nopcsa, 1928:22). Boulenger realized that the type specimen of *Pachypleura edwardsii* (fig. 6.1) had to be one collected in the Viggiù area, since the (geologically older) specimens from Besano that were included in Cornalia's sample he thought to possibly represent the genus *Neusticosaurus*. In his paper of 1863, Curioni compared *Pachypleura edwardsii* with *Lariosaurus balsami*, confirming their difference, but he also declared *Macromerosaurus Plinj* identical with *Lariosaurus balsami*. According to Peyer, *Lariosaurus balsami* must take priority since the specimen that became the holotype had already been described and figured, although not named, by Balsamo-Crivelli in 1839 (1933:6). His view has prevailed, although *Macromirosaurus Plinj* takes page priority to *Lariosaurus balsami* in Curioni's 1847 publication (see Zittel, 1887–1890:484; Arthaber, 1924:489). In another development, Lydekker in his *Catalogue of the Fossil Reptilia and Amphibia in the British Museum* (*Natural History*) found the name *Pachypleura* to be already in use for a coleopteran beetle; he drew no nomenclatural consequences, however, as he thought the specimen so named should possibly be referred to the genus *Neusticosaurus* (Lydekker, 1889:285). *Pachypleura* was independently renamed *Pachypleurosaurus* by Broili (1927) and Nopcsa (1928).

The confusion concerning the genera *Lariosaurus* and *Pachypleurosaurus* resulted from the fact that Curioni claimed in his 1863 paper that all fossil saurians from the Perledo locality are to be referred to a single species, *Lariosaurus balsami*, whereas Cornalia had claimed that *Pachypleura* also occurs in the Perledo deposits (Cornalia, 1854). If both of these authors were right, this would imply that *Pachypleura* (*Pachypleurosaurus*) would be a junior synonym of *Lariosaurus*. Curioni's claim was decidedly refuted by Peyer, however, who unequivocally established the presence of pachypleurosaurs in the Perledo material (1933). In fact, *Lariosaurus* is a nothosaurid, and hence quite far removed from *Pachypleurosaurus* in terms of its phylogenetic relationships (Rieppel, 2000a).

The first pachypleurosaurs from the Grenzbitumenzone of Monte San Giorgio were described and illustrated by Peyer under the name *Pachypleurosaurus edwardsii* (1932). This was a preliminary account based on 11 specimens selected because of their completeness and excellent preservation. This preliminary report was followed by the publication in 1935 of Rainer Zangerl's (1912–2004) PhD thesis that was based on 105 prepared specimens. At that time, a total of approximately 500 specimens of pachypleurosaurs had already been collected at Monte San Giorgio. According to Zangerl the specimens he prepared and described came from three different horizons (the Cava Inferiore beds, the Cava Superiore beds, and the Alla Cascina beds) of the Meride Limestone (Ladinian) (1935). However, two specimens described by Zangerl are labeled as coming from Val Porina (1935, text-figs. 32, 36, and 44). The latter yielded fossils from two time horizons: the Grenzbitumenzone accessible in the Val Porina mine, and the Cava Inferiore horizon at an outcrop in the lower Val Porina (Sander, 1989a:572). As a result, Zangerl's sample did include specimens from the Grenzbitumenzone (Rieppel, 1989b:3). As is known today, pachypleurosaurs do not occur in the Grenzbitumenzone below bed 139 of the 190 beds that have been recognized (Rieber, 1973a; Sander, 1989a:571). This is the reason for Zangerl's claim that *Mixosaurus*, a common reptile in the Grenzbitumenzone, and pachypleurosaurs do not stratigraphically overlap, but that pachypleurosaurs follow *Mixosaurus* in their temporal distribution at Monte San Giorgio (1935:5; see also Peyer, 1932). Drawing on specimens from four different time horizons, Zangerl had to deal with remarkable variability in his material, all of which he referred to as *Pachypleurosaurus edwardsi*.

In 1959, Kuhn-Schnyder described a new species of *Pachypleurosaurus*, that is, *P. staubi*, from the Prosanto Formation (Ladinian, Middle Triassic) of Stulseralp near Bergün, Canton of Graubünden, Switzerland (Kuhn-Schnyder, 1959c:656; Furrer, 1995). In order to assess the taxonomic status of the new find, Kuhn-Schnyder compared it to the rich material of *Pachypleurosaurus edwardsii* from Monte San Giorgio that had in the meantime accumulated in the Zurich collections. Noting even more extensive morphological variation than had been described by Zangerl, he concluded: "It seems likely that another species or perhaps even several species of pachypleurosaurs occur in the Triassic of the southern Alps, which have hitherto been referred to *Pachypleurosaurus edwardsi* (CORNALIA)" (Kuhn-Schnyder, 1959c:652). He declared the reexamination of the type specimen of *Pachypleurosaurus edwardsii* a most essential step in any attempt to gain further insight into the systematics of Monte San Giorgio pachypleurosaurs. This was exactly the step taken by Carroll and Gaskill in the next publication that dealt with this vast material (1985).

During their visit at the Paleontological Institute of the University of Zürich, Carroll and Gaskill concentrated on well-preserved, relatively large specimens of pachypleurosaurs, which previously had informally been referred to as "*Grosser Pachypleurosaurus*." These, they contrasted with the vast material of smaller pachypleurosaurs, which they recognized as generically different from the *Grosser Pachypleurosaurus*. Differences became evident in the phalangeal formula of manus and pes, where the *Grosser Pachypleurosaurus* invariably shows one and two phalanges in the first and second digit, respectively, whereas the smaller pachypleurosaurs invariably show two and three phalanges in the first and second digit, respectively. Other than size and phalangeal counts, differences of relative proportions were noted in features concerning the skull and the appendicular skeleton (pectoral and pelvic girdles and limbs). Comparison with the type specimen of *Pachypleurosaurus edwardsii* (see fig. 6.1) led Carroll and Gaskill to conclude that the latter represents a juvenile specimen of the *Grosser Pachypleurosaurus*. Carroll and Gaskill referred the multitude of small pachypleurosaurs from Monte San Giorgio to the genus *Neusticosaurus* (1985). In 1881 Oscar Fraas described small, lizard-like fossils from the Lettenkohle (lower Keuper, Ladinian, Middle Triassic) of Ludwigsburg near Stuttgart, Germany, under the name of *Simosaurus pusillus*. Seeley immediately recognized that the structure of the skull in this latter species differs markedly from that of the previously described *Simosaurus gaillardoti*, and hence referred the species *pusillus* to a new genus, which he named *Neusticosaurus*; *N. pusillus* thus became the type species for its genus (Seeley, 1882; Meyer, 1842; Rieppel, 1994a). Carroll and Gaskill did not attempt to clarify the taxonomic status of the small pachypleurosaurs from Monte San Giorgio, referring those simply to *Neusticosaurus* sp. (1985). With respect to *Pachypleurosaurus staubi* described by Kuhn-Schnyder, Carroll and Gaskill concluded: "It resembles closely the multitude of small pachypleurosaurid specimens from

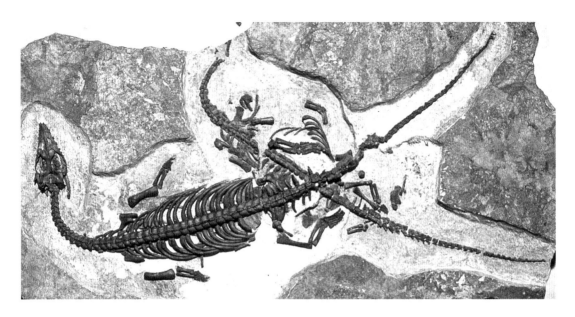

6.2. Two immature specimens of *Serpianosaurus mirigiolensis* (Paleontological Institute and Museum, University of Zurich T1071), from Mirigioli, Point 902. The fossil-bearing slab measures 36 cm in width (Photo © Heinz Lanz/ Paleontological Institute and Museum, University of Zurich).

Monte San Giorgio and it would certainly be placed in the same species if it had been found at that locality" (1985:348). Indeed, "*Pachypleurosaurus staubi* cannot be distinguished from *N. pusillus*" (Stockar and Renesto, 2011:S175; see also Rieppel, 2000a). The same would probably apply to the two specimens from a locality near Lake Lugano, northern Italy, described by Mateer as *Pachypleurosaurus* cf. *staubi*, specimens which in Carroll and Gaskill's judgment reinforce the distinction of *Pachypleurosaurus edwardsii* (the *Grosser Pachypleurosaurus*) from the smaller form (*Neusticosaurus*) (Mateer, 1976; Carroll and Gaskill, 1985:349).

The mystery of the small pachypleurosaurs (*Neusticosaurus*) was finally resolved by P. Martin Sander (1989a) through his graduate research. Sander took an essentially stratigraphic approach when trying to sort out the taxonomy of *Neusticosaurus* of Monte San Giorgio. Conducting fieldwork in order to stratigraphically correlate with one another the various localities at which pachypleurosaurs had been collected, he noted that pachypleurosaurs occur in four different horizons, each of these characterized by a different pachypleurosaur species. The earliest occurrence of pachypleurosaurs is in the upper part of the Grenzbitumenzone (Cava Tree Fontana, Val Porina mine, P. 902 excavation). It yielded medium-sized pachypleurosaurs—larger than *Neusticosaurus*, but smaller than *Pachypleurosaurus edwardsii*—which had originally, and informally, been referred to the genus *Phygosaurus*. The latter genus was based on a pachypleurosaur from the Grigna mountains (Perledo), kept at the Institut de Géologie, of the Université Louis Pasteur de Strasbourg and first described by Deecke under the name *Lariosaurus balsami* (Deecke, 1886; see also Rieppel, 1987a). The specimen can no longer be located today, as it most probably was destroyed in a fire in 1967. It consisted of a trunk skeleton and a partial neck, pectoral girdle, and incompletely preserved fore limbs, and elements of the pelvic girdle, all exposed in ventral view. Arthaber noted a number of differences between the Strasbourg specimen

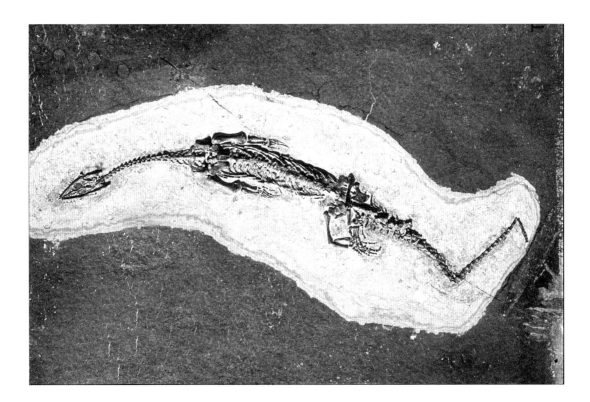

6.3. A well-preserved specimen of *Neusticosaurus pusillus* (Paleontological Institute and Museum, University of Zurich 3943) from the Cava Inferiore horizon at Monte San Giorgio. The specimen measures 21 cm in length (Photo © Heinz Lanz/ Paleontological Institute and Museum, University of Zurich).

and *Lariosaurus*, such as the absence of rib pachyostosis and the presence of only three sacral ribs (as opposed to four or more sacral ribs in *Lariosaurus*) (1924). Arthaber consequently referred the Strasbourg specimen to a new genus and species, *Phygosaurus perledicus* (1924:439). The specimen was subsequently re-prepared and re-described by Peyer, who provided an excellent illustration and a formal diagnosis for the taxon (1933:120, and pls. 38, 39). When describing the pachypleurosaurs from the upper part of the Grenzbitumenzone, Rieppel noted that all the features listed as diagnostic for *Phyosaurus perledicus* are plesiomorphic, and hence considered the name a *"nomen dubium"* (Rieppel, 1989b; the name "is not certainly applicable to any known taxon": Blackwelder, 1967:398). Rieppel consequently referred the Grenzbitumenzone pachypleurosaurs to a new genus and species, *Serpianosaurus mirigiolensis* (fig. 6.2) (1989b).

The next level up to yield pachypleurosaurs is the Cava Inferiore beds of the Lower Meride Limestone, Ladinian (Cascinello, lower Val Porina, Val Serrata, Acqua del Ghiffo, Acqua Ferruginosa, and a few other spots). Here, the typical small pachypleurosaurs have been collected, which Sander recognized as being con-specific with *Neusticosaurus pusillus* (fig. 6.3) (Sander, 1989a:576; Fraas, 1881; Seeley, 1882). Above that horizon, small pachypleurosaurs have been collected from the Cava Superiore beds, Lower Meride Limestone, Ladinian (Acqua del Ghiffo, Ca' del Monte), which Sander found to represent a different, new species that he named *Neusticosaurus peyeri* (1989a:601). It differs from *N. pusillus* in being of slightly larger overall size, with a wedge-shaped skull that displays very small, slit-like upper temporal fenestrae, a rounded anterior margin of the shoulder girdle, a short scapular blade, and a somewhat more distinctly expressed sexual dimorphism in the humerus. The neural spines on the posterior dorsal vertebrae are of somewhat increased height, a feature that is more distinctly expressed in *Pachypleurosaurus edwardsii*. The latter taxon comes from the youngest horizon, the Alla Cascina beds, Lower Meride Limestone, Ladinian (Alla Cascina, Val Serrata, lower Val Porina, road Crocifisso-Serpiano) (Carroll and Gaskill; 1985). There are then, following Sander, four pachypleurosaur horizons at Monte San Giorgio, each one characterized by its own pachypleurosaur species, originally thought to be without any taxonomic overlap between successive horizons (1989a). A phylogenetic analysis revealed the pachypleurosaurs from Monte San Giorgio to form a monophyletic group, within which *Pachypleurosaurus edwardsii* nested inside the *Neusticosaurus* clade (Sander; 1989a). This result necessitated the referral of the species *edwardsii* to the genus *Neusticosaurus* as well (Kuhn-Schnyer, 1994, objected to that conclusion). Having sorted out the pachypleurosaur taxonomy, Sander could identify *Serpianosaurus mirigiolensis*, *Neusticosaurus pusillus*, and N. *peyeri* in the sample described by Peyer, as well as *Pachypleurosaurus edwardsii* along with the other three species in the sample described by Zangerl (Sander, 1989a:603; Peyer, 1932; Zangerl, 1935). This, of course, explains the perplexingly broad range of variation recognized by Zangerl (1935).

Box 6.1. *Neusticosaurus peyeri* from the lower Meride Limestone of Monte San Giorgio

Neusticosaurus peyeri is a small species of pachypleurosaur, adults usually reaching an overall length of 45 cm. The group is named after the pachyostosis of the ribs and vertebrae, a thickening of bone that adds ballast and so helps counteract positive buoyancy. This reduces energy requirement during diving excursions. Pachypleurosaurs are rather generalized, lizard-like reptiles that show few specific adaptations to life in the sea other than pachyostosis. Adults are subject to sexual dimorphism, the males being a bit more robustly built than the females. Pachypleurosaurs were viviparous and did not crawl on land to deposit eggs in burrows like extant sea turtles do. The deposits at Monte San Giorgio did not yield any gravid pachypleurosaur females, such as are known from Triassic marine deposits of southwestern China, but an isolated embryo of *Neusticosaurus peyeri* has been described with no fragments of a calcified eggshell associated with it (Sander, 1988). Pachypleurosaurs, especially the smaller species *Neusticosaurus pusillus* and *Neusticosaurus peyeri*, are known from abundant material that allows investigation of individual as well as ontogenetic variation. They have also been portrayed as an example of anagenetic evolution, that is, evolutionary transformation and speciation within a single discrete evolutionary lineage. Their taxonomy is not easily sorted out, however, because of the uncertain provenance and taxonomic status (it is an immature individual) of the holotype of *"Pachypleuro-saurus edwardsii,"* the latter one of the earliest fossil marine reptiles collected in the Middle Triassic of Northern Italy around the middle of the nineteenth century.

© Beat Scheffold / PIMUZ)

The Oldest Pachypleurosaur from the Besano Formation

At Monte San Giorgio, the earliest occurrence of pachypleurosaurs is that of *Serpianosaurus mirigiolensis* in the upper part of the Grenzbitu-menzone, the latter geologically equivalent to the Besano Formation. In 2011, a team from the Museo Civico di Storia Naturale in Milano pursued fieldwork targeting the Besano Formation at the Rio Vallone locality near Besano. Sifting through mining debris they found a slab containing a new pachypleurosaur. Lithological peculiarities of the slab that contained the new pachypleurosaur indicated that it came from the middle part of the Besano Formation. Hence the new pachypleurosaur, named *Odoiporosaurus teruzzii* (fig. 6.4), is geologically somewhat older than *Serpianosaurus*, and indeed the earliest occurrence of a pachypleu-rosaur in the Besano Formation/Grenzbitumenzone (Renesto, Binelli et al., 2014). The new taxon is similar in size to *Serpianosaurus*, but distinct from the latter by a relatively smaller skull, a relatively larger upper temporal fenestra, a broadened ulna, and the dentition. The relatively large upper temporal fenestra is a plesiomorphic feature which *Odoipo-rosaurus* shares with the *Anarosaurus–Dactylosaurus* clade from the Ger-manic Muschelkalk (Anisian; Rieppel and Lin, 1995). The dentition of *Odoiporosaurus* is of particular interest as it consists of characteristic leaf-shaped teeth similar to those observed in *Anarosaurus heterodontus* from the Lower Muschelkalk (Anisian) of Germany and the Nether-lands (Rieppel and Lin, 1995; Klein, 2009). The phylogenetic analysis showed *Odoiporosaurus* to be the sister taxon to the *Serpianosaurus–Neusticosaurus* clade, with the *Anarosaurus–Dactylosaurus* clade being

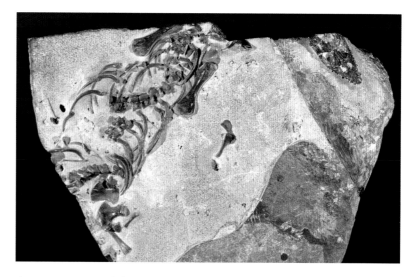

the sister group of the pachypleurosaurs from the southern Alpine Triassic: ((*Anarosaurus, Dactylosaurus*) (*Odoiporosaurus* (*Serpianosaurus, Neusticosaurus*))).

The Anatomy of Monte San Giorgio Pachypleurosaurs

The fact that it proved so difficult to untangle the taxonomy of the pachypleurosaurs from Monte San Giorgio indicates the close morphological similarities that prevail amongst the four species that have been recognized. Differences between the species concern mainly numerical and metric characters, such as tooth and vertebral counts, as well as phalangeal counts, and proportional relations such as relative skull size, relative height of neural spines, and limb proportions. The pachypleurosaurs retain a generalized, lizard-like habitus with a relatively small skull, a distinct neck, an elongated trunk, and a long tail. Across the four species, the number of cervical vertebrae ranges from 13 to 21, the number of dorsal vertebrae ranges from 17 to 23/24. The number of trunk vertebrae increases from stratigraphically older to younger species. The neuro-central suture generally remains open. The sacral ribs, as well as the caudal ribs present in the proximal tail region, do not fuse with the respective vertebrae on which they articulate. The caudal transverse processes are elongated in *Neusticosaurus pusillus* and *N. peyeri*. The name of the clade refers to the pachyostotic ribs (prominent in *Neusticosaurus pusillus* and *N. peyeri*), and pachyostosis also affects the centra and neural arches of the vertebrae. In the skull, the orbits are enlarged, the upper temporal fenestrae distinctly reduced in size, the upper temporal arch is broad, and the ventral margin of the cheek region is embayed. The snout may be more slender or more rounded. The lower jaw is delicately built, the coronoid process weakly developed. The teeth are numerous, small, and peg-like. Pachypleurosaurs probably preyed on small nektonic invertebrates and fish in the coastal surface waters (Sander, 1989a:653). In the pectoral girdle the clavicles and interclavicle form a solid bar anteriorly, behind which is located the pectoral fenestra, an opening bounded laterally by the medial margins of scapula and coracoid. The interclavicle underlies the clavicular blades, a sauropterygian synapomorphy, as is also the relation of the clavicle to the scapula. Unlike other tetrapods, the dorsal part of the clavicle is received on the morphologically medial surface of the scapula. In the pelvic girdle, a thyroid fenestra is present. The obturator foramen in the pubis may be slit-like, that is, not enclosed by bone, as in immature specimens or also in adult *Neusticosaurus edwardsii*. Both pubis and ischium are biconcave, whereas the ilium is a rather small, stubby ossification with a blunt dorsal process and a ventrally expanded acetabular portion. *Neusticosaurus* and *Serpianosaurus* have three sacral ribs articulating with the ilium. The humerus is always more robustly built than the femur, and distinctly longer than the femur in *Neusticosaurus edwardsii*. The ratio of humerus to femur length, as well as the robustness of the humerus, is sexually dimorphic in *Serpianosaurus* and *Neusticosaurus*. In analogy to the Triassic pachypleurosaur *Keichousaurus* from southwestern China, of which gravid specimens are known, it can be concluded that the humerus is relatively shorter and less robustly built in females (Y.-N. Cheng, Wu, and Ji, 2004). In fact, the humerus in females often resembles the immature humerus in males. It is possible that the more robustly built fore limbs of males in these sexually dimorphic pachypleurosaurs played some role in mating behavior. The radius again is longer and more robust than the ulna, the rounded or elongate rectangular ulnare typically located distal to the ulna and posterior to the distal end of the radius. Ossifications in the manus (carpus and digits) are subject to ontogenetic variation and skeletal paedomorphosis. The phalangeal formula in the manus is variably reduced by comparison to the primitive condition, or else distalmost phalanges failed to ossify. The femur is straight and more lightly built than the humerus. The tibia again is more robust and somewhat longer than the fibula. The tarsal ossifications typically comprise the astragalus and the calcaneum, the first larger than the latter and located in an intermedium position. The phalangeal formula is either primitive reptilian or slightly reduced in the pes. Phalangeal reduction appears to follow a gradient from stratigraphically older to younger species (Hugi et al., 2011).

An embryo of *Neusticosaurus peyeri* was described by Sander; it is the smallest specimen known of *N. peyeri*, clearly immature to judge from the degree of ossification, with a relatively large skull and equally relatively large orbits (Sander, 1988; see also Sander, 1989a). The scleral ossicles are well ossified, however, forming a closed ring of overlapping bony plates. This indicates that the specimen represents a late stage embryo or indeed a neonate; in the turtle *Chelydra serpentina*, for example, scleral ossicles overlap to form a closed ring only at the time of hatching (Franz-Odendaal, 2006). Quite generally in vertebrates, the timing of the onset of ossification of these ocular elements is highly variable, but overall they develop late in ontogeny (Franz-Odendaal and Vickaryous,

The Life History of the Pachypleurosaurs from Monte San Giorgio

6.5. The holotype of *Serpianosaurus mirigiolensis* (Paleontological Institute and Museum, University of Zurich T3931), from Mirigioli, Point 902. The specimen measures 67 cm in length (Photo © Heinz Lanz/Paleontological Institute and Museum, University of Zurich).

2006). The curled-up position of the specimen suggests that the egg membrane must have been intact or ruptured only very briefly before the carcass became embedded in sediment. This speaks for viviparity in the genus *Neusticosaurus* and most probably other pachypleurosaurs as well. It is inconceivable that *Neusticosaurus* would have buried eggs with a calcified eggshell on land, and that one of these eggs would somehow have been washed into the sea, getting preserved as it is—an immature skeleton in an embryonic posture without any trace of a calcified eggshell remaining. Viviparity has also been documented by way of gravid females for the pachypleurosaur *Keichousaurus hui* from the Middle to early Upper Triassic of southwestern China (Y.-N. Cheng, Wu, and Ji, 2004).

Given the abundance of the material available, skeletochronological studies on pachypleurosaurs from Monte San Giorgio could be used to determine when individuals of the different species would have reached sexual maturity and how long they would have lived. It seems that the greater the individual size, the longer the longevity. The largest specimen of *Serpianosaurus mirigiolensis* (fig. 6.5). is estimated to have been in its 14th year at the time of death. It would have reached sexual maturity during its second or third year (Hugi et al., 2011:421). *Neusticosaurus pusillus* and *N. peyeri* reached sexual maturity at three to four years old, and are estimated to have lived to an age of nine to ten years (Sander, 1989a). The largest specimen of the *Grosser Pachypleurosaurus—Neusticosaurus edwardsii*—was in its 23rd year at the time of death and had reached sexual maturity "in its sixth to seventh year" (Hugi et al., 2011:423).

Sedimentary analysis of the layers yielding pachypleurosaurs at Monte San Giorgio indicate a seasonal subtropical climate subject to rhythmic monsoons (Röhl et al., 2001). Data from sedimentary analysis also "support a habitat with more open marine influences for *S. mirigiolensis*, but more restricted lagoonal habitats for *N. pusillus*, *N. peyeri*, and *N. edwardsii*" (Hugi et al., 2011:423). There also seems to prevail a trend to increasing aquatic adaptation from stratigraphically older to younger species, with *Neusticosaurus edwardsii* being the most pelagic one as revealed by the higher vascularization of the bone tissue (Hugi et al., 2011:425). Indeed, looking at the ontogenetic sequence of ossification in the skeleton of these pachypleurosaurs, and associated changes in bone histology, it has been shown that the stratigraphically oldest species, *Serpianosaurus mirigiolensis*, most closely approaches the patterns discerned in the presumed terrestrial ancestral condition. In contrast, the stratigraphically youngest species, *Neusticosaurus edwardsii*, departs from that ancestral pattern to the largest degree due to heterochronic shifts in the ontogeny of skeleton formation (Hugi and Scheyer, 2012). This then raises the question of the evolutionary relationships of the species of pachypleurosaurs that occur in the Monte San Giorgio basin, as discussed below.

The Evolution of the Pachypleurosaurs from Monte San Giorgio

All four species in the *Serpianosaurus–Neusticosaurus* clade are marked out by autapomorphy. There is also cladistic structure within the clade: (*Serpianosaurus* (*N. pusillus* (*N. peyeri*, *N. edwardsii*))) (Sander, 1989a). Such cladistic structure is meant to indicate relative degrees of relationships, not to represent the process of phylogeny, i.e., a sequence of ancestors and descendants. Cladistic structure is therefore not necessarily tied to the stratigraphic succession of the four species. Some have nevertheless considered the consequences that obtain when the dichotomies in the cladogram of the *Serpianosaurus–Neusticosaurus* clade are taken to represent speciation events. If the sister taxa *N. peyeri* and *N. edwardsii* had originated from the same speciation event, one would expect both taxa to temporally overlap. Such, it was claimed, is not observed, in spite of the very rich and dense fossil record. *N. edwardsii* remains a ghost lineage in the Cava Superiore beds which was claimed to yield exclusively *N. peyeri* (O'Keefe and Sander, 1999). The state of knowledge at the time indicated a strict temporal succession with no overlap between the four species of the *Serpianosaurus–Neusticosaurus* clade that occur in the Monte San Giorgio basin, which shows open marine influence in early stages of its formation (Grenzbitumenzone), but signs of progressive restriction in later phases of its evolution (Meridekalke) (Furrer, 1995). Sander even recognized a subpopulation of *Neusticosaurus pusillus*, which is "slightly younger than its ancestor, *N. pusillus*, and somewhat older than its descendant, *N. peyeri*," and that is morphometrically intermediate between the latter two species (1989a:658). This suggests an anagenetic transition between *N. pusillus* and *N. peyeri*, which would violate the cladistic structure mentioned above. Indeed, O'Keefe and Sander make a case for considering the whole *Serpianosaurus–Neusticosaurus* clade as a single anagenetic lineage, as *N. peyeri* was found to be morphometrically intermediate between *N. pusillus* and *N. edwardsii* (1999). The four species would thus constitute an anagenetic lineage of ancestors and descendants, which would require the evolutionary reversal of the autapomorphic characters that mark out ancestral species (O'Keefe and Sander, 1989). O'Keefe and Sander further recognized that morphometric analysis "discriminates between *Serpianosaurus* and the members of *Neusticosaurus* more strongly than between the species of the later genus" (1999: 526). This observation renders the transition from *S. mirigiolensis* to *N. pusillus* a more complex event, one that requires a decrease of overall body size coupled with an increase in vertebral count (O'Keefe et al., 1999). The transition from *N. pusillus* to *N. peyeri* is documented by the intermediate population identified by Sander, whereas the transition from *N. peyeri* to *N. edwardsii* (the *Grosser Pachypleurosaurus*) involved a marked heterochronic shift: "we advance the hypothesis that *N. edwardsii* arose from *N. peyeri* via hypermorphosis" (O'Keefe et al., 1999:514).

More recently, this hypothesis of an anagenetic evolution of the *Serpianosaurus–Neusticosaurus* lineage in the Monte San Giorgio basin has been challenged by new fossil finds. The hypothesis was based on a presumed lack of overlap in the stratigraphic succession of the species in the *Serpianosaurus–Neusticosaurus* clade. Recent fieldwork, however, has unearthed a specimen of *Neusticosaurus peyeri* in the Alla Cascina beds (Stockar and Renesto, 2011). In another campaign, a specimen of *N. edwardsii* was collected in the Cava Superiore beds (Furrer, in Stockar and Renesto, 2011:S176). There is, thus, stratigraphic overlap between these two species after all.

The Problem of the Type Specimen of *Neusticosaurus edwardsii*

One of the surprising findings of the morphometric analysis of the species in the *Serpianosaurus–Neusticosaurus* clade was not just the fact that all four species were well demarcated, but that the type specimen of *Neusticosaurus edwardsii* (see fig. 6.1.) did "not group with the other specimens referred to that taxon," thus raising the question of whether it was "incorrectly classified" (O'Keefe and Sander, 1999:526). This is particularly interesting when keeping in mind that the type specimen was found at a locality (Ca' del Frate near Viggiù) and possibly in a stratigraphic level different from those of the specimens collected from the Alla Cascina beds at Monte San Giorgio (Carroll and Gaskill, 1985). The literature concerning the type specimen of *N. edwardsii* is consequently rather confused. According to Kuhn Schnyder, the type specimen comes from "the region around Besano"; its precise stratigraphic provenance is unknown (Kuhn-Schnyder, 1974:65, 67). Carroll and Gaskill believed the type specimen to come "from near Besano," not far from the Alla Cascina locality at Monte San Giorgio: "the fossil-bearing beds can be traced without break between the two localities" (1985:346). However, the exact stratigraphic provenance of the type specimen could no longer be established, and there exist "lithological differences" between the type specimen and the Monte San Giorgio material (Carroll and Gaskill, 1985:346). In addition, the small size of the type specimen of *N. edwardsii* renders it a juvenile individual, as evidenced by the ill-defined ends of the long-bones (Carroll and Gaskill, 1985:346). According to Sander, the fossil bearing beds at Ca' del Frate that yielded the type specimen are of the same age as the Alla Cascina beds at Monte San Giorgio (1989a:572, 617).

Carroll and Gaskill established the presence of *Neusticosaurus edwardsii* in the Alla Cascina beds of Monte San Giorgio on the basis of the type specimen selected from Cornalia's sample by Boulenger (1898:7, n. 1). Recall that Cornalia's sample comprised material from "near Besano," that is, the Besano Formation, as well as from "near Viggiù," that is, the Ca' del Frate locality (Cornalia, 1854; see also Pinna and Teruzzi, 1991:6). Recognizing the Besano sample as probably representing the genus *Neusticosaurus*, Boulenger selected a Viggiù specimen as holotype for *Pachypleura (Pachypleurosaurus) edwardsii*. In that he was followed by Broili (1927) and Nopcsa (1928). The latter in particular obtained permission to further prepare the type specimen, which he re-described in detail, and illustrated (Nopcsa, 1928, pl. II). *N. edwardsii* carries neural spines of increased height on the posterior dorsal vertebrae, and it has "one phalanx on digit 1, and two phalanges on digit 2 in manus and pes" (Carroll and Gaskill, 1985:348). The height of the neural spines on the posterior trunk vertebrae cannot be assessed in the type specimen, and its "digital formula can no more be made out" (Nopcsa, 1928:24). Sander (1989a:617) lists the relatively small head as a diagnostic character of *N. edwardsii*, the skull:trunk ratio varying from 0.23 to 0.27. There are two specimens of *N. edwardsii* identified by Carroll and Gaskill that approach the skull:trunk ratio of the type specimen with a ratio of 36% and 45%, respectively (1985:358, table 3). These two specimens are the smallest among the Alla Cascina material; the specimen with the 45% ratio is of about the same size as the type specimen. The relatively high skull:trunk ratio in small specimens thus reflects the negative allometric growth of the head, which raises the issue of ontogenetic variation impacting potentially diagnostic features in an immature type specimen. Indeed, and as stated by Stockar and Renesto, body proportions of small specimens of *N. edwardsii* "to some extent overlap with those of an adult *N. peyeri*, thus rendering them of little use for species identification" (2011:S176). Furthermore, the sample described by Carroll and Gaskill comprises a specimen from Aqua del Ghiffo, and two specimens from the lower Val Porina, where the Cava Superiore and the Cava Inferiore beds are exposed (Carroll and Gaskill, 1985:355; Sander, 1989a:572; see also Furrer, in Stockar and Renesto, 2011:S176). All of those are referred to *N. edwardsii*, as is also a specimen (not the type specimen) from Ca' del Frate (Sander, 1989a:618). Conversely, a specimen from the Alla Cascina beds was identified as *N. peyeri* (Stockar and Renesto, 2011). In summary, the question arises whether the Viggiù specimen selected by Boulenger from Cornalia's sample as type specimen for *Pachypleura (Pachypleurosaurus) edwardsii* does, indeed, unequivocally diagnose the Monte San Giorgio material from the Alla Cascina beds in the Lower Meride Limestone (Boulenger, 1898; Cornalia, 1854).

Fossils near Viggiù, Varese, were historically as well as in recent years collected at the Ca' del Frate locality. Fossils, both historical and more recent collections, from that locality have been attributed to the upper part of the Kalkschieferzone (Tintori, Muscio et al., 1985:202). The Kalkschieferzone also crops out in the Monte San Giorgio area, where it corresponds to the uppermost part of the Meride Limestone (uppermost Ladinian; Furrer, 1995:832). The Kalkschieferzone deposits are thus younger in age than the pachypleurosaur yielding horizons in the Lower Meride Limestone recognized by Sander (1989a). The Kalkschieferzone both at Ca' del Frate and in the area of Monte San Giorgio have yielded actinopterygian fishes as well as *Lariosaurus*, the latter also known from Perledo (Peyer, 1933; Kuhn-Schnyder, 1987; Tschanz, 1989; Tintori and Renesto, 1990; Renesto, 1993). Recent fieldwork has signaled the occurrence of pachypleurosaurs in central levels of the Upper Meride Limestone of late Ladinian age; the single specimen recovered is too poorly preserved, however, to allow identification to the species level (Renesto and Felber, 2007). Finally, there remains the possibility that the type specimen of *Neusticosaurus edwardsii* comes from a lower, geologically older horizon of the Ca' del Frate deposits. Heinz Furrer communicated that while coming from the Ca' del Frate locality, the type specimen must have come from a horizon that is of the same age as the Alla Cascina beds at Monte San Giorgio, and not from younger deposits targeted by modern fieldwork (letter dated April 24, 1997; also in Sander, 1989a)

Comparing the occurrence of actinopterygians at the different localities (i.e., Besano/Monte San Giorgio, Perledo, and Ca' del Frate), Tintori and Lombardo noted that actinopterygian species known from the Grenzbitumenzone or Lower Meride Limestone are also found in the lower part of the Perledo-Varenna Formation (the Varenna Limestone Member), whereas actinopterygian species known from Ca' del Frate (Kalkschieferzone) also occur in the upper part of the Perledo-Varenna Formation (the Perledo Member) (1999:495). So if the historic finds of fossil reptiles at the Perledo locality came from the Perledo Member, they are likely to be geologically younger than the pachypleurosaurs from the Lower Meride Limestone (Tintori, Muscio et al., 1985). Peyer listed the historical material of pachypleurosaurs from Perledo, previously erroneously referred to *Lariosaurus* by Mariani, but the material is rather fragmentary (Mariani, 1923; Peyer, 1933). A specimen kept at the Museo di Scienze Naurali "E. Caffi," Bergamo, has a femur that is longer than the humerus, and thus must represent a *Neusticosaurus* different from *N. edwardsii* (Carroll and Gaskill, 1985:348, table 1). A poorly preserved specimen kept at the Museo Civico di Storia Naturale in Milan has a humerus length of 19 mm, and a femur length of 15.5 mm, such that the humerus length is 122% of the femur length (specimen c in Peyer, 1933:106). This is close to the type specimen for *Neusticosaurus edwardsii*, at the top of the range of variation of *Neusticosaurus* other than *N. edwardsii*, and somewhat below the range of variation of *N. edwardsii* (Carroll and Gaskill, 1985, table 1).

In summary it can be concluded that the type specimen of *Neusticosaurus edwardsii*, collected in the neighborhood of Viggiù (Varese), most probably came from the Ca' del Frate locality, but its precise stratigraphic provenance remains unknown. Morphometrically, the type specimen of *N. edwardsii* does not cluster with the specimens collected from the Alla Cascina beds at Monte San Giorgio that are referred to that same species (O'Keefe and Sander, 1999). It would seem, then, that the applicability of the name *Pachypleurosaurus* (*Neusticosaurus*) *edwardsii*, tied as it is to an immature specimen that has been designated as holotype, requires further taxonomic scrutiny. Of the type specimen of *N. edwardsii*, Stockar and Renesto write that it "obviously raises serious taxonomic questions about its correct identity" (2011:S176).

7.1. A large individual of *Ceresiosaurus calcagnii* (Paleontological Institute and Museum, University of Zurich T4836) from Acqua Ferruginosa. The specimen measures 230 cm in length. It is surrounded by a number of small pachypleurosaurs (*Neusticosaurus pusillus*) (Photo © Heinz Lanz/ Paleontological Institute and Museum, University of Zurich).

Lariosaurs and Nothosaurs

7

The pachypleurosaurs are commonly referred to the family Pachypleurosauridae that was first erected by Nopcsa (1928). Nothosaurs and lariosaurs have been referred to the families Nothosauridae (first introduced by Baur without diagnosis or content), and Lariosauridae, respectively (Baur, 1889:312; Lydekker, 1889:284). Subsequent workers assigned different sauropterygian taxa to the Pachypleurosauridae and Nothosauridae, until Peyer clarified the situation and for the first time validly diagnosed the family Pachypleurosauridae (Peyer, 1933; Rieppel, 2000a). In hindsight, the early discussion and debates about the validity and identity of the genera *Pachypleura* (*Pachypleurosaurus*) and *Lariosaurus*, sketched in chapter 6, are perhaps difficult to understand given the rather distinct morphology that characterizes pachypleurosaurs and lariosaurs, especially in skull structure, where lariosaurs closely resemble nothosaurs instead (Rieppel, 2000a). However, confusion could arise because the first specimen of *Lariosaurus balsami* that was described, and later became the holotype of its species, approaches pachypleurosaurs in size, showing cervical vertebrae with pachyostotic neural arches that carry a very low neural spine (the dorsal neural arches are abraded), and the skull is missing (Balsamo-Crivelli, 1839; Curioni, 1847; Rieppel, 2000a:90, fig. 66). The skull was again only partially and poorly preserved in the second specimen of *Lariosaurus balsami* to become known, described under the name of *Macromirosaurus Plinj* by Curioni (1847; see comments in Peyer, 1933:84, and pl. 41, fig. 1). Some similarity of these specimens to pachypleurosaurs must therefore have been apparent to these early workers.

Pachypleurosaurs retain rather generalized skull proportions. The preorbital region is generally longer than the postorbital region, but not by much. The snout is rounded, the dentition in all but one species is homodont, the jaws being furnished with numerous small, peg-like teeth. There is a broad parietal skull table with a centrally located pineal foramen. The upper temporal fossa is small, distinctly smaller than the orbit, one of the hallmark characteristics of pachypleurosaurs. The lower temporal arch is absent, the cheek region embayed ventrally. The skull structure in lariosaurs and nothosaurs is strikingly different from that in pachypleurosaurs. The skull is more elongate and narrowly shaped than in pachypleurosaurs. The preorbital region is elongated to a variable degree but is always shorter than the elongated postorbital region. The dentition is heterodont. The premaxilla and the anterior end of the dentary bear elongated, strongly procumbent fang-like teeth. A couple of fang-like teeth in the maxilla are separated from the premaxillary fangs

The Monophyly of *Lariosaurus* and *Nothosaurus*

by a set of distinctly smaller teeth. The parietal skull table is narrow, the pineal foramen displaced posteriorly. The upper temporal fenestra is large, elongated, with a longitudinal diameter that equals or in most species exceeds the longitudinal diameter of the orbit. The snout is constricted, matching the distinctly reinforced, "spoon-shaped" mandibular symphysis.

The genus *Lariosaurus* has been distinguished from the genus *Nothosaurus* on the basis of several characteristics. In *Lariosaurus*, the moderate elongation of the rostrum matches that of basal nothosaurs but is never as pronounced as it is in derived nothosaurids. The same is true of the elongation of the postorbital region of the skull and the size of the upper temporal fenestra relative to the orbit. The rostral constriction, the constriction of the parietal skull table, as well as the posterior displacement of the pineal foramen may be less pronounced in some species of *Lariosaurus* than in nothosaurs. It is largely these features that most recently have been recognized to blur the generic distinction of *Nothosaurus* and *Lariosaurus* (J. Liu, Hu et al., 2014; Klein et al., 2016). On the basis of the postcranial skeleton, *Lariosaurus* is readily recognized by its distinctly broadened ulna. The humerus likewise has a very characteristic shape, evenly curved and broader distally than proximally, but without a central constriction of the shaft as is seen in *Nothosaurus*. The neural arches in the vertebral column are pachyostotic in *Lariosaurus*, and there are four or five sacral ribs. Lariosaurs may develop hyperphalangy in manus and pes. *Lariosaurus balsami*, for example, shows hyperphalangy in the manus, the maximal phalangeal count being 4-5-5-4(5)-3 (Tschanz, 1989:157). As this account shows, there is overlap in general skull characteristics between lariosaurs and basal nothosaurs, whereas *Lariosaurus* is well diagnosed on the basis of its postcranial skeleton (Rieppel, 2000a; see also Rieppel, 1998b). Historically, the taxonomy of the genera *Nothosaurus* and *Lariosaurus* was based on European material. The genera occur in two different depositional environments. One is the Germanic basin (Muschelkalk, Gipskeuper; Anisian–Ladinian), the other is the Alpine Triassic (intraplatform basins). Fossil reptiles in the Germanic basin, in particular in the Muschelkalk, tend to be disarticulated but more or less three dimensionally preserved. In contrast, fossil reptiles from the southern Alps, in particular at localities such as Monte San Giorgio, Besano, Perledo, and Vigiù (Ca' del Frate), are mostly articulated, but mostly strongly compressed. *Nothosaurus* is a very frequently found fossil in the Germanic basin, whereas *Lariosaurus* is exceedingly rare (Rieppel and Hagdorn, 1997:137; Klein et al., 2016). The reverse is true for the southern Alpine Triassic, where *Lariosaurus* is relatively common, *Nothosaurus* very rare. Given the distribution of the fossil material, it seems that *Nothosaurus* diversified as it spread through the Muschelkalk basin from east to west, the same direction in which the marine ingression progressed. In contrast, *Lariosaurus* seems to have diversified along the western shoreline of the Tethys, from where it only rarely ventured into the Germanic basin (Rieppel and Hagdorn, 1997). The result of these taphonomic and paleobiogeographic biases is that the taxonomy of *Nothosaurus* is largely one based on more or less well-preserved skulls, their morphological and proportional characteristics. Sutural patterns are very hard to discern in the mostly strongly compressed skulls of *Lariosaurus*, such that the systematics of *Lariosaurus* are largely based on skull proportions and characters in the postcranial skeleton.

Such traditional distinction of the genera *Lariosaurus* and *Nothosaurus* has become blurred as articulated specimens of nothosaurids became known from the Middle and early Upper Triassic of southwestern China (J. Liu, Hu et al., 2014), and as a small skull from the middle Anisian of the Muschelkalk of Winterswijk, the Netherlands, was attributed to the genus *Lariosaurus* (Klein et al., 2016). In their discussion of *Lariosaurus xingyiensis* from the Ladinian of Xingyi, southwestern China, Rieppel, J.-L. Li et al. noted characters that this taxon shares with *Nothosaurus* (2003:631). Similarly, in their description of a new specimen of *Nothosaurus youngi*, again from the Ladinian of Xingyi, Ji, Jiang, Rieppel et al. noted a "confusing mixture of features typical of *Nothosaurus* and *Lariosaurus*" (2014:469). The genera *Lariosaurus* and *Nothosaurus* are clearly not unequivocally diagnosed in the material from the Middle Triassic of southwestern China.

The same situation has in the meantime been recognized in the Germanic basin (Klein et al., 2016). This insight was brought about by the discovery of a small skull and a few associated postcranial elements in the lower Muschelkalk (Bithynian, Anisian, early Middle Triassic) of Winterswjik, the Netherlands, which are referred to a new species of *Lariosaurus*, namely, *L. vosseveldensis* (Klein et al., 2016). This species, together with *Lariosaurus hongguoensis* from the Pelsonian (Anisian, Middle Triassic) of Panxian, southwestern China, mark the earliest occurrences of *Lariosaurus* in the fossil record (Jiang, Maisch, Sun et al., 2006). *L. vosseveldensis* again "exhibits a mosaic of nothosaurian and lariosaurian skull characters" (Klein et al., 2016:e1163712–5). A phylogenetic analysis showed three species commonly referred to the genus *Nothosaurus*, namely, *N. juvenilis*, *N. winkelhorsti*, and *N. youngi*, to nest inside the genus *Lariosaurus*. To render the latter taxon monophyletic, these three nothosaur species would have to be referred to *Lariosaurus*. Formal taxonomic and nomenclatural conclusions have not yet been drawn but warrant further scrutiny (Klein et al., 2016).

Ceresiosaurus (*Lariosaurus*) *calcagnii* and *C. lanzi*

During the 1928 field season Peyer dug through layers in the lower Meride Limestone (Cava Superiore beds, Ladinian) at the Acqua del Ghiffo locality in the Valle Nera near Serpiano. These efforts were rewarded with the collection of a hitherto unknown sauropterygian, one complete skeleton, and two less well-preserved specimens (plus some isolated bones: Hänni, 2004: 14). Peyer compared these new fossils with *Lariosaurus balsami* but found enough differences to erect a new genus and species, *Ceresiosaurus calcagnii* (Peyer, 1931a). Curioni introduced the name

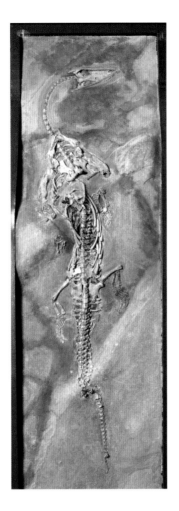

7.2. The holotype of *Ceresiosaurus lanzi* (Paleontological Institute and Museum, University of Zurich T2464) from the Alla Cascina beds. The specimen measures 172 cm in length (Photo © Heinz Lanz/Paleontological Institute and Museum, University of Zurich).

Lariosaurus to designate a saurian collected near Perledo above Varenna on Lake Como. The name he derived from Lago di Lario, the local Italian name for Lake Como (Curioni, 1834). Since Peyer's new specimen came from Monte San Giorgio above Porto Ceresio on Lake Lugano, he derived the name for the new genus from the local Italian name for Lake Lugano, that is, Lago Ceresio. The specific epithet derives from the name of Emilio Calcagni, the landowner who authorized Peyer's excavation activities at the Acqua del Ghiffo locality. Five years later, in 1933, Peyer collected another two specimens of *Ceresiosaurus* in the Alla Cascina beds. A spectacular, large specimen, 2.3 meters long, was collected in the Cava Inferiore beds at the Acqua Ferruginosa locality in 1937. This adult specimen of *Ceresiosaurus* is surrounded by several much smaller specimens of *Neusticosaurus pusillus*, which may have been part of its prey (fig. 7.1). A coprolite preserved on the same slab comprises skeletal elements of *Neusticosaurus* (Kuhn-Schnyder, 1974:63). Skeletal elements of pachypleurosaurs have also been found in the abdominal cavity of *Ceresiosaurus* (Brinkmann, 1994:69). The largest specimen of *Ceresiosaurus* was collected from the Alla Cascina beds in September 1973 (Hänni, 2004:8). A specimen collected in 1933 in the Alla Cascina beds was described by Hänni as the holotype of a new species, *Ceresiosaurus lanzi* (2004) (fig. 7.2).

Box 7.1. *Ceresiosaurus calcagnii* preying on *Neusticosaurus pusillus* from the lower Meride Limestone at Monte San Giorgio

The large lariosaur *Ceresiosaurus calcagnii* is here shown preying on the small pachypleurosaur *Neusticosaurus pusillus*. *Ceresiosaurus calcagnii* is by far the largest lariosaur known, adults reaching a total length of 2.3 meters. Other than in certain proportions, the skull of lariosaurs is closely similar to that of the related nothosaurs. The fore limbs in *Ceresiosaurus calcagnii* are longer and more massively built than the hind limbs, clearly in support of locomotion. In addition, the hands and feet in that species show the addition of supernumerary phalanges (hyperphalangy), another distinct adaptation to an aquatic lifestyle. Also, in *Ceresiosaurus calcagnii*, the neural spines in the proximal caudal vertebrae are distinctly enhanced, rendering the tail a more effective propulsive organ. *Ceresiosaurus calcagnii* and *Neusticosaurus pusillus* are known to have coexisted in time and space. A coprolite referable to *Ceresiosaurus calcagnii* comprised bones of *Neusticosaurus*, indicating that these smaller reptiles were part of the diet of *Ceresiosaurus*, along with fish and cephalopods.

© Beat Scheffold

© Beat Scheffold

Ceresiosaurus is by far the largest lariosaur known. In the adult, the front limbs are much more powerfully developed than the hind limbs. This speaks to their importance in locomotion, which was analyzed on the basis of the otariid-model, introduced by Godfrey in an analysis of plesiosaur locomotion, and adopted for *Ceresiosaurus* by Hänni (Godfrey, 1984; Hänni, 2004; see also Araújo and Correia, 2015). *Ceresiosaurus* is also known to have developed hyperphalangy in manus and pes. The largest number of phalanges known for the manus is 3-5-4-5-3 (left) and 3-5-5-3-3 (right), respectively; the largest number of phalanges known for the pes is 2-3-5-6-8 (Hänni, 2004:87). In mature specimens, the carpus comprises six ossifications: the intermedium, the smaller fibulare, and distal carpals one through four, where the fourth is the largest, and the second is the smallest. The adult tarsus comprises three ossifications: the large astragalus located in an intermedium position, the smaller calcaneum, and the fourth distal tarsal. The fore and hind limbs thus differ not only in relative size but also in degree of ossification (see also Rieppel, 1989a:140, and fig. 10).

As is the case with pachypleurosaurs, the two species of *Ceresiosaurus* recognized at Monte San Giorgio succeed one another stratigraphically, with no overlap yet known. *Ceresiosaurus calcagnii* comes from the Cava Inferiore and Cava Superiore beds in the lower Meride Limestone, whereas *Ceresiosaurus lanzi* comes from the somewhat younger Alla Cascina beds, again in the lower Meride Limestone (Peyer, 1931a; Hänni, 2004). Given their temporal succession, the characters that differentiate the two species are unlikely to reflect sexual dimorphism. *Ceresiosaurus calcagnii* is characterized by the presence of 50 presacral vertebrae, of which 24–25 are cervicals; there are 70 caudal vertebrae with heightened neural spines in the proximal caudal region; the surface of the neural spines is not striated; the dorsal ribs are pachyostotic; the sacral ribs are closely juxtaposed; and there is no posteromedial process on the interclavicle. *C. calcagnii* also differs from *C. lanzi* in some skull proportions, such as the diameter of the upper temporal fossa relative to the longitudinal diameter of the orbit. In addition, *C. lanzi* is characterized by the presence of 47 presacral vertebrae, of which 22–24 are cervicals; there are 55 caudal vertebrae without heightened neural spines in the proximal tail region; the neural spines are vertically striated; the dorsal ribs are not pachyostotic; the sacral ribs are widely spaced; the interclavicle carries a short posteromedial process. The species is named after Heinz Lanz, a longtime preparator and photographer at the Paleontological Institute in Zurich.

The genus name *Ceresiosaurus* has in the past been treated as synonymous with the much older name *Lariosaurus*. Pursuing a cladistic analysis of lariosaur interrelationships, Rieppel found *Ceresiosaurus* to nest inside the tree comprising the species that are generally referred to the genus *Lariosaurus* (1998b). In order to maintain the monophyly of the genus *Lariosaurus*, the name *Ceresiosaurus* was declared a subjective junior synonym, and the species re-named as *Lariosaurus calcagnii*. A similar result was obtained by Storrs (1993b; see below for a discussion of the synonymy of *Lariosaurus* and *Silvestrosaurus*). However, Hänni found the two species of *Ceresiosaurus* she recognized to form the sister clade to a clade that comprises all the species commonly referred to the genus *Lariosaurus* (2004). She consequently retained both generic names as valid and referred the two genera to the family Lariosauridae (Lydekker, 1889:284).

In his survey of the fossil biota of Monte San Giorgio, Kuhn-Schnyder illustrated a small lariosaur that was collected in 1971 in Val Mare near Meride, which he referred to *Lariosaurus balsami* (1974:65, fig. 42). The specimen came from the top of the upper Meride Limestone, that is, the late Ladinian Kalkschieferzone, and thus qualifies as the geologically youngest fossil reptile so far collected at Monte San Giorgio. In his monographic description of the specimen, Kuhn-Schnyder referred it to a new species, *Lariosaurus lavizzarii*, named after Luigi Lavizzari (1814–1875), an influential naturalist and politician of Canton Ticino, the Italian-speaking part of Switzerland where Monte San Giorgio is located (1987:19). The species is described as differing from *Lariosaurus balsami* in the contours of the skull (a slightly offset rostrum), and a relatively shorter humerus. With a glenoid-acetabular length of only 60 to 62 mm, the specimen was recognized as a juvenile individual by Kuhn-Schnyder (1987:7). This raises questions regarding the diagnostic value of these characters, in particular relative humerus length, although Sanz had characterized the relation of humerus to femur growth as isometric in *Lariosaurus* (1983).

The specimen was later revisited by Tschanz, who declared *Lariosaurus lavizzarii* a subjective junior synonym of *L. balsami* (1989). He noted that the rostrum is even more distinctly set off from the orbital region of the skull in the Munich specimen of *L. balsami*, which was designated the neotype of its species by Kuhn-Schnyder (1987:19; see Peyer, 1933, pl. 32, fig. 1; Rieppel, 1998b:12, fig. 10A). The relative humerus length in *L. lavizzarii* corresponds to that of small individuals of *L. balsami* (Tschanz, 1989:157). However, in 1993, Renesto described a juvenile specimen of *Lariosaurus* from the Kalkschieferzone at Ca' del Frate near Viggiù that he found to be not diagnostic at the species level and which therefore cannot be referred to any species of *Lariosaurus* with certainty. Comparing this new specimen with *Lariosaurus lavizzarii* from the Kalkschieferzone of a nearby locality, he drew the same conclusion with respect to the latter specimen (Renesto, 1993). Otherwise, the Kalkschieferzone near Ca' del Frate has yielded a beautifully preserved lariosaur described as *Lariosaurus valceresii* by Tintori and Renesto (1990). Renesto's conclusion was that it is impossible to decide whether the juvenile specimens from the Kalkschieferzone should be referred to the latter species or to *Lariosaurus balsami* instead (1993). *Lariosaurus lavizzarii* thus becomes a *nomen dubium* (Rieppel, 2000a:91).

In September 1961, a small nothosaurian reptile was collected in layer 97 at Point 902. This renders the specimen of uppermost Anisian age and hence, as was later found, the earliest record of lariosaurs at Monte

San Giorgio (Brack and Rieber, 1986, 1993; Tschanz, 1989; Furrer, 2003).
Kuhn-Schnyder illustrated the specimen, which he referred to as a "small
nothosaurid" (1974:31, figs. 9 through 12; see also Kuhn-Schnyder, 1963b).
At the time, the description of this new find did not constitute "one of the
urgent tasks of the Paleontological Institute," of which he was director
(Kuhn-Schnyder, 1990:314). The specimen was eventually described by
Tschanz as a new species, *Lariosaurus buzzii* (fig. 7.3), named after Giorgio Buzzi, a mine worker who participated in Peyer's fieldwork until he
was killed in an accident in the Val Porina mine on August 20, 1931 (1989).
The longitudinal diameter of the upper temporal fossa equals that of the
orbit in *L. buzzii*, the rostrum is slightly constricted in front of the external
nares, the number of cervical (less than 15) and dorsal vertebrae appears
to be reduced, and the species is characterized by additional details in
the morphology of the pectoral girdle and humerus. As noted in chapter
5, the specimen is intriguing as it contains fragments of tooth-bearing elements of a juvenile placodont *Cyamodus hildegardis* as stomach content.
In his cladistic analysis, Tschanz found the genus *Lariosaurus* to be the
sister taxon to the genus *Ceresiosaurus*; he included both genera in the
family Lariosauridae (1989).

Tschanz's description of *Lariosaurus buzzii*, and the systematic
conclusions he drew, were criticized by Kuhn-Schnyder (1990). In his

critique, Kuhn-Schnyder rejected the referral of the new species to the genus *Lariosaurus* and instead introduced a new genus, named *Silvestrosaurus*, to receive *S. buzzii* (1990:315). This, he claimed, removes the occurrence of *Lariosaurus* in the Anisian of Monte San Giorgio. Kuhn-Schnyder also rejected a close relationship between *Lariosaurus* and *Ceresiosaurus*, and hence did not allow for the inclusion of both genera in the same family, Lariosauridae. The cladistic analysis of Tschanz he characterized as "going off on the wrong track," and he concluded his critique with a word about Linné. Himself in search of a comprehensive natural system of plants, Linné is said to have declared the researcher who would eventually fulfill that goal as the Olympian deity Apollo for botany. In Kuhn-Schnyder's assessment, "the sauropterygians are presently still awaiting their Apollo" (1990:316). Kuhn-Schnyder's critique of Tschanz's work on lariosaurs thus provides a beautiful example of a clash of paradigms: the traditional taxonomist versus the cladist seeking monophyly (see also Kuhn-Schnyder and Rieber, 1984). The treatment of *Lariosaurus buzzii* as a lariosaur has stood the test of time, however (see Storrs, 1993b; Rieppel, 1998b).

Nothosaurus giganteus (Paranothosaurus amsleri)

In 1932, Peyer was able to retrieve from the roof of the Fortuna mine at the Cava Tre Fontane locality the complete skeleton of one of the largest nothosaurs known at that time, embedded in a thick dolomite layer, itself intercalated with bituminous layers as is characteristic of the Grenzbitumenzone. The fossil was discovered when blasting exposed, and partially destroyed, the distal end of a hind limb as well as the distal-most part of the tail. The specimen has a total length of approximately 3.8 meters, and it was described in a monograph published by Peyer in 1939 (Peyer, 1939a). He considered it to represent a new genus and species, which he named *Paranothosaurus amsleri* (fig. 7.4). The specific epithet honors Alfred J. Amsler from Schaffhausen, who for years had funded the position of one preparator to work on Monte San Giorgio material. At the time, the discovery of the specimen—the complete articulated skeleton of a nothosaur—was of greatest importance, as only one partial articulated nothosaur skeleton was then known from the Germanic Muschelkalk, the holotype of *Nothosaurus raabi*. The latter is a rather small, incompletely and, in part, poorly preserved skeleton of a basal nothosaur, which was collected 1896 in a mine pushing through the lower middle Muschelkalk (*orbicularis* beds, Anisian) near Rüdersdorf close to Berlin. It was described in detail by Schroeder and mounted for exhibit at the Natural History Museum in Berlin (1914). The missing parts of the skeleton were reconstructed using woodcarvings. Today, the species is considered a subjective junior synonym of *Nothosaurus marchicus*, based on an incomplete skull from the same locality and horizon (Koken, 1893; Rieppel and Wild, 1996). Clearly, Peyer's new, larger, and much more complete nothosaur skeleton promised to offer a lot more anatomical detail than these historic finds (1939a).

7.4. The holotype of *Paranothosaurus amsleri* (Paleontological Institute and Museum, University of Zurich T4829; now considered a junior synonym of *Nothosaurus giganteus*), from the Cava Tre Fontane. The specimen measures 380 cm in length (Photo © Heinz Lanz/Paleontological Institute and Museum, University of Zurich).

Box 7.2. "*Paranothosaurus amsleri*" from the Grenzbitumenzone of Monte San Giorgio

Paranothosaurus amsleri is today considered a junior synonym of *Nothosaurus giganteus*, the latter known from skulls and isolated vertebrae from the Germanic Muschelkalk. When collected in 1932, "*Paranothosaurus*" was not only one of the largest nothosaurs known, reaching a total length of 3.8 meters. It was at the time also the only known complete and articulated skeleton of *Nothosaurus*. Today, many well-preserved, complete and articulated skeletons of nothosaurs are known from marine Middle and Upper Triassic deposits in southwestern China. The skeleton of "*Paranotosaurus amsleri*" was strongly compressed during fossilization, however, which obscures some morphological details, especially in the skull. The skull of nothosaurs was elongate but very low. The procumbent fang-like teeth in the anterior part of the upper and lower jaw were suitable for catching fish and cephalopods with a quick sideways snapping bite. As in *Ceresiosaurus*, the fore limbs are again more robustly built than the hind limbs in support of locomotion. It remains unknown whether lariosaurs and nothosaurs were viviparous, or whether they deposited their eggs on the shore.

© Beat Scheffold / PIMUZ

Peyer found himself unable to discern sutural details in the skull of his new find but otherwise noted its close similarity to the skull of *Nothosaurus*: "The similarity the skull of the Ticino nothosaur shares with *Nothosaurus* is so close, that if only the skull had been found, it would without any doubt have been referred to the genus *Nothosaurus*" (Peyer, 1939a:69). As mentioned above, *Nothosaurus* is a frequently found faunal element in the Germanic Muschelkalk, and Peyer felt justified to erect a new genus and species on the basis of the striking differences in the structure of the posterior cervical and dorsal vertebrae. The neural arches are distinctly broader than long, with "swollen" and "domed" (i.e., pachyostotic) prae- and postzygapophyses. This contrasts with nothosaur neural arches commonly found in the Germanic Muschelkalk, which are smaller, not distinctly broader than long, without pachyostosis of the pre- and postzygapophyses, and carry a distinctly heightened neural spine. These neural arches were referred to the genotypical species *Nothosaurus mirabilis* by H. v. Meyer (1847–1855, pl. 23). Peyer did note, however, the rare occurrence of isolated neural arches closely comparable to those of *Paranothosaurus* in the Germanic Muschelkalk (1939a:70).

Following a renewed preparation of the skull, Kuhn-Schnyder was able to decipher the sutural pattern in the skull of *Paranothosaurus*: "The skull of *Paranothosaurus amsleri* PEYER most closely compares to that of *Nothosaurus* not only in its shape, but also in the sutural pattern" (Kuhn-Schnyder, 1974:61; see also Kuhn-Schnyder, 1966:534). Kuhn-Schnyder considered it probable that the low neural spines in *Paranothosaurus* could have resulted from compression during fossilization, but he left the question of a possible synonymy of *Nothosaurus* and *Paranothosaurus* undecided (1966:535–536). That question, he argued, could not be answered until such time as dorsal vertebrae of *Paranothosaurus* were found that were embedded and preserved in lateral view (Kuhn-Schnyder, 1974:61).

A revision of the genus *Nothosaurus* from the Germanic Muschelkalk showed that a low neural spine in the dorsal vertebrae is the plesiomorphic condition (Rieppel and Wild, 1996). That the vertebral structure of *Paranothosaurus* resulted from compression during fossilization is rendered unlikely by the occurrence of perfectly preserved, large dorsal vertebrae with a low neural spine and pachyostotic pre- and postzygapophyses in the Ladinian of Fusea, Province Udine, northeastern Italy (Dalla Veccia, 1993, fig. 9). Neural arches most closely similar to those of *Paranothosaurus* both in size and structure are also found in the Upper Muschelkalk of the Germanic basin, which on the basis of their size can be attributed to *Nothosaurus giganteus*. Given the similarities in size and morphology, as well as the occurrence of both genera in the late Anisian, it appears justified to consider *Paranothosaurus amsleri* a subjective junior synonym of *Nothosaurus giganteus* (Münster, 1834; Rieppel and Wild, 1996).

The total length of the single specimen of "*Paranothosaurus amsleri*" approximates four meters. The skull itself is approximately 48.5 cm long.

This size places *Paranothosaurus* at the apex of the trophic pyramid in the Monte San Giorgio biota (Brinkmann and Mutter, 1999). The dentition is typically nothosaurian, with procumbent fangs inserted in the premaxillae and in the reinforced mandibular symphysis, separated by a set of smaller teeth from the maxillary fangs located just in front of the orbit. The postorbital region of the skull is distinctly elongated, the longitudinal diameter of the upper temporal fenestra twice that of the orbit. Powerful jaw adductor muscles effected a snapping bite during a sideways strike effected by the long neck (Rieppel, 2002a). There are 19 cervical, 26 dorsal, 5 sacral, and 33 caudal vertebrae; the distal tip of the tail is missing. The humerus (24.5 cm) is slightly shorter than the femur (26 cm) but more robustly built and characteristically curved. The femur is more delicately built and straight. The carpus comprises four, the tarsus three ossifications. Indeed, the fore limb is generally more robustly built than the hind limb, indicating its relevance in locomotion. The phalangeal formula in the pes is plesiomorphic (2-3-4-5-4); that of the manus is unknown. A well-developed ventral gastral rib basket must have stiffened the trunk to some degree, leaving the limbs and the tail to do most of the work during locomotion.

Since Peyer's find, very little articulated, or even only associated nothosaur material has been collected, or at least published, in the western Tethyan faunal province (including the Germanic basin). One such find is *Nothosaurus jagisteus* (Rieppel, 2001b), from the upper Muschelkalk (lower Ladinian) of Berlichingen, Hohenlohe area, southern Germany, which comprises the skull, neck, anterior trunk region, pectoral girdle, and remains of the fore limbs. The other specimen concerns an articulated, partially preserved trunk with associated elements of the pectoral girdle and fore limbs from the Lower Muschelkalk (lower Anisian) of Winterswijk, the Netherlands, a locality that has also yielded isolated nothosaur skulls (Bickelmann and Sander, 2008; Klein and Albers, 2009; see also the discussion in Albers, 2011). This situation changed dramatically with the collection of articulated skeletons of *Nothosaurus* in southwestern China (of which more later). One of the earliest occurrences of *Nothosaurus* in the eastern Tethyan faunal province is *Nothosaurus yangjuanensis* from the Pelsonian (middle Anisian) of Panxian County, Guizhou Province (Jiang, Maisch, Hao et al., 2006).

8.1. The holotype of
Askeptosaurus italicus Nopcsa,
kept at the Museo Civico
di Storia Naturale di Milano
(Photo © Giorgio Teruzzi/
Museo Civico di Storia
Naturale di Milano).

Thalattosaurs

The recognition of the Thalattosauria as a monophyletic clade and the understanding of its relationship among reptiles in general, follows a tortuous path of confusing research that continued through many decades, and in many ways has not quite come to a conclusion even today (Nicholls, 1999; Müller, 2005). The first representative of the group to have been described was *Thalattosaurus alexandrae* from the Hosselkus Limestone (Carnian, Upper Triassic) of Shasta County, northern California (Merriam, 1904). The following year, Merriam described a second thalattosaur of the same provenance, which he named *Nectosaurus halius* (1905). He included these taxa in his order Thalattosauria, characterized as marine diapsid reptiles with an upper and lower temporal fenestra, a distinct pineal foramen, and a specialized (durophagous) palatal and marginal dentition (Merriam, 1904:420; Nicholls, 1999:1). In the first edition of his textbook of 1933, Romer placed *Thalattosaurus* with its paddle-like limbs "with very great doubt" in the Eosuchia, a loosely defined assemblage of varied diapsid reptiles (1933:157). In the synoptic classification of reptiles he appended to his textbook, the Thalattosauridae is formally listed as part of the Eosuchia. Along with thalattosaurs, Romer's eosuchians at the time included, among others, the Permian younginids and the rhynchocephalians that first appear in the Triassic, the latter at the time including the fossil relatives of the tuatara (*Sphenodon*) along with the herbivorous rhynchosaurs. The Eosuchia together with the Squamata constituted the Lepidosauria, a subclass that Romer at the time considered of doubtful validity (1933:438f). With his influential textbook, Romer set the framework for subsequent discussions of thalattosaur interrelationships. In the second edition of his textbook of 1945, Romer included the younginiforms (inclusive of prolacertiforms) and choristoderes along with thalattosaurs in his Eosuchia, which together with the Rhynchocephalia (including the rhynchosaurs) and Squamata would constitute the Lepidosauria. By that time, two thalattosaurs had been described from the Middle Triassic of Monte San Giorgio, namely, *Clarazia* and *Hescheleria* (Peyer, 1936a, 1936b). Romer followed Peyer's assignment of the two genera to a separate family, the Claraziidae, which he included in the Rhynchocephalia, the latter comprising the sphenodontids and their fossil relatives, as well as the rhynchosaurs (Peyer, 1936b:41; Romer, 1945:595, 1966:368). The Claraziidae were thus not considered to be thalattosaurs, the Thalattosauria comprising *Thalattosaurus* and *Nectosaurus* only (Merriam, 1904, 1905).

Thalattosaurs, an Enigmatic Clade of Triassic Marine Reptiles

The thalattosaur *Askeptosaurus italicus* was first described in 1925 on the basis of very fragmentary material from the Besano locality. Since then, several specimens have been collected both at Besano and at Monte San Giorgio. *Askeptosaurus* is most closely related to *Anshunsaurus*, a thalattosaur from the Middle and early Upper Triassic of southwestern China. Clearly a marine predator, the skull is elongate and flat in *Askeptosaurus*, the jaws extended to form a prominent rostrum carrying distinct pointed teeth. Prey was secured with a rapid sideways bite, the range of which was expanded through an elongation of the neck. Body and tail are elongate and slender, the animal reaching a total length of 2.5 meters. The limbs are well developed in support of locomotion, which was probably predominantly driven by lateral undulation of body and tail. The mode of reproduction in thalattosaurs remains unknown.

© Beat Scheffold / PIMUZ

Peyer assigned the description of the new specimens of *Askeptosaurus* as a graduate student research project to Kuhn-Schnyder, who published his PhD thesis in 1952. He had three specimens at his disposal: two from the Besano locality, which given their provenance had eventually to be returned to the Milan Natural History Museum, and one from the Grenzbitumenzone of Valle Stelle at Monte San Giorgio. He recognized a saurian of up to 2.5 meters length, with a skull characterized by an elongate rostrum and retracted external nares, betraying aquatic, or at least amphibious habits (fig. 8.2). There are approximately 14 cervical and 25 dorsal vertebrae; the long tail, comprising approximately 70 vertebrae, is laterally compressed and supported locomotion through lateral undulation. The zeugopodium is distinctly shorter than the stylopodium in both fore and hind limbs, carpus and manus are well ossified, and the phalangeal

8.2. A specimen of *Askeptosaurus italicus* (Museo Civico di Storia Naturale di Milano V 456) from the Vallone mine, Besano. The specimen measures 180 cm in length (Photo © Giorgio Teruzzi/Museo Civico di Storia Naturale di Milano).

formula in the manus is 2-3-3-4-3, in the pes it is 2-3-4-4-4. There is, thus, a slight reduction in phalangeal numbers compared to the plesiomorphic condition. At the time of his PhD thesis research, Kuhn-Schnyder was able to offer a detailed description of the postcranial skeleton of *Askeptosaurus*, but he had to admit that the skull structure remained far less well known due to the strong dorsoventral compression of the material (1952). He nonetheless concluded that there is an upper and lower temporal fenestra, but the lower temporal arch remained incomplete, as in squamates and *Prolacerta*. The rostrum of *Askeptosaurus* he found to resemble that of *Thalattosaurus*, although the limbs of *Thalattosaurus* he found to be more highly adapted to an aquatic lifestyle than those of *Askeptosaurus*. In summary, he concluded that *Askeptosaurus italicus* represents a specialized representative of early squamates (Kuhn-Schnyder, 1952:66).

An additional specimen of *Askeptosaurus italicus* had already been collected in 1937, in the Galeria Arnaldo Superiore at the Tre Fontane locality at Monte San Giorgio, but because of what promised to be a difficult job of preparation, it had not been prioritized for investigation (Kuhn-Schnyder, 1974: 43). Eventually, however, its partially disarticulated skull, exposed in ventral view, became the subject of a separate description by Kuhn-Schnyder (1971). The new specimen showed similarities to *Thalattosaurus* with regard to the differentiation of the rostrum, of the temporal fenestrae, and in the deeply embayed occiput, but revealed rather profound differences in the dentition. Although Kuhn-Schnyder stipulated a common ancestry of *Thalattosaurus* and *Askeptosaurus*, he nevertheless concluded that "they evolved at different rates and in different directions to become marine predators" (1971:96). He referred the taxa to the Askeptosauroidea and Thalattosauroidea, respectively, both representing specialized offshoots of the Eosuchia. Squamates and rhynchocephalians he considered to have had a separate origin among "primitive eosuchians" (Kuhn-Schnyder, 1971:96).

A modern revision of all the *Askeptosaurus* material then available, and an analysis of thalattosaur interrelationships, was offered by Johannes Müller in 2005. By that time, a number of older assumptions of reptile interrelationships had been abandoned on cladistics principles. The Eosuchia, for example, had been dissolved as a paraphyletic assemblage; similarly, the archosauromorph rhynchosaurs had been separated from the lepidosauromorph rhynchocephalians. The implications of these and other changes in reptile classification, and their relevance for thalattosaur interrelationships, will be discussed after all thalattosaur taxa from Monte San Giorgio have been introduced in their historical context.

Müller found *Askeptosaurus* to be a marine reptile of approximately three meters total length (2005). An upper temporal fossa in the slightly depressed skull is present, but small and slit-like—the result of a secondary reduction. The supratemporal is elongate, slender, intercalated between parietal and squamosal, and closely approaches the posterior corner of the upper temporal fenestra. The pineal foramen is large and somewhat displaced anteriorly. The cheek region is wide open, and the lower temporal arch is incomplete. The jugal carries a prominent posterior process, but a quadratojugal is absent. The orbit is relatively large, the scleral ossicles are well developed, and the external nares are retracted. The rostrum is elongate and parallel-sided. In contrast to other thalattosaurs, the dentition is homodont, the marginal teeth of conical, pointed, and slightly recurved shape. There is no diastema separating the premaxillary from the maxillary teeth. The dermal palate is toothless.

The pectoral girdle is noteworthy for its interclavicle, which carries a significantly expanded posterior stem. The humerus is more sturdily built than the femur, but somewhat shorter. There are two ossifications in the

8.3. The holotype and only known specimen of *Clarazia schinzi* (Paleontological Institute and Museum, University of Zurich T4778) from the Val Porina. The specimen measures approximately 100 cm in length (Photo © Heinz Lanz/ Paleontological Institute and Museum, University of Zurich).

proximal carpus (intermedium and ulnare), and four in the distal carpus (distal carpals one through four); to these may be added a centrale in some individuals. The phalangeal formula in the manus is again subject to individual variation, the count being 2-3-3(4)-4-3. In the pelvic girdle, a thyroid fenestra is not differentiated. The fifth metatarsal is straight. The tarsus comprises six ossifications, the astragalus, the calcaneum, and the distal tarsals one through four. The complete phalangeal formula of the pes remains unknown, but the ungual phalanges represent moderately developed claws, as is also the case in the manus.

Clarazia, Hescheleria, and the Tocosauria

Askeptosaurus was not the first thalattosaur to be collected in the Besano Formation/Grenzbitumenzone. Two other taxa had previously been collected in the Grenzbitumenzone of Monte San Giorgio, both of which have so far not been recorded from the Besano locality (Peyer, 1936a, 1936b).

The holotype and only known specimen of *Clarazia schinzi* (fig. 8.3) was collected in 1933 after blasting in the industrially exploited bituminous shales (Anisian) in the Val Porina mine (Peyer, 1936a; Kuhn-Schnyder, 1974:45). The genus name honors Georges Claraz (1832–1930), a Swiss naturalist and explorer who made a considerable fortune with farming and cattle ranching in Argentina (Schinz and Wolfer, 1931:489). The significant endowment he established, administered by the director of the Zoological Institute of the University of Zurich, generously supported Peyer's Monte San Giorgio project through the years.

The specimen as preserved, with the tip of the tail missing, measures 90 cm in length; it may have reached one meter, or even a bit more in total length (Rieppel, 1987b; approx. 116 cm in Peyer, 1944:76). The skull was originally prepared in ventral view, exposing the durophagous dentition. The premaxilla is slightly deflected ventrally, and carries four, the maxilla five blunt, bulbous crushing teeth. The vomers are fused, and again carry globular crushing teeth of which five are exposed. The transverse pterygoid flange underlaps the movable palatobasal articulation and again is furnished with a patch of crushing teeth. The mandibles are heavily built, each dentary furnished with a set of nine crushing teeth the size of which increases in an antero-posterior gradient. Peyer deplored the fact that due to the ventral exposure of the skull, he could not ascertain the configuration of its temporal region: "probably there was only a single temporal opening" (Peyer, 1936a:46). The significance of the presence of a single upper temporal opening was at the time not at all clear since a debate was raging as to whether the squamates had lost the lower temporal arch otherwise characteristic of diapsids, or whether the cheek region of the lizard skull had become excavated from below. Nevertheless, and not wanting to overemphasize the significance of the configuration of the temporal region of the skull, Peyer concluded: "The character of the entire skeleton suggests that *Clarazia* belongs to the *Tocosauria*, i.e., to the group of reptiles that comprises the squamates, the rhynchocephalians, and all the intermediate forms between those two" (1936a:46). More specifically, Peyer found Merriam's Thalattosauria to share characteristics both with squamates and with rhynchocephalians and noted that the configuration of the jaws and dentition in particular rendered a common ancestry of *Thalattosaurus* and *Clarazia*, and hence a referral of *Clarazia* to the Thalattosauria "highly probable" (Peyer, 1936a:47).

Box 8.2. *Clarazia schinzi* from the Grenzbitumenzone of Monte San Giorgio

Clarazia is a more derived thalattosaur, known only from a single but complete and articulated specimen from the Grenzbitumenzone of

© Beat Scheffold / PIMUZ

Monte San Giorgio. Characterized by a long tail, the animal reached a total length of approximately one meter. *Clarazia* is most closely related to another thalattosaur from Monte San Giorgio, *Hescheleria*, the two included in the family Claraziidae. In *Clarazia*, the jaws form a relatively short, pointed, and slightly downward deflected rostrum. The lower jaw is very massively built. The dentition was durophagous, which makes hard-shelled invertebrates the likely prey. Picking invertebrate prey off the substrate, *Clarazia* must have foraged in the marginal areas of the Grenzbitumenzone basin, as the deeper waters were oxygen depleted. *Hescheleria* is a durophagous thalattosaur as well but differs from *Clarazia* in details of the dentition and by the fact that in *Hescheleria*, the rostrum is much more markedly downturned, assuming an essentially vertical position suitable to sift through bottom sediments in search of invertebrate prey.

Following on the heels of his monograph on *Clarazia*, Peyer published the description of a second unique specimen he thought to be closely related, namely, *Hescheleria ruebeli* (1936b). The disarticulated skeleton was collected during the 1929 field season in the Grenzbitumenzone (Anisian) of Val Porina (fig. 8.4). Peyer estimated the reptile to have reached one meter in length (Peyer, 1944:77). The genus name honors Karl Hescheler (1868–1940), who as the director of the Zoological Institute of the University of Zurich administered the "Georges and Antoine Claraz endowment," using it to generously support Peyer's activities at Monte San Giorgio, the significance of which he certainly appreciated. Hescheler himself pursued interests in paleontology, mostly concerning Pleistocene mammals, which he researched with a paleobiological rather than stratigraphic focus. Indeed, Hescheler was the first to be named professor of paleontology at the University of Zurich in 1903. He became a founding member of the Swiss Archaeological as well as the Swiss Paleontological Society, and he established paleontology as a research program at the University of Zurich (Strohl, 1940).

Given the disarticulated nature of the specimen, the configuration of the temporal region of the skull once again defied precise analysis. What instead caught Peyer's attention were the specializations for durophagy evident in the jaws of *Hescheleria*, which are much different from those of *Clarazia*, however. A distinct facet at the anterior end of the short maxilla for the reception of a posterior process on the premaxilla, the latter defining the anterior ventral margin of the external naris, allowed the inference that the alveolar margin of the premaxilla, furnished with four or five small, peg-like teeth, is deflected downward nearly vertically relative to the alveolar margin of the maxilla, the latter again carrying small, peg-like teeth. That way, the premaxillary rostrum overhangs the anterior end

8.4. Radiograph of *Hescheleria ruebeli* (PMUZ T2469) from the Val Porina. The fossil-bearing slab measures 346 mm in width (radiograph University Hospital Zurich, director Prof. Dr. H. R. Schinz, 1937; photographic rendition John Weinstein, the Field Museum, Chicago).

of the lower jaw, its teeth pointing backward with no occlusal dentition to work against. As in *Clarazia* the mandible is heavily built, with a set of four or five small, peg-like teeth restricted to its anterior end. However, there rises in the symphyseal region of the lower jaw a very prominent bump, to which both mandibles contribute. This bump, or cusp, vastly protrudes beyond the dentary teeth, and must have worked against the upper inner surface of the premaxillary rostrum, perhaps also against the dermal palate (vomers) in an effort to crush hard-shelled invertebrate prey that was dug up from the substrate using the ventrally projecting rostrum (Peyer, 1936b:37; see also Rieppel, Müller et al., 2005). A palatal dentition was otherwise absent in *Hescheleria*, at least on the palatine and pterygoid. Peyer noted these differences in dentition between *Hescheleria* and *Clarazia*, but otherwise found the two species close enough to unite them in a separate family, the Claraziidae, which he characterized as "marine *Toscosauria* [*sic*] (in the sense of M. Fürbringer)" (1936b:41). Referencing his monograph on *Clarazia*, he reiterated his belief that the Tocosauria *sensu* Fürbringer (1900) are a natural group encompassing squamates and rhynchocephalians as well as intermediate forms, one to which the Claraziidae had also to be referred. In summary, Peyer speculated that these marine reptiles had evolved from ancestral forms close to squamates or rhynchocephalians at different times and in different directions, a radiation represented by, among others, *Thalattosaurus*, *Clarazia*, and *Hescheleria* (1944:76).

The term Tocosauria first appears in the second volume of Ernst Haeckel's (1834–1919) *Generelle Morphologie* (Haeckel, 1866:cxxxiii). The name referred to a heterogeneous group of Paleozoic reptiles, one that certainly no longer captures our current understanding of reptile phylogeny and classification. But Peyer wanted the name Tocosauria to be understood in the sense of Fürbringer, who characterized them as a group comprising the "Streptostylia s. Squamata with the two well-known orders Lacertilia and Ophidia . . . the monimostylic order (super-order?) Rhynchocephalia with the suborders (orders?) Protorosauria and Rhynchocephalia vera, and the monimostylic order Ichthyopterygia" (Fürbringer, 1900:76; Peyer, 1936b). With the ichthyosaurs included in the Tocosauria, Peyer may have seen no problem deriving other marine reptiles from that "ancestral stock" as well, but the systematic arrangement he thereby arrived at certainly no longer matches current ideas of reptile phylogeny (1936b).

Clarazia and *Hescheleria* were re-described by Rieppel, following the preparation of the skull of *Clarazia* in dorsal view (1987b). The latter immediately revealed the close relationship of *Clarazia* with *Thalattosaurus* as indicated by a number of shared derived characters: the premaxillaries form a slightly ventrally deflected rostrum furnished with a durophagous dentition, the posteromedial processes of the premaxillaries extend backward to contact the frontals, thus separating the nasals from one another, the upper temporal fossa is obliterated (secondarily closed), and—perhaps most important—the supratemporal is an elongated element intercalated between parietal and squamosal, anteriorly contacting the posterolateral process of the frontal. *Clarazia* was thus recovered as sister taxon of *Thalattosaurus*, with *Askeptosaurus* the sister taxon of those two (see also Benton, 1985, on a close relationship of *Askeptosaurus* with *Thalattosaurus*, both to be included in the Thalattosauria *sensu* Merriam, 1904). Given the incomplete preservation of *Hescheleria*, its precise placement remained unresolved beyond the relationship it was postulated to share with *Clarazia* by Peyer (1936b). Diapsid classification had at the time come under close scrutiny, providing a powerful framework for the discussion of thalattosaur relationships either with the archosauromorph, or with the lepidosauromorph lineage of diapsids (Sauria) (Gauthier, 1984; Evans, 1984; Benton, 1985). In Rieppel's assessment, "a fair amount of character incongruence renders the assignment of the Thalattosauria either to the Lepidosauromorpha or to the Archosauromorpha equivocal" (1987b:128).

In his critical comments on thalattosaur relationships, Kuhn-Schnyder rejected his earlier interpretation of *Askeptosaurus* as a specialized squamate (1988:882). He had previously characterized *Askeptosaurus* as a diapsid, noting, however, that "the upper temporal opening is reduced to a narrow slit or even entirely closed" (Kuhn-Schnyder, 1971:94). He further argued that "the skull of *Askeptosaurus* resembles that of *Thalattosaurus* in the differentiation of the rostrum, in the configuration of the temporal openings, and in the deeply embayed occiput," although stark differences are apparent in the palatal dentition or the

Thalattosaur Interrelationships

absence thereof (Kuhn-Schnyder, 1971:95). In his later evaluation he instead argued that *Askeptosaurus*, just as *Thalattosaurus* and *Clarazia*, was characterized by the presence of but a single, ventrally positioned temporal opening, and a monimostylic quadrate. He further argued that there is no evidence to support the conclusion that the absence of an upper temporal fenestra in thalattosaurs is due to a secondary closure (Kuhn-Schnyder, 1988:884). He dismissed Peyer's warning that the configuration of the temporal region of the skull should not be overrated in reptile classification, and on the basis of the presence of a single lower temporal opening denied any close relationship of thalattosaurs with diapsids; concerns on which branch of diapsid (saurian) reptiles the thalattosaurs should be placed he consequently declared irrelevant (Peyer, 1936a; Kuhn-Schnyder, 1988:881). *Hescheleria* he found too incompletely preserved to allow the assessment of its relationships with any certainty.

In 1984, a new lepidosaurian reptile was described from the Zorzino Limestone (Norian, Late Triassic) of Endenna, near Zogno, in the Bergamo Prealps, under the name of *Endennasaurus acutirostris* (Renesto, 1984). A more detailed study revealed its nature as an edentulous thalattosaur that constitutes the sister taxon of the (*Askeptosaurus* (*Clarazia*, *Thalattosaurus*)) clade (Renesto, 1991). The first comprehensive analysis of thalattosaur interrelationships was offered by Nicholls, based on a reexamination of the fossil remains of *Thalattosaurus* and *Nectosaurus* from the Hosselkus Limestone (Carnian, Upper Triassic) of Shasta County, northern California (1999). In addition to the Monte San Giorgio thalattosaurs and *Endennasaurus*, she also included the relatively poorly known genera *Agkistrognathus* and *Paralonectes* from the Lower to Middle Triassic Sulphur Mountain Formation of eastern British Columbia in her analysis (Nicholls and Brinkmann, 1993b). A nested hierarchy for all thalattosaurs then known forms a monophyletic clade named Thalattosauriformes by Nicholls (1999: 25). The relationships of the Thalattosauriformes to other diapsid reptiles was not the objective of her investigation, however.

A reappraisal of the anatomy and phylogenetic relationships of *Endennasaurus* by Johannes Müller and colleagues again did not address thalattosaur relationships within reptiles in general, but it did include the first thalattosaur that had in the meantime been described from southwestern China, namely, *Anshunsaurus huangguoshuensis* from the Xiaowa Formation (formerly Wayao Member of the Falang Formation) of Carnian age, in the Guanling Biota, Guizhou Province (Rieppel, Liu, and Bucher, 2000; Müller et al., 2005). A milestone in the analysis of thalattosaur interrelationships in terms of numbers of characters considered and taxa included is Müller's work published in 2005, based on a monographic treatment of all known material of *Askeptosaurus*. The phylogenetic analysis included all thalattosaur taxa known at the time, among them the second thalattosaur that had by then been described from southwestern China, namely, *Xinpusaurus suni*, again from the

Xiaowa Formation of Carnian age, Guanling Biota (J. Liu and Rieppel, 2001). Some of the highlights of Müller's analysis in the ever-changing landscape of thalattosaur relationships deserve to be mentioned (2005). *Clarazia* was recovered as sister taxon of *Hescheleria*, with *Thalattosaurus* the sister taxon of the two. This appears to be a rather robust grouping, indicating an affinity of western Tethyan with eastern Pacific taxa. Further, the Chinese genus *Anshunsaurus* was recovered as sister taxon of *Askeptosaurus*, a relationship that indicates affinities of western Tethyan with eastern Tethyan thalattosaurs (see also Rieppel, Liu, and Bucher, 2000; L. Cheng, Chen, Zhang et al., 2011). *Endennasaurus* was found to be the sister taxon of the *Askeptosaurus-Anshunsaurus* clade (see also J. Liu and Rieppel, 2005). Finally, and significantly, the Chinese thalattosaur *Xinpusaurus* was found to be the sister taxon of *Nectosaurus* from northern California, indicating trans-Pacific relationships of these two genera that are notoriously difficult to explain for non-pelagic organisms (Rieppel, 1999a; Bardet et al., 2014). But once again, "a problem of any hypothesis on the biogeographic origin of thalattosaurs is that there is currently no consensus on the identity of their sister group" among diapsid reptiles in general (Müller, 2005:1363f). Since Müller's work, additional Triassic marine reptiles have been described from southwestern China and have become subject to phylogenetic and biogeographic analysis, most notably two new genera of the enigmatic saurosphargids (see chap. 5): *Sinosaurosphargis,* and *Largocephalosaurus* (C. Li, Rieppel, Wu et al., 2011; C. Li, Jiang, Cheng et al., 2014). Analysis of the phylogenetic relationships of these two genera, and by implication of the saurosphargids generally, showed the latter to group with thalattosaurs, on the one hand, and with sauropterygians, on the other. This three-taxon statement remains poorly resolved, however: saurosphargids have been recovered as sister group of thalattosaurs, with the sauropterygians as sister of these two (C. Li, Rieppel, Wu et al., 2011). Alternatively, saurosphargids were recovered as sister group of sauropterygians, with thalattosaurs as sister of these two (C. Li, Jiang, Cheng et al., 2014). Either way, the three clades would appear to be part of the same marine radiation. In contrast, earlier investigations into diapsid interrelationships recovered a signal that groups thalattosaurs with ichthyopterygians (Müller, 2003). At the present time it therefore remains unknown how many times in the course of the early Triassic reptiles invaded the sea. One thing that is known, however, is that thalattosaurs are certainly not close to lepidosaur or even squamate origins.

Before moving on to the next, and last, major clade of marine reptiles represented in the Middle Triassic of Monte San Giorgio, another look at the extraordinary structure of the rostrum in *Hescheleria* seems to be in order. Three basic rostral morphologies can be identified among thalattosaurs (Rieppel, Müller et al., 2005). The presumed plesiomorphic condition is observed in *Askeptosaurus* and *Anshunsaurus*. The elongate rostrum is parallel-sided, terminating in a rounded tip; the external nares are retracted; the dentition is isodont; a diastema separating the premaxillary from the maxillary dentition is absent. The second, relatively derived rostral structure is observed in *Clarazia*, *Thalattosaurus*, and *Miodentosaurus*, the latter most probably from the Xiaowa Formation (Carnian) of Guanling, Guizhou Province, southwestern China (Y.-N. Cheng, Wu and Sato, 2007). In these taxa, the rostrum is moderately elongate, its lateral margins converging toward a narrow, pointed tip. The anterior tip of the rostrum is moderately deflected ventrally, and a distinct diastema separates the premaxillary from the maxillary teeth. *Xinpusaurus* approaches the claraziid rostral structure, although the premaxillaries are in general much more delicately built. Among known thalattosaurs, *Xinpusaurus* is autapomorphic in that the alveolar margin shows a dorsal curvature at the anterior end of the maxilla, matched by a corresponding dorsal curvature of the alveolar margin of the dentary, and a diastema separating the premaxillary from the maxillary teeth is absent.

Highly derived is the rostral structure in *Hescheleria*, with an almost vertically deflected premaxillary rostrum (Peyer, 1936b). The maxilla is relatively short, with a truncated anterior end and a prominent, vertically ascending process. The premaxillary rostrum is suspended from the anterior end of the maxilla laterally on either side, and from the anterior end of the frontal(s) dorsomedially; details of the contact with the vomer(s) remain unknown. The premaxillary rostrum overhangs the anterior end of the lower jaw; its teeth point backward with no occlusal dentition to work against. A distinct diastema separates the premaxillary from the maxillary teeth. A reinvestigation of the incompletely, yet very well preserved and acid-prepared rostrum of *Nectosaurus* from the Carnian (Upper Triassic) of Shasta County, northern California, allows the reconstruction of a similarly almost vertically deflected premaxillary rostrum overhanging the anterior end of the lower jaw (Rieppel, Müller et al., 2005). Such a reconstruction of *Nectosaurus* is corroborated by a complete, extraordinarily well-preserved and acid-prepared thalattosaur skull that was collected several decades ago from the Upper Triassic (Carnian) Natchez Pass Formation of the Humboldt Rage (Buffalo Mountains) in northwestern Nevada (Nicholas Hotton III, personal communication 1991; see also Storrs, 1991b:2066). Unfortunately, the specimen, which is kept in the collections of the National Museum of Natural History, Smithsonian Institution, remains undescribed to the present day. It would not be surprising if that specimen represented the genus *Nectosaurus* as well. Finally, an almost vertically deflected premaxillary rostrum overhanging the anterior end of the lower jaw was tentatively also reconstructed for *Paralonectes* from the Lower to Middle Triassic Sulphur Mountain Formation of eastern British Columbia, although preservation in this case is much more problematical (Nicholls and Brinkman, 1993b; Rieppel, Müller et al., 2005). Nevertheless, it seems intuitively likely that such a highly derived rostral structure is indicative of close phylogenetic relationships of *Hescheleria* with *Nectosaurus* (and possibly *Paralonectes*), although such relationships based on rostral structure have not yet been more comprehensively tested through phylogenetic analysis. If confirmed in future work, such relationships would indicate faunal affinities of the western Tethyan and the eastern Pacific faunal provinces with interesting paleobiogeographical implications (Rieppel, 1999a; see also Bardet et al., 2014).

9.1. A specimen of *Macrocnemus bassanii* (Paleontological Institute and Museum, University of Zurich T2472) from the Cava Tre Fontane. As preserved (most of the tail missing), the specimen measures 50 cm in length (Photo © Heinz Lanz/ Paleontological Institute and Museum, University of Zurich).

Protorosaurs

The protorosaurs are a diverse group of Permo-Triassic reptiles that have been recovered from continental as well as marine deposits of Laurasian and Gondwanan origin. Their diverse skeletal morphology testifies to their varied adaptations to vastly different modes of life, ranging from marine to terrestrial to arboreal (but see Pritchard and Nesbitt, 2017), but at the same time renders the analysis of their phylogenetic relationships difficult, indeed controversial.

The Protorosauria was introduced by Huxley in 1872 to include *Protorosaurus speneri* from the Kupferschiefer Formation (Wuchipingian, Upper Permian) of Germany and England (Huxley, 1872:195; Proterosauria in Lydekker, 1888:301; see also Meyer, 1830, 1832; Gottmann-Quesada and Sander, 2009). A pivotal discovery for the discussion of diapsid reptile interrelationships was *Prolacerta broomi* from the Early Triassic of the Karroo Basin, South Africa, and from Antarctica (Parrington, 1935; Colbert, 1945a, 1987; see also Modesto and Sues, 2004). The description of a second skull of *Prolacerta* led to the erection of the Prolacertiformes (Camp, 1945:97). The prolacertiforms constituted a subgroup of diapsids that over time became co-extensive with Huxley's Protorosauria, the latter taking priority (Chatterjee, 1986; see also Evans, 1988). The discovery of *Prolacerta* proved so important because, as is implied by its name, it was considered a form close to the origin of the Squamata (lizards)—a point of view that heavily colored not only the interpretation of thalattosaurs (*Askeptosaurus*, see chap. 8), but also influenced the interpretation of the protorosaurs from Monte San Giorgio (Kuhn-Schnyder, 1971). A crucial role in this debate was played by *Araeoscelis*, a gracile reptile from the Lower Permian of Texas, which was again originally described as an ancestral lizard (Williston, 1910, 1914a). But *Araeoscelis* differs fundamentally from *Prolacerta* in the configuration of the temporal region of its skull (Vaughn, 1955). In *Araeoscelis*, there is a single upper temporal fenestra, below which the temporal region is entirely covered by dermal bone, a cover that comprises a broadly expanded squamosal. *Prolacerta* likewise has an upper temporal opening, but the cheek region is widely open. The jugal carries a distinct posterior process but fails to contact the quadratojugal, leaving the lower temporal arch incomplete (Modesto and Sues, 2004). Whereas Camp considered *Prolacerta* a diapsid with an incomplete lower temporal arch, he included *Araeoscelis*—along with *Protorosaurus* and sauropterygians (nothosaurs, plesiosaurs, and placodonts)—in his Euryapsida, a group he diagnosed as "reptiles with very

The Problem of Protorosaur Monophyly

broad arches below the upper temporal opening" (Colbert, 1945b:148; see also Camp, 1945).

With *Araeoscelis* and *Prolacerta*, two reptiles with a different configuration of the temporal region of the skull had both originally been described as ancestral, or close to the ancestry, of squamates (lizards) (Williston, 1910, 1914a; Parrington, 1935). Generalized squamates (basal lizards) are characterized by the presence of an upper temporal fenestra, bounded laterally (ventrally) by an upper temporal arch. The cheek region is wide open, and although the jugal may carry a rudimentary posterior process, a lower temporal arch is absent (as also is the quadratojugal). There were, then, two alternative ways to derive the generalized lizard skull from the alternative ancestral conditions. If squamates were descended from a form like *Araeoscelis*, the dermal covering of the cheek region would have been reduced by an embayment from below, a scenario championed by D. M. S. Watson (1914:89) and Williston (1914b:138). The alternative, and today generally accepted scenario, is that squamates are of diapsid origin, hence ancestrally had a lower temporal fenestra as is present in the tuatara (*Sphenodon*), but have lost the lower temporal arch (Parrington, 1935:204).

It is ironic that this latter scenario was motivated, in large part, by Parrington's description of *Prolacerta* as providing an intermediate condition of form, when today *Prolacerta* is recognized as an archosauromorph taxon quite unrelated to lepidosaurs. Indeed, *Protorosaurus* and *Prolacerta* have for a long time been considered the standard-bearers of the Protorosauria (Prolacertiformes), which in current phylogenetic analyses nest at the base of the archosauromorph clade (see Rieppel, Fraser et al., 2003, for a review). More recently, doubts have been rising about the monophyly of protorosaurs (prolacertiforms), especially concerning their basal members, such as *Protorosaurus* and *Prolacerta*. In particular, *Prolacerta* was found to be more closely related to basal Archosauriformes such as *Proterosuchus* and *Euparkeria* than to other protorosaurs, and under some data combinations, *Protorosaurus* was found to join *Prolacerta* in these alternative relationships (Dilkes, 1998; Rieppel, Fraser et al., 2003; for more recent analyses of basal archosauromorph interrelationships see Nesbitt, 2011; Pritchard et al., 2015; Ezcurra, 2016; Pritchard and Nesbitt, 2017). Such uncertainties of basal protorosaur/prolacertiform interrelationships need not concern the discussion of the protorosaurs from the Middle Triassic of Monte San Giorgio, however, which clearly are part of a monophyletic clade diagnosed by a suite of intriguing characteristics.

Monte San Giorgio protorosaurs comprise two genera, *Macrocnemus* and *Tanystropheus*. A close relationship of the two genera is generally accepted, as indicated by the shared character of hollow long-bones in the limbs (Dilkes, 1998; Rieppel, Fraser et al., 2003). *Tanystropheus*, in turn, is part of a tanystropheid clade that is again diagnosed by a highly peculiar feature. In the generalized tetrapod hand or foot, the metacarpals and metatarsals, respectively, are elongate elements that are distinctly longer than the more distal phalanges of the digits. Such is not the case *Tanystropheus*, where in the foot metatarsal five is at least weakly hooked (the ancestral condition in saurians, i.e., lepidosauromorphs and archosauromorphs) but much reduced in length. The reduction in length of the fifth metatarsal is compensated for by an elongation of the first proximal phalanx in the fifth toe, which consequently takes on a metatarsal-like appearance. *Tanystropheus* shares this peculiar character with *Amotosaurus rotfeldensis* from the terrestrial upper Buntsandstein of Baden-Württemberg, southern Germany; with *Langobardisaurus pandolfii* from the Zorzino Limestone Formation (middle Norian, Upper Triassic) of Cene, Bergamasque Prealps, northern Italy, and from the Seefeld Formation (Norian) near Innsbruck, Austria; and with *Tanytrachelos ahynis*, a freshwater form from the Late Triassic of eastern North America (Olsen, 1979; Bizzarini and Muscio, 1994; Renesto, 1994; Fraser, Grimaldi et al., 1996; Fraser and Rieppel, 2006; Casey et al., 2007; Saller et al., 2013). This part of the protorosaur tree can be accepted with at least some confidence. But even this small part of the protorosaur tree shows a broad array of adaptations, from terrestrial (*Macrocnemus, Langobardisaurus*), to freshwater aquatic (*Tanytrachelos*), to marine (*Tanystropheus*) conditions.

Following the paleontological excavation campaigns organized by the Museo Civico di Storia Naturale di Milano at the Besano locality in the years 1863 and 1878, Francesco Bassani in 1886 published a preliminary account of the material thus obtained: "The work was long lasting and difficult, as one had to deal with specimens that often were fragmentary, distorted, or mixed up, requiring repeated observation, patience, and a lot of caution" (1886:20). It must have been during these investigations that Bassani was dealing with a fragmentary specimen of a reptile that he duly labeled but did not include in his published account. During a brief visit to the Milan museum, Franz Nopcsa noticed the fossil and proceeded to describe it in a brief publication of 1930 under the name *Macrochemus bassanii*. Given its poor preservation, as well as the poor contrast the fossil bones provided against the black matrix, Nopcsa published neither a drawing, nor a photograph of the specimen, but a diagrammatic reconstruction instead (1930:254). The latter appeared suspicious to Bernhard Peyer, who was also proficient in classic Greek and found Nopcsa's spelling of the genus name linguistically misconstrued. Going back to the original specimen, Peyer found Bassani's original label attached to it, which read: "This reptile appears to me to belong to a new genus, one which is also different from *Tribelesodon*. I have imposed on it the genus name *Macrocnemus* (derived from classic Greek, in reference to the large 'fibia' [meaning the tibia]). However, I have preferred not to talk of it in my memoir [of 1886], because the fossil does not allow a detailed examination of all the [relevant] osteological characteristics. BASSANI" (Peyer, 1931f:191). The name was thus meant to refer to the elongate zeugopodium of the hind limb (tibia), and hence had to read

Macrocnemus not
Macrochemus

Macrocnemus, not *Macrochemus* (Peyer, 1931f; Nopcsa, 1931). What had happened was that the label Bassani had written had been folded, the fold running right through the "n" in *Macrocnemus*; later restoration of the label turned the "n" into an "h," which accounts for Nopcsa's misspelling (Peyer, 1937:4). Nopcsa noted the characteristically elongated cervical vertebrae in *Macrocnemus*, which prompted him to compare the taxon with *Trachelosaurus*, *Protorosaurus*, and *Araeoscelis*. With *Trachelosaurus* and *Protorosaurus* as possible allies, Nopcsa hit on the protorosaur affinities of *Macrocnemus* (Broili and Fischer, 1917; Nopcsa, 1930). Given the characteristic limb proportions, Nopcsa interpreted *Macrocnemus* as a terrestrial form, although it was collected from marine sediments (1930).

Alerted by Nopcsa's brief description of the poorly preserved holotype, Peyer hoped to secure better specimens of this terrestrial reptile from the marine sediments of Monte San Giorgio (1944). His expectations were, again, duly fulfilled. In 1937, Peyer was able to publish an extensive monograph on *Macrocnemus bassanii*, based on material that in addition to the holotype comprised six specimens, of which four were collected at Monte San Giorgio, two at the Besano locality (fig. 9.1). The Monte San Giorgio specimens mostly derived from the Anisian of the Grenzbitumenzone (Val Porina mine and Cava Tre Fontane) except for one, the latter representing the geologically youngest occurrence of *Macrocnemus* in the Meride Limestone (Ladinian) of the Alla Cascina beds (Peyer, 1937:6; Kuhn-Schnyder, 1971:112; Furrer, 2003:46). Following the detailed description of the material at his disposal, which unequivocally established a close relationship of *Macrocnemus* with *Tanystropheus*, Peyer conducted extensive morphological comparisons in an effort to decipher the broader phylogenetic relationships of *Macrocnemus* (1937). *Protorosaurus* and *Araeoscelis* figured prominently in that discussion, but so did other Permo-Triassic reptiles, as well as rhynchocephalians (*Sphenodon*) and squamates. Peyer found the general habitus of *Macrocnemus* to be lizard-like (lepidosaurian), and based on the absence (incompleteness) of the lower temporal bar concluded that *Macrocnemus* "is to be classified along with *Araeoscelis* amongst the 'ancient lizards' in the sense of S. W. Williston" (Peyer, 1937:120). Such a conclusion was predicated on the hypothesis, however, that the open cheek region of the lizard skull evolved through ventral embayment from a condition exemplified by *Araeoscelis* (Williston, 1917:418). "I have no hesitation in saying that the skull and skeleton of *Araeoscelis* present distinctly primitive characters of the *Squamata*, to such an extent indeed that I believe the genus has a definite phylogenetic relationship with the order. In fact, as far as the skull is concerned, all that seems necessary to convert *Araeoscelis* into a primitive lizard is the erosion of the lower part of the squamosal bone. . . . [T]he development of streptostyly in the quadrate . . . would necessarily ensue with the loss of the support of the squamosal" (Williston, 1914a:392).

Commenting on the reconstruction of the skeleton of *Tanystropheus*, Peyer once again noted a similar elongation of the cervical vertebrae, albeit to a far less dramatic degree, in *Macrocnemus*, *Protorosaurus*, and

Araeoscelis (1939b:209). This characteristic was taken to indicate phylo-genetic affinities that mark out a group Kuhn-Schnyder identified as the Protorosauria (1954a:220). However, in the *Traîté de Paléontologie*, Peyer and Kuhn-Schnyder instead classified *Macrocnemus* and *Tanystropheus*, along with *Askeptosaurus* (see chap. 8) as Triassic squamates, noting the profound difference in the configuration of the temporal region of the skull in these genera as opposed to *Araeoscelis* (1955b:578). They further commented on the incomplete knowledge of the skull anatomy in *Proto-rosaurus*, and expressed doubts as to the validity of the Protorosauria: "It does not seem appropriate to unite these two genera [viz. *Araeoscelis and Protorosaurus*] in a single order. . . .[I]t does neither appear practical to us to choose as the banner-bearer of an order a genus as poorly known and controversial as *Protorosaurus*" (Peyer and Kuhn-Schnyder, 1955b: 600). What had happened? The interpretation of the origin of the squamate skull had obviously shifted from the Williston model based on *Araeoscelis*, to the Parrington model, the latter based on *Prolacerta* and deriving the lizard skull from an ancestral diapsid condition through loss of the lower temporal arch (Williston, 1914, 1917; Parrington, 1935).

Given persisting uncertainties in the interpretation of the temporal region of the skull of *Macrocnemus*, Peyer in his original monograph had conceded: "Should it be found, against all expectations, that *Mac-rocnemus* does after all have a [complete] lower temporal arch, then the genus would have to be classified with, or nearby the rhynchocephalians" (1937:120). In 1954, Kuhn-Schnyder published an essay on the origin of squamates (lizards) that triggered the feud between him and Peyer sketched in the first chapter (1954a, 1954b). In that essay, Kuhn-Schnyder offered a revised reconstruction of the skull of *Macrocnemus*, based on a specimen that was collected in 1938 at the Cava Tre Fontane locality of Monte San Giorgio (Anisian of the Grenzbitumenzone) (1954a:220; see also Kuhn-Schnyder, 1971) (see fig. 9.1). On the basis of these new insights he drew the following systematic conclusions: "*Macrocnemus* is characterized by two temporal openings; the lower temporal arch is incomplete; the quadrate is streptostylic; a supratemporal is present. *Macrocnemus* is to be classified with ancient lizards. There exists no close relationship [of *Macrocnemus*] with *Araeoscelis*. The structure of the skull of *Protorosaurus* remains unknown" (Kuhn-Schnyder, 1954a: 220). Based on his later detailed description of the skull of the 1938 Cava Tre Fontane specimen, Kuhn-Schnyder continued to classify *Macrocnemus* with squamates ("lacertilians"), but on account of the elongation of the cervical vertebrae, and given that the presence of a supratemporal he now considered to be uncertain, he concluded: "*Macrocnemus* cannot be considered a direct ancestor of lacertilians" (Kuhn-Schnyder, 1962b:124; see also Kuhn-Schnyder, 1974:46). He further specified that "it is a secure fact that *Macrocnemus*, *Tanystropheus* and *Askeptosaurus* did no longer possess a lower temporal arch. All three genera can be derived from diapsid forms. However, today their relationship does no longer appear as close as had previously been assumed" (Kuhn-Schnyder, 1962b:126). In summary, Kuhn-Schnyder classified *Macrocnemus* as a representative of the Prolacertilia, more closely related to *Prolacerta* than to *Tanystropheus* and *Askeptosaurus* (1962b:131). For Kuhn-Schnyder, the name "Prolacertilia" in that context meant "ancestral lacertilians," and thus had a quite different connotation than it has today (see also Kuhn-Schnyder, 1974).

Box 9.1. *Macrocnemus bassanii* from the Middle Triassic of Monte San Giorgio

Several specimens of *Macrocnemus* have been collected, both at the Besano locality and at Monte San Giorgio. The gracile animal appears lizard-like, reaching a total length of approximately 120 cm. Earlier theories that *Macrocnemus* might be related to lizards are no longer held, the taxon now classified as a protorosaur of archosauromorph affinities. The skull is relatively lightly built, with a pointed snout. Protorosaurs are generally characterized by an elongation of the neck vertebrae. Such elongation is relatively moderate in the neck of *Macrocnemus*, which comprises eight cervical vertebrae. In view of its slender body, elongate, slender limbs, and long tail, *Macrocnemus* is thought to have been a predominantly terrestrial organism. Remarkable are the much-elongated hind limbs, after which the genus was named (*Macrocnemus*, derived from Ancient Greek, meaning "large tibia"). It is likely that the animal switched to bipedal locomotion when running at high speed in pursuit of insect prey. Remains of the scaly skin are preserved in the sacral region and along the tail in a juvenile specimen of *Macrocnemus* collected at the Besano locality. The mode of reproduction in *Macrocnemus* remains unknown, although a marine protorosaur from the Middle Triassic of southwestern China is known to have been viviparous.

© Beat Scheffold / PIMUZ

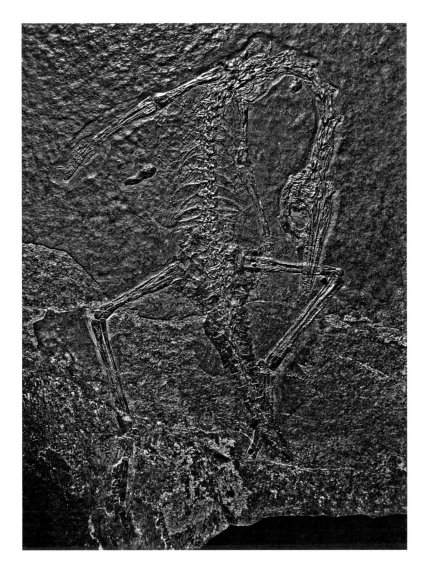

9.2. A juvenile specimen of *Macrocnemus bassanii* (Museo Civico di Storia Naturale di Milano BES SC 111) from Besano. As preserved (most of the tail is missing), the specimen measures 30 cm in length (Photo © Giorgio Teruzzi/Museo Civico di Storia Naturale di Milano).

In fact, *Macrocnemus* is a typical example of a protorosaur (prolacertiform) as currently understood, that is, part of clade that represents a basal branch of archosauromorph reptiles. As a basal archosauromorph *Macrocnemus* is far removed from lepidosaurian, or even squamate relationships. In view of these earlier discussions, it is interesting that skin preserved in the sacral region and along the tail of a juvenile *Macrocnemus* collected at the Besano locality (fig. 9.2) should display a squamation recalling the squamate condition (Renesto and Avanzini, 2002:42). *Macrocnemus* is a gracile reptile with a long tail, estimated to have reached a total length of 110 to 120 cm. The telltale protorosaur character is the elongation of the cervical vertebrae (of which there are eight in *Macrocnemus*), their length increasing toward the middle of the cervical region, then decreasing again. The cervical vertebrae are a little more than twice as long as the dorsal vertebrae, and characterized by a low neural spine, or rather, a low neural ridge. Another protorosaur character relates to the cervical ribs that are elongate and slender, and aligned parallel to the cervical vertebrae. The anterior (morphologically proximal) articular head carries a distinct free-ending anterior process.

The skull is relatively lightly built, with large orbits and a pointed snout. The preorbital region of the skull is distinctly longer than the postorbital region. A characteristic feature shared with *Tanystropheus* and other protorosaurs is an elongated narial groove that extends from the external naris posteriorly on the posterior (nasal) process of the premaxilla up on to the nasal (Jiang, Rieppel, Fraser et al., 2011). It might have accommodated a salt gland. The dentition is homodont, the jaws being furnished with numerous small, peg-like, monocuspid teeth. The upper temporal fossa is well delimited, bounded posteriorly by the characteristically tetra-radiate squamosal as described by Peyer (1937). The cheek region is wide open, revealing a wide anteromedial flange extending anteriorly from the shaft of the quadrate. Although Kuhn-Schnyder interpreted the quadrate of *Macrocnemus* as streptostylic, the ventrally descending process of the tetraradiate squamosal, overlapping the quadrate anterolaterally, would seem to have prevented such movement (Kuhn-Schnyder, 1962b; Rieppel and Gronowski, 1981). The jugal is a triradiate ossification, its posterior process ending in a pointed tip at about the midpoint of the lower temporal fenestra. The lower temporal arch thus remains incomplete, and the lower temporal fenestra is open ventrally. The presence of a quadratojugal has not been confirmed for *Macrocnemus*.

The dorsal vertebral column of *Macrocnemus* comprises 16 to 17 vertebrae, followed by two sacrals, and at least 40 caudals (Rieppel, 1989c; Renesto and Avanzini, 2002). The second pleurapophysis in the sacral region is characteristically bifurcated in *Macrocnemus*. The dorsal ribs are slender, and the gastral rib basket is delicately built, features that indicate an agile organism that seeks no bone ballast to counter buoyancy. The pectoral girdle is of a generalized structure, with slender clavicles and a cruciform interclavicle carrying a prominent posterior stem. The scapula and coracoid are broad and plate-like. The pelvic girdle is characterized by a distinct thyroid fenestra in its ventral part; the ilium carries a distinct, laterally deflected preacetabular process that serves as the site of attachment of the iliopubic ligament. High agility of *Macrocnemus* is also indicated by its limb proportions (Rieppel, 1989c). This is particularly apparent in the hind limb, the proportions of which gave the taxon its name (Nopcsa, 1930; Peyer, 1931f). In the Besano/Monte San Giorgio specimens the length of the tibia is typically longer than the femur, but in a Chinese specimen it is just under the length of the femur (Rieppel, 1989c; Jiang, Rieppel, Fraser et al., 2011). The tarsus is well ossified, comprising two proximal (an astragalus located in an intermedium position that is smaller than the calcaneum) and four distal ossifications in the mature animal; a centrale may variably be present. The metatarsus is strongly asymmetrical, the fourth metatarsal being the longest and most sturdily built in the series; the fifth metatarsal is relatively short and distinctly hooked.

Until recently, only a single species of *Macrocnemus* had been recognized in the Middle Triassic Monte San Giorgio/Besano biota, namely, *Macrocnemus bassanii* (Nopcsa, 1930). A second species, *Macrocnemus obristi*, is now known from the Prosanto Formation (early Ladinian) of the Ducan area, Grisons, Switzerland (Fraser and Furrer, 2013). A third species in the genus, *Macrocnemus fuyuanensis*, was described from the Middle Triassic (Ladinian) of southwestern China, differing from *M. bassanii* only in skeletal proportions (humerus/radius) in the fore limb (Li, Zhao, and Wang, 2007; Jiang, Rieppel, Fraser et al., 2007). Interestingly, a newly prepared specimen from the upper Grenzbitumenzone (late Anisian) of Mirigioli/Point 902 was identified, based on similar fore limb proportions, as *Macrocnemus* aff. *M. fuyuanensis* (Jaquier et al., 2017). This testifies to close faunal connections of the western and eastern Tethyan province, an inference that is further corroborated by the occurrence of *Tanystropheus* in the Middle Triassic (Ladinian) of southwestern China (see chap. 11 for further discussion).

The whole skeletal habitus of *Macrocnemus* suggests a gracile, predominantly terrestrial lizard-like animal, with limb proportions that indicate it may well have been bipedal in rapid locomotion (Rieppel, 1989c). This would not have involved an erect posture and a digitigrade stance as reconstructed for *Macrocnemus* by Nopcsa, a reconstruction that raised Peyer's initial suspicions (Nopcsa, 1930; Peyer, 1931f, 1937). Neck elongation as is characteristic of *Macrocnemus* might seem incompatible with facultative bipedalism at high speed, but the weight of the long neck might well have been compensated for by the long tail. The proximal tail region of *Macrocnemus* is characterized by very prominent, horizontally oriented caudal ribs, which indicates a fleshy base in a tail that was not laterally compressed. It thus appears that a heavy tail counterbalanced the head, neck, and body in bipedal locomotion, rather than being used in lateral undulation to provide propulsion in an aquatic environment (Snyder, 1954, 1962; see discussion in Rieppel, 1989c; Renesto and Avanzini, 2002). In summary, albeit collected from marine sediments in considerable numbers, skeletal correlates indicate that *Macrocnemus bassanii* was a predominantly terrestrial, insectivorous or more generally carnivorous species. Terrestrial habits are easily explained, since neighboring reef structures are known to have existed during the deposition of the Grenzbitumenzone in the Monte San Giorgio–Besano basin (Kuhn-Schnyder, 1974).

Tanystropheus, the "giraffe-neck saurian" is unquestionably the most iconic fossil from Monte San Giorgio, although it has been reported from various other localities as well (Peyer, 1939b; F. Drevermann in Peyer, 1944:65). And once again, it was the well-preserved material collected by Bernhard Peyer at Monte San Giorgio that solved the riddle surrounding

Macroscelosaurus, Zanclodon, Tribelesodon and *Tanystropheus*

this taxon, the history of which reaches back to pioneer times in vertebrate paleontology.

Count Georg Münster (1776–1844) had collected fossils in the Muschelkalk around the town of Bayreuth, Bavaria (Germany), both in its
upper and lower parts, for more than 25 years. Chancing upon a partial
skeleton, its elements preserved in association, when prospecting in the
upper Muschelkalk (mo1, Anisian) on the Oscherberg near Laineck east
of Bayreuth, he finally put pen to paper and reported his findings in the
*Neues Jahrbuch für Mineralogie, Geognosie, Geologie und Petrefakten-
kunde* (Münster, 1834). This most respectable find he named, following
its brief description, *Nothosaurus mirabilis*; the specimen—the holotype
of its species—was eventually figured by Hermann von Meyer, along
with additional material, which validated the species name according to
modern standards (1847–1855. pl. 23). Münster mentioned and named
a variety of other fossil remains he had collected, but he concluded his
preliminary report with an allusion to many more different species yet
to be described and named: "I will refrain from a more detailed description of these saurians until I have had the time to investigate them
properly" (Münster, 1834:527). He must have informally communicated
his observations about strange, elongated elements from the upper Muschelkalk of the surroundings of Bayreuth as those of a reptile he called
Macroscelosaurus, however (Peyer and Kuhn-Schnyder, 1955b:578). As
indicated by this name, Münster believed these elements to represent
the elongated limb-bones of an "exceptionally long-legged saurian," but
never formally published the name or the material to which the name
referred (Wild, 1976:124). The name appears in print for the first time
in Hermann von Meyer's account of the *Saurians of the Muschelkalk*,
who recognized that the putative limb bones of *Macroscelosaurus* in fact
represent what he thought to be caudal vertebrae of a reptile that H. v.
Meyer named *Tanystropheus conspicuus*; no specific epithet was ever
associated with the genus name *Macroscelosaurus* (Meyer, 1847–1855
[1852:42[1]]). Nonetheless, the published record thus established the genus
name *Macroscelosaurus* as a potential senior synonym of *Tanystropheus*.
Huene treated *Macroscelosaurus* as a junior synonym of *Tanystropheus
conspicuus* (1907–1908:22). Broili noted the technical priority of *Macroscelosaurus* over *Tanystropheus*, but continued to use the latter name, given
its general entrenchment (1915:51). Citing Broili, O. Kuhn followed the
International Code of Zoological Nomenclature (ICZN), and treated
Tanystropheus as a subjective junior synonym of *Macroscelosaurus*, cred-
iting Münster as the author of the latter name (1934:118). In fact, the
names *Macroscelosaurus* and *Tanystropheus* do not appear in Münster
(1834). The genus name *Macroscelosaurus* was eventually suppressed by
the International Commission on Zoological Nomenclature, which exer-
cised its plenary power to conserve the genus name *Tanystropheus*, with
Tanystropheus conspicuus as genotypical species, by monotypy (Wild,
1976:125; see also Wild, 1975). Since *Tanystropheus conspicuus* from the
Germanic Muschelkalk is known exclusively from cervical vertebrae,

Peyer proposed to designate the much more completely known *Tanystropheus longobardicus* from Monte San Giorgio as "neogenotype," a proposal that is not sanctioned by the ICZN, however (Peyer, 1931c:92; Kuhn, 1934:118).

To add to the confusion, Plieninger described and illustrated an upper jaw fragment, isolated vertebrae, phalanges, and osteoderms from the Lettenkeuper (Ladinian) of Gaildorf (Württemberg, southern Germany) under the name of *Smilodon laevis* (1847a:152). Realizing that the genus name *Smilodon* was preoccupied by the saber-toothed cat, he replaced that name by *Zanclodon*, with *Z. laevis* as genotypical species (Plieninger, 1847b:248; Wild, 1976:125). Huene considered some of the vertebrae among that material as those of *Tanystropheus* (1931). All the material originally referred to *Zanclodon laevis* by Plieninger is now lost except for the upper jaw fragment, which is distinctly different from *Tanystropheus* and was designated as lectotype for the species *Zanclodon laevis* by Wild, in order to clearly demarcate *Zanclodon* from *Tanystropheus* (1976:125). This he considered necessary in order to fix the meaning of the genus name *Tanystropheus*, as some of the vertebrae published by Plieninger under the name *Zanclodon laevis* may well have represented *T. conspicuus* (Wild, 1973:149). As noted by Huene, *Zanclodon* would take priority over *Tanystropheus* if a synonymy were established (1931:83).

In his report on the paleontological excavations of 1863 and 1878 in the bituminous shales at the Besano locality, Bassani reported on a relatively small reptile fossil "of great importance, as it signals in all probability the presence of pterosaurs" in the Besano Formation (1886:25). He called the fossil *Tribelesodon longobardicus*, the genus name referring to the tricuspid teeth identifiable in the jaws of a skull some 30 mm long (Bassani, 1886:29). But in his synoptic classification of the fossils from Besano, Bassani only tentatively referred *Tribelesodon* to the Pterosauria (1886:48). Nevertheless, he was followed in this assessment by several subsequent authors (e.g., Arthaber, 1922; Nopcsa, 1923; see chap. 1 for a more detailed account). It was not until Peyer's discovery of a complete and articulated juvenile specimen at the Val Porina locality at Monte San Giorgio (figs. 9.3, 9.4) that the true nature of *Tanystropheus* was revealed (Peyer, 1931c, 1944; Kuhn-Schnyder, 1974:50). Peyer immediately recognized the close similarity of the cervical vertebrae of the Val Porina specimen with those of *Tanystropheus conspicuus* from the Germanic Muschelkalk, which refuted the interpretation of the Muschelkalk vertebrae as caudal ones, as well as the pterosaur nature of *Tribelesodon*. The latter name thus was recognized as a subjective junior synonym of *Tanystropheus*, the proper species binomen for the Besano/Monte San Giorgio taxon thus being *Tanystropheus longobardicus*. As was soon recognized, tricuspid teeth in the posterior part of the maxillary and dentary tooth row are characteristic of small specimens of *T. longobardicus* only, which were interpreted as terrestrial and insectivorous juveniles by Wild—a conclusion that will require further discussion below (1973).

9.3. Radiograph of the juvenile specimen of *Tanystropheus longobardicus* (Paleontological Institute and Museum, University of Zurich T2791), collected in 1929 in the Val Porina. The skull measures 46 mm in length (radiograph University Hospital Zurich, director Prof. Dr. H. R. Schinz, 1931; photographic rendition John Weinstein, the Field Museum, Chicago).

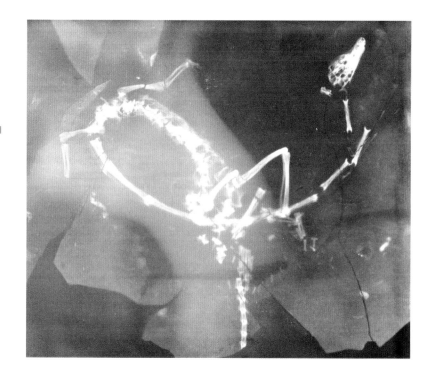

9.4. The juvenile specimen of *Tanystropheus longobardicus* (Paleontological Institute and Museum, University of Zurich T2791), collected in 1929 in the Val Porina. The skull measures 46 mm in length (Photo © Nicholas C. Fraser/ National Museums Scotland).

At the time when Peyer clarified the nature of *Tanystropheus longobardicus*, two other species of *Tanystropheus* had already been described, both from the Triassic of the Germanic basin (1931c). The genotypical species, by monotypy, was *Tanystropheus conspicuus* originally described by H. v. Meyer; the holotype comprises nine cervical vertebrae from the upper Muschelkalk (mo1, Anisian) of the Bindlacher Berg near Bayreuth. Wild estimated the species, which had a widespread occurrence in the upper Muschelkalk of the Germanic basin, to have reached a total length of approximately six meters (1973: 149).

Tanystropheus antiquus was originally described by Huene on the basis of 10 cervical vertebrae from the lower Muschelkalk (about 2 meters above the *Dadocrinus* beds, Anisian) of Krappitz, Upper Silesia (now Krapkowice, Poland) (1907–1908:223). The taxon had also been collected in the lower Muschelkalk of Jena, and Rüdersdorf near Berlin. In contrast to *Tanystropheus conspicuus* and *T. longobardicus*, the cervical vertebrae are not as pronouncedly elongated in *T. antiquus*, and the neural spines are somewhat less reduced.

In 1967, Ortlam described a puzzling array of incomplete fossil material from the terrestrial upper Buntsandstein of the northern Black Forest, Germany, which he referred to *Tanystropheus longobardicus*, and *Macrocnemus bassanii* (Ortlam, 1967:518, 522), respectively. These latter two taxa are exclusively known from late Anisian and Ladinian marine deposits. The upper Buntsandstein, however, is generally considered to represent the late Lower Triassic, extending upward to straddle the Spathian–Anisian boundary. The Black Forest fossils are thus of a stratigraphically older age than the taxa to which they have been referred, reason enough to subject them to a critical evaluation, along with the description of much more complete, newly collected specimens (Wild, 1980a). The conclusion of that revision was that the specimens referred to *Macrocnemus bassanii* and *Tanystropheus longobardicus*, respectively, by Ortlam, as also the new material, all represent *Tanystropheus antiquus*, which seemed to make sense given its stratigraphic provenance (Wild, 1980a). However, Wild noted their relatively small size, which to him indicated that the Buntsandstein material represents juvenile specimens which, as in *Tanystropheus longobardicus*, would pursue terrestrial habits, and become marine as adults only (1980a:204). More puzzling still was the presence of eight cervical vertebrae only in the Black Forest taxon, a number clearly lower than the 12 cervicals then considered to be characteristic of *Tanystropheus longobardicus* (Wild, 1973).

Langobardisaurus is another undisputed tanystropheid known from articulated specimens that again confirm the presence of 12 cervical vertebrae (Renesto, 1994). The low number of cervicals consequently raised doubts as to the taxonomic identity of the Black Forest protorosaur, which Wild eventually recognized as a separate genus, one he did not formally name but which he found to be "closest to *Tanystropheus* and *Macrocnemus*" (Wild, 1987:37; see also Evans, 1988). This new taxon was eventually described under the name of *Amotosaurus rotfeldensis*, for which one of the more complete specimens, collected from the upper Buntsandstein near Rotfelden, in the Calw district of Württemberg (southern Germany) was selected as the holotype (Fraser and Rieppel, 2006). Given the incompleteness of the material, *Tanystropheus antiquus* was found to be distinguished from *T. conspicuus* only by its smaller size, and its earlier stratigraphic occurrence (lower Muschelkalk of Krappitz, Upper Silesia). Insufficiently diagnosed as it thus is, referral of the species to the genus *Tanystropheus* remains equivocal (Fraser and Rieppel, 2006). The species has indeed been referred to a separate genus, *Protanystropheus antiquus* (Sennikov, 2011). Specimens corresponding to *Protanystropheus antiquus* in size and stratigraphic occurrence have also been signaled from the lower Muschelkalk of Rüdersdorf (Berlin), Jena, Bonnhof (southwestern Germany), and Winterswijk, the Netherlands (Wild and Oosterink, 1984). Other than in the Muschelkalk of the Germanic basin and in the Alpine Triassic, *Tanystropheus* has been reported from Spain, eastern Europe, the Middle East, and China, proving it to be a pan-Tethyan faunal element (Jurcsak, 1978; Wild, 1987; Vickers-Rich et al., 1999; Rieppel, 2001c; C. Li, 2007; Rieppel, Jiang et al., 2010;).

Box 9.2. *Tanystropheus longobardicus* from the Middle Triassic of Monte San Giorgio

Tanystropheus, another protorosaur, is a signature fossil from Monte San Giorgio. Several specimens have been collected both at Monte San Giorgio and at Besano. The largest individuals attained a total length of five to six meters. Small individuals with tricuspid teeth have alternatively been interpreted as juveniles, or as representatives of a separate, smaller species. The most conspicuous feature is the extreme elongation of the neck, which comprises a total of only 13 cervical vertebrae. Such elongation of the neck has motivated a number of competing physiological and functional interpretations. It seems, however, that the neck elongation is a consequence of the positive allometric growth of the cervical vertebrae rather than a specific adaptation. Conflicting hypotheses have also sought to explain the lifestyle of *Tanystropheus*. Given the extreme elongation of the neck, and the fact that the low neural spines on the "thoracic" vertebrae do not suggest a particularly strong neck musculature, most authors consider *Tanystropheus* a fully aquatic (marine) animal. There is the alternative interpretation, however, of *Tanystropheus* as a predominantly terrestrial organism, sitting on the shore while using its long neck to catch prey items under water. Stomach contents reveal fish and cephalopods as prey items of *Tanystropheus*.

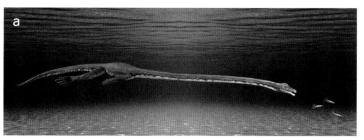

© Beat Scheffold

© Beat Scheffold / PIMUZ

In spite of such intensive research, the taxonomy of the genus *Tanystropheus* in the Middle Triassic of Monte San Giorgio/Besano (figs. 9.5, 9.6) remains to some degree unresolved (Wild, 1973, 1980b; Nosotti, 2007). The crux of the matter lies in the interpretation of the relatively small specimens, which show tricuspid teeth in the posterior maxillary and dentary tooth row. In his revision of *Tanystropheus* from the Grenzbitumenzone of Monte San Giorgio, Wild treated the larger specimens with monocuspid teeth as adults with marine habits, preying on fish and cephalopods as indicated by their stomach contents (1973). The smaller specimens with tricuspid teeth he treated as insectivorous juveniles with terrestrial habits. Later, Wild described a second species from stratigraphically younger deposits of Monte San Giorgio, *Tanystropheus meridensis* from the Alla Cascina beds (Meride Limestone, Ladinian), based on a skull and a string of six cervical vertebrae (1980b). *Tanystropheus meridensis* is of relatively small size, and again characterized by tricuspid teeth in the posterior maxillary and dentary tooth row, motivating Wild to specify in the diagnosis of the new species that it is "currently known from a juvenile specimen only" (1980b:5). In a critical revision of the Monte San Giorgio material, Fraser, Nosotti, and Rieppel noted that "*T. meridensis* cannot be distinguished from the smallest specimens of *T. longobardicus*" (2004:60A). Wild's (1980b) motivation to consider a geologically younger occurrence of *Tanystropheus* to represent a different species were stratigraphic considerations, which in view of the potential longevity of vertebrate taxa might be considered problematical (Wild, 1980a). The taxonomic issue that obtains in this case is rather the question whether the smaller specimens with tricuspid teeth in the posterior maxillary and dentary tooth row are, indeed, juveniles of *Tanystropheus longobardicus*, or whether they represent a different species of *Tanystropheus* altogether (see also Rieppel, 1994b; Renesto, 2005). Add to this conundrum the possibility that *Tanystropheus longobardicus* might be synonymous

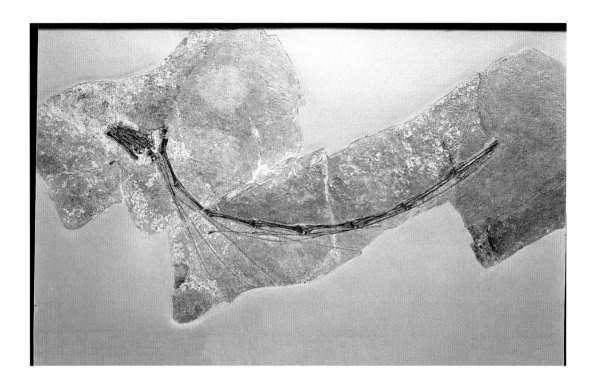

with *T. conspicuus*, a question that presently cannot be settled because skull material representing the latter species remains unknown (Wild, 1980a:204; Renesto, 2005:386). The expert judgment on these questions, and the taxonomic and nomenclatorial consequences that might obtain, have at this time not yet been settled and formalized. Given such taxonomic uncertainties, Renesto referred a new specimen of *Tanystropheus* from the Cava Inferiore beds (Meride Limestone, Ladinian) of Monte San Giorgio, and hence stratigraphically intermediate between *T. longobardicus* and *T. meridensis*, to *Tanystropheus* cf. *T. longobardicus* (2005). Another species of similarly uncertain taxonomic status that has been reported from the southern Alpine Triassic is *Tanystropheus fossai*, based on a few isolated cervical vertebrae from the Argillite di Riva di Solto (late Norian, Upper Triassic), Val Brembana, Bergamasque Prealps, northern Italy (Wild, 1980b). While this would be the stratigraphically youngest occurrence of the genus, Renesto again notes that the vertebrae in question, while unquestionably protorosaurian, "lack unequivocal characters for assignment to *Tanystropheus*" (2005:386).

Tanystropheus longobardicus could reach a total length of up to six meters, with a neck the length of which equals that of the trunk and tail. Its anatomy is well known from specimens collected in the Grenzbitumenzone and Besano Formation at Monte San Giorgio, and at the Besano locality, respectively (Wild, 1973, 1980b; Nosotti, 2007) (see figs. 9.5, 9.6). The skull is that of a diapsid reptile with an incomplete lower temporal arch. The orbit is large, and the preorbital region of the skull is longer than the postorbital region. The external naris opens into a narial trough that extends backward on the dorsal surface of the snout, as is also

9.6. A specimen of *Tanystropheus longobardicus* (Museo Civico di Storia Naturale di Milano BES SC 1018) from the Sasso Caldo quarry, Besano. The estimated overall length of the specimen is 140 cm (Nosotti, 2007:20; photo © Giorgio Teruzzi/ Museo Civico di Storia Naturale di Milano).

observed in *Macrocnemus*. The frontal forms a peculiar, laterally projecting shelf that overhangs the dorsal margin of the orbit (Nosotti, 2007, fig. 37). The most remarkable feature of *Tanystropheus* is unquestionably its neck, comprising much-elongated cervical vertebrae of a highly specialized structure, hollow, and characterized by a very low neural crest. The neck had previously been thought to comprise 12 cervical vertebrae, but is now known to be composed of 13 cervicals, of which the last two are distinctly shorter that the preceding ones (Nosotti, 2007; Rieppel, Jiang et al., 2010). In general, the length of the cervical vertebrae increases up to the ninth and tenth elements, behind which it decreases gain. Equally characteristic are the much-elongated cervical ribs, bridging several intervertebral joints and thus clustering to form bundles of slender bony rods aligned parallel to the ventrolateral aspect of the cervical vertebral column. The dichocephalous articular head is oriented at right angles to the longitudinal axis of the shaft of the rib, and articulates low on the anterior aspect of the cervical vertebra. A distinct free-ending anterior process projects beyond the articular head of each cervical rib. The cervical region is followed by 12 dorsal vertebrae, 2 sacrals, and in excess of 50 caudals (Wild, 1973:62). The occurrence of fracture planes in anterior caudal vertebrae, indicative of caudal autotomy in *Tanystropheus*, remains controversial, although the available evidence weighs against it (Wild, 1973, 1987; Renesto, 2005; Nosotti, 2007:70). The dorsal ribs are holocephalous, and show no sign of pachyostosis, at best a moderate posterior fan-shaped expansion of their shaft (Rieppel, Jiang et al., 2010). The laterally projecting caudal ribs (transverse processes) are prominent in the proximal tail region. The gastral rib basket is of a delicate structure. A peculiarity of *Tanystropheus* is the occurrence, in some specimens, of heterotopic ossifications in the base of the tail (Wild, 1973; similar elements are also known in *Tanytrachelos*: Olsen, 1979). Present only in some specimens, and hence indicative of sexual dimorphism, their function remains unknown, however (their interpretation as support structures of paired male copulatory organs was predicated on the now abandoned classification of *Tanystropheus* as an early squamate: Wild, 1973; Kuhn-Schnyder, 1974).

In the shoulder girdle, the scapula and coracoid are plate-like, kidney-shaped elements that meet in the formation of the glenoid facet. The clavicles are slender, curved elements, and the cruciform interclavicle is robustly built, carrying a prominent posterior stem. In the pelvic girdle, a distinct thyroid fenestra separates the ventral parts of pubis and ischium. The ilium carries a distinct, posteriorly projecting process. The limbs are rather long and slender, the fore limbs distinctly shorter than the hind limbs. The femur is characteristically sigmoidally curved. Carpus and tarsus show a greater degree of skeletal paedomorphosis than is evident in *Macrocnemus*, indicating a more prominent aquatic lifestyle. The best-preserved carpus comprises four ossifications: the radiale, ulnare, and two distal carpals (Wild, 1973:109). Four ossifications likewise characterize the mature tarsus, comprising the astragalus, calcaneum, and the distal tarsals three and four (Wild, 1973, fig. 119; Nosotti, 2007:71, and fig. 61). The perforating foramen is located on the astragalo-calcaneal suture. The short and sturdy fifth metatarsal shows traces of the ancestrally hooked structure, partially obscured by skeletal paedomorphosis. The phalangeal formula in *Tanystropheus longobardicus* is 2-3-4-4-3 in the manus, 2-3-4-5-4 in the pes (Nosotti, 2007, table 9).

The ecology of *Tanystropheus longobardicus* has been a topic of intense debate. Much of that discussion turns on the almost grotesquely elongated neck of *Tanystropheus*, which is something of a conundrum of its own. Although the neck equals trunk and tail in length (up to 60% of the total length of the animal), it is composed of only 13 much elongated cervical vertebrae. Add to such a low number of vertebrae the elongated cervical ribs, which form bundles of bony rods bridging multiple intervertebral joints along the ventrolateral aspect of the cervical vertebral column. Such a morphology does not support the inference of great mobility or flexibility of the elongated neck in *Tanystropheus*. Another issue in the reconstruction of neck mobility in *Tanystropheus* is the tendency to infer the function of intervertebral joints on the basis of bony structures alone, without taking the possible structure of cartilaginous components into account.

Peyer pictured *Tanystropheus* as a terrestrial animal, patrolling the shoreline snapping up fish or squid that happened to be in reach of its long neck (1931c). Stomach contents comprising fish scales and cephalopod hooklets prove that at least large specimens of *Tanystropheus* foraged in the marine environment. Wild placed the adult *Tanystropheus* as a

fully aquatic (marine) predator, which ventured on land only to repro-
duce, that is, to deposit eggs (1973). The small specimens with tricuspid
teeth in the posterior maxillary and dentary tooth row he considered to
be juveniles, adapted to an insectivorous life on land. A heterodont dentition need not indicate terrestrial and insectivorous habits, however, as
was argued by Cox (1975). The putative juveniles could also have been
grazing on algae on the seafloor along the margins of the Grenzbitumen basin, in analogy to the marine iguana *Amblyrhynchus*. Kummer
placed *Tanystropheus* on land again, but stipulated a swan-like S-curve
as the resting position for the long neck so as to prevent the animal from
toppling over due to the imbalanced weight of head and neck (1975).
Tschanz based his functional interpretation of the neck of *Tanystropheus*
on a comparative study of the epaxial neck musculature in extant iguanid
and varanid lizards, and concluded that *Tanystropheus* was unable to raise
its head significantly above ground with the neck pulled up to a position
higher than the shoulder region, as in the postures suggested by Wild or
Kummer (Tschanz, 1986; Wild, 1973; Kummer, 1975). This conclusion
obtained from the absence of distinct and robust neural spines in the
pectoral and anterior dorsal region of the vertebral column, which would
have been necessary to provide a site of origin for strong neck muscles.
Tschanz consequently reconstructed *Tanystropheus* as an obligatory
aquatic (marine) organism, which given its relatively stiff yet elastic neck
would have been an axial-subundulatory swimmer, providing propulsion
mainly through lateral undulation of the trunk and tail (1986, 1988).

Renesto in contrast, emphasized the elongate, horizontally oriented
caudal ribs (transverse processes) that articulate on the proximal caudal
vertebrae (2005). In his judgment, these would have prevented efficient
lateral undulation, which would have been more effective in a laterally
compressed tail. The specimen of *Tanystropheus* he described from the
Cava Inferiore beds shows preservation of skin and other organic material
in the pelvic and proximal tail region, in a pattern similar to that seen
in *Macrocnemus* (Renesto and Avanzini, 2002). His conclusion was that
a significant muscle mass was associated with the caudal ribs in the
proximal tail region, thus providing a balancing counterweight to the
head and neck. He consequently reconstructed *Tanystropheus* as a terrestrial organism. Renesto found limb morphology to further support a
terrestrial lifestyle of *Tanystropheus*, as had originally been sketched by
Peyer, placing the animal in a terrestrial, near-shore habitat together with
Macrocnemus (Peyer, 1931c; Renesto, 2005; for taphonomic clues pointing in the same direction see Beard and Furrer, 2017). Rejecting these
arguments as unconvincing, Nosotti cited the higher degree of skeletal
paedomorphosis observed in the limbs of *Tanystropheus* as compared to
Macrocnemus, interpreting *Tanystropheus* "as an aquatic protorosaur with
close terrestrial ancestors, living in shallow waters . . . and in all probability returning to land for reproduction" (2007:77). Most recently, bone
histology was investigated in an attempt to assess the lifestyle of *Tanystropheus*. A purely aquatic lifestyle found no support, but neither would the

elongated neck be suitable for fishing from the shoreline. Instead, bone density analysis suggested amphibious habits for *Tanystropheus* (Jaquier and Scheyer, 2017).

Some degree of relative elongation of the cervical vertebrae is a diagnostic feature of the Protorosauria, among which *Tanystropheus longobardicus*—along with *Dinocephalosaurus orientalis* from the Anisian (Middle Triassic) of southwestern China—is among the largest representatives, growing to up to six meters in total length (Wild, 1973; C. Li, Rieppel, and LaBarbera, 2004; Rieppel, C. Li et al., 2008). Although the individual cervical vertebrae are again elongated in *Dinocephalosaurus*, they are so to a lesser degree than in *Tanystropheus*. Neck elongation in *Dinocephalosaurus* is effected not only by elongation of the cervical vertebrae, but also by increasing their number up to 33 (C. Li, Rieppel and Fraser, unpubl. obs.). This contrasts with only 13 cervical vertebrae in *Tanystropheus longobardicus*, which are the relatively longest of any observed among all protorosaurs. Their extreme elongation may have resulted from simple positive allometric growth in a protorosaur that grows to large overall adult body size, rather than having developed as a specific adaptation to either an aquatic (marine), or a terrestrial (near shore) habitat (Tschanz, 1988; Taylor, 1989; Nosotti, 2007). The extreme body proportions of *Tanystropheus* may thus have resulted from developmental constraints in an animal selected for large overall body size for reasons other than neck elongation, for example, sexual competition or defense (Taylor, 1989). This would explain why the explanation of neck elongation per se as a specific adaptation remains so elusive. Wild has argued that an increase in overall body size, along with increasing specialization in the cervical region of the vertebral column, and coupled with a high juvenile mortality rate, eventually led to the demise of the evolutionary lineage represented by *Tanystropheus* in the Middle to Late Triassic (1987). In Taylor's estimation, *"T. longobardicus* had a long neck simply because it was a big animal, and it survived in spite of it rather than because of it" (1989:689).

Note   1. The date of publication of the relevant fascicle as part of H. v. Meyer (1847–1855) is 1852: Quenstedt, 1963:70.

10.1. The holotype of *Ticinosuchus ferox* (Paleontological Institute and Museum, University of Zurich T4779) from the Val Porina. The specimen measures 250 cm in length (Photo © Heinz Lanz/Paleontological Institute and Museum, University of Zurich).

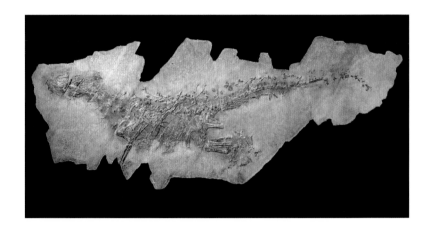

A Dinosaur Lookalike from Monte San Giorgio

10

It is not a dinosaur; there are no dinosaurs that are part of the marine Middle Triassic Monte San Giorgio biota. But if you found an isolated tooth, you might be tempted to think otherwise. The tooth crown is distinctly set off from the root, the latter anchored in the deep alveolus of the tooth-bearing element, the dentary or the maxilla, in the living animal. The tooth crown is tall, slender, pointed, somewhat recurved and slightly laterally compressed. The serrated anterior and posterior cutting edges are distinct, revealing the carnivorous habits of an apex predator of its time. It was aptly named *Ticinosuchus ferox* by Bernhard Krebs, the "ferocious archosaurian from the Canton Ticino" (1965). It is known from three specimens only, two collected in the Grenzbitumenzone of Monte San Giorgio, and one coming from the Besano locality (Krebs, 1965; Pinna and Arduini, 1978; but see comments in Nesbitt, 2011:26, who does not accept this taxonomic assignment of the Besano find). *Ticinosuchus* is traditionally pictured as a terrestrial carnivorous predator, the carcass of which was washed into the marine deposits of the Monte San Giorgio basin from nearby shores. Setting aside the ongoing debate about the ecology of *Tanystropheus*, *Macrocnemus* is the only other reptile of the Grenzbitumenzone believed to have been predominantly terrestrial (Rieppel, 1989c; Renesto and Avanzini, 2002). Not a frequent fossil find, *Macrocnemus* is known from at least nine reasonably complete specimens, as opposed to three of *Ticinosuchus*, two of which are very incomplete. This makes sense given the interpretation of *Macrocnemus* as a predominantly insectivorous animal, as opposed to *Ticinosuchus*, which was the top carnivore of its terrestrial biota and must have been a much less common faunal element. However, although isolated fish scales are strewn all across the slab on which the holotype is preserved, Nesbitt observed a notable concentration of unidentified fish scales at the base of the tail of *Ticinosuchus*, possibly indicating piscivorous habits in a predator living close to the shoreline (2011: 26). The skull and mandible of *Ticinosuchus* show no special adaptation to piscivory, however, as is also not the case in two potentially semi-aquatic (marine) archosaurs from the Middle Triassic of China (Nesbitt, Brusatte et al., 2013:261; C. Li, Wu, Cheng et al., 2006; C. Li, Wu, Zhao et al., 2016; see chap. 11 for further discussion of the latter taxa).

Ticinosuchus ferox—A Ferocious Predator

Ticinosuchus is unquestionably a terrestrial carnivore, whose carcass must have been washed into the Grenzbitumenzone basin. It sports the typical archosaurian dentition composed of large teeth with a slightly recurved crown that is somewhat laterally compressed and carries an anterior and posterior serrated cutting edge. The girdle and limb elements indicate an erect gait for *Ticinosuchus*, as is also characteristic of the distantly related dinosaurs, but different from the sprawling gait that characterizes such forms as lizards. Characteristic of the taxon is a double row of osteoderms running along the dorsal midline of the neck and trunk, while a single row of osteoderms runs along the dorsal and ventral midline of the tail. *Ticinosuchus* is here shown roaming the margins of the Grenzbitumenzone basin, amid a vegetation that is known from plant remains that had been washed into the basin. The dominant land plant remains found in the Middle Triassic of Monte San Giorgio comprise the conifer *Voltzia*, and the horsetail *Equisetites*. Flowering plants (angiosperms) had not yet evolved in the Triassic. The reconstruction of the pattern of locomotion revealed *Ticinosuchus* and its close allies as possible makers of the *Chirotherium* tracks of Triassic age, first described under that name in 1835.

© Beat Scheffold

Following the initial assessment of Krebs, as well as the judgment of later workers, *Ticinosuchus* belongs to the Rauisuchidae, a family included in the rauisuchioids that are early and basal representatives of crown-group archosaurs, more closely related to crocodiles and their fossil relatives than to dinosaurs and their descendants, the birds (Huene, 1948; Krebs, 1965; Lautenschlager and Desojo, 2011). The first rauisuchid to have been described was *Teratosaurus suevicus* from the Triassic Stubensandstein (Löwenstein Formation, Norian), discovered in 1860 in the surroundings of Stuttgart and described by H. v. Meyer in 1861. It was not until 1948, however, that the family Rauisuchidae was formally recognized by Huene based on carnivorous archosaurs (i.e., the pseudosuchians *Prestosuchus* and *Rauisuchus*) he had described from the Triassic of southern Brazil, and which at that time he classified in the subfamily Rauisuchinae within the family Stagonolepidae (Huene, 1942:243; 1948:87). A wide geographic distribution of rauisuchians thus had become apparent early on in the course of their investigation. Today, rauisuchians are characterized as "an assemblage of mostly large terrestrial Triassic top predators, whose skulls are vaguely reminiscent of theropod dinosaurs" (Scheyer and Desojo, 2011:1289). The term "assemblage" reflects the fact that the monophyly or non-monophyly of rauisuchians continues to be debated, as also are many issues concerning the species-level taxonomy within the group (Nesbitt et al., 2013). The reason is that many taxa are based on very incomplete, fragmentary, or disarticulated material. A low-level general consensus regarding rauisuchians has nevertheless emerged in recent research. Rauisuchians are generally loosely diagnosed as "extinct, Triassic crurotarsan archosaurs with a well-defined rotary joint between astragalus and calcaneum very similar to that seen in crocodilians, and paired, paramedian dorsal osteoderms along the presacral vertebral column" (Gower, 2000:456). It was Krebs's original description of *Ticinosuchus* that highlighted the complex crurotarsal joint between the astragalus and calcaneum in the proximal tarsus as a crocodilian feature that *Ticinosuchus* shares with *Rauisuchus*, thus revealing rauisuchians to be more closely related to the crocodyliform lineage (the pseudosuchian clade) than to the pterosaur-dinosaur-bird lineage (the ornithosuchian clade) of crown-group archosaurs (1965). Rauisuchians are restricted to the Middle to Upper Triassic, having achieved a cosmopolitan distribution early during their evolutionary diversification: they are known from North (USA) and South America (Argentina, Brazil), from Africa (Morocco, Tanzania), from Europe (including Russia) and from Asia (including India and China), but are apparently absent from Australia and Antarctica (Gower, 2000:457; see also Nesbitt et al., 2013). They represented the top carnivores of their time, some growing to a total length of nearly six meters. During the late Carnian a major global faunal turnover took place due to a collapse of the terrestrial ecosystem that led to the extinction of then dominant herbivores such as dicynodonts and rhynchosaurs (Benton, 1986, 1994). Following their demise, they were replaced by early herbivorous dinosaurs that first appeared during the Norian. Throughout this major biotic crisis, rauisuchians continued to fill the role of top carnivores worldwide, before going extinct toward the end of the Triassic. They were replaced by theropod dinosaurs that first appeared in the Late Triassic (Gower, 2000:457).

Two morphotypes have been identified among quadrupedal rauisuchians: relatively small and slender forms with a long neck, such as *Ticinosuchus*; and relatively large and more stoutly built forms with a short neck, such as *Stagonosuchus* (Lautenschlager and Desojo, 2011:379). *Ticinosuchus ferox* is known from three specimens, the first a complete yet strongly compressed skeleton collected in 1933 in the Grenzbitumenzone at the Val Porina locality of Monte San Giorgio (fig. 10.1). The second specimen was collected in 1943 in debris that had resulted from industrial exploitation of the Cava Tre Fontane deposits; it consists of the neural spines of five caudal vertebrae capped by a single row of overlapping arrow-head-shaped osteoderms that are characterized by a prominent dorso-medial longitudinal keel. This second find immediately signaled the presence of an archosaur in the Middle Triassic of Monte San Giorgio biota, and thus motivated prioritization of the difficult and protracted preparation of the first, principal find from Val Porina of 1933. The third specimen of *Ticinosuchus ferox*, collected at the Besano locality in 1975 by a party from the Museo Civico di Storia Naturale di Milano, comprises an incomplete right mandible, left and right radius and ulna, and four osteoderms (Pinna and Arduini, 1978). *Ticinosuchus* had also been reported from the Upper Triassic (Carnian) of the Argana Basin ("*Couloire d'Argana*") in the western High Atlas of Morocco, but that material is now recognized as representing a different taxon, *Arganasuchus dutuiti* (Dutuit, 1979; Jalil and Peyer, 2007).

Ticinosuchus ferox* reached a total length of some 2.5 meters. The skull is that of a typical archosaur, with a prominent antorbital opening located in the elongated preorbital region in front of the orbit. The pointed and slightly recurved teeth are relatively widely spaced as is required for their cutting function, the latter revealed by an anterior and posterior serrated cutting edge. Tooth implantation is thecodont. Krebs (1965) established the presence of approximately 24 presacral, 2 sacral, and approximately 55 caudal vertebrae. Krebs reconstructed the neck with short dichocephalous cervical ribs pointing backward, the distal ends of their shafts overlapping (1965). The trunk is characterized by a distinct lumbar region, with the posteriormost dorsal ribs shortened, and the last few absent altogether. The neural spines of the vertebrae in the cervical and dorsal region are capped by two parallel rows of osteoderms; a single row of osteoderms runs along the dorsal and ventral midline of the tail. The gastral rib basket is well developed. The pelvis is distinctly triradiate, with an ilium characterized by a horizontal ridge on the postacetabular process, a feature *Ticinosuchus* shares with other rauisuchians (Lautenschlager and Desojo, 2011). The limbs are gracile and elongate, the hind limbs somewhat longer and more strongly built than the fore limbs. The erect gait and an initial degree of digitigrade locomotion betrays a terrestrial, quadrupedal predator. The phalangeal formula is 2-3-4-4(5?)-2(3?) in the manus, 2-3-4-5-3 in the pes. Geometrically, the third digit is the longest in the hand and foot.

Aside from *Ticinosuchus* from the Grenzbitumenzone/Besano Formation of Monte San Giorgio/Besano, rauisuchids are only very scarcely documented from the southern Alpine Triassic, as by an isolated tooth from the Carnian of the Bergamasque Prealps, or trackways from the late Carnian of the Dogna Valley, Friulia, Italy (Dalla Vecchia, 1996; Renesto, Confortini et al., 2003).

Chirotherium— Mysterious Tracks in the Buntsandstein

Strange pentadactyl tetrapod footprints and trackways were discovered in 1833 by Friedrich Sickler (1773–1836), grammar school teacher in Hildburghausen (Thuringia, Germany), on a slab of sandstone he had ordered from a quarry located in Heβberg near his hometown, to be used for construction in his garden. Recognizing the potential significance of the find, Sickler offered the quarry workers remuneration for alerting him to further discoveries of similar kind. When more tracks were exposed in the quarry in situ in 1834, Sickler reported their occurrence in the same year in an open letter to the famous anatomist and zoologist from the University of Göttingen, Johann Friedrich Blumenbach (1752–1840): "Missive to the honorable Dr. J. F. Blumenbach on the most extraordinary tracks of primeval, large and unknown animals, discovered only a few months ago in the sandstone quarries of Heβberg near the town of Hildburghausen"[1] (summarized in Bronn, 1835). Additional tetrapod tracks were reported, again in 1834, from standstone quarries in Weikersrode near Hildburghausen, first discovered by Carl Barth (1787–1853), a copper engraver from Hildburghausen. The writer recognized larger and smaller footprints, which "on first cursory inspection seemed to share some similarity with a human hand, in which the thumb is somewhat set off" (Bernhardi, 1834:641). The sandstone on which these tracks occurred was further identified as belonging to the Buntsandstein of Lower to early Middle Triassic age. The leading paleontologist Heinrich Georg Bronn (1800–1862) concurred: "We will have to most likely ascribe these tracks to some ape" (Bronn, 1835:233). In 1835, the paleontologist and zoologist Johann Jakob Kaup (1803–1873) from Darmstadt commented on these "*Quadrumanen*-tracks," and, assuming their mammalian, indeed marsupial origin, proceeded to formally name them *Chirotherium Barthii*, after Carl Barth, the engraver. He nonetheless reserved the right to rename the trace fossil *Chirosaurus* should it prove to be of "amphibious origin" (Kaup, 1835:328). This was the first ever formally named trace fossil, consisting of terrestrial vertebrate footprints with the fifth digit—but at the time identified as the "thumb"—strangely set off from the remaining four. The only problem was that the putative thumb in the tracks appeared on the wrong side of hand and foot. This in turn meant that whatever animal was the originator of the footprints, it would have had to cross its legs when walking. Based on ample British finds of Triassic *Chirotherium* tracks, on sedimentary rock that also yielded an isolated tooth with strangely infolded dentine and enamel, the eminent paleontologist and comparative anatomist Sir Richard Owen (1804–1892) identified labyrinthodont amphibians as originators of the *Chirotherium* tracks (Owen, 1841, 1842). The leading British geologist of the time, Charles Lyell (1797–1875) concurred, and in 1855 published the sketch he had obtained from Owen of a bizarre frog-like creature with a labyrinthodont skull, arguing that "only an amphibian would have been able to walk in this extraordinary way" (Bowden et al., 2010:217), that is, with crossed legs.

Following their discovery near Hildburghausen in 1833, many more *Chirotherium* tracks were found in the Germanic Buntsandstein Formation, especially in the upper part of its middle section (of Anisian, Middle Triassic, age). But similar tracks were also recorded from abroad, including Britain (here reported even before the Hildburghausen discovery), France, Italy, and Spain, and, in the twentieth century, from Arizona and Argentina (Bowden et al., 2010:215). A rich para- (ichno-) taxonomy of

Chirotherium footprints started to spring up, but some voices urged caution. The feet of terrestrial animals crossing the slick surface of wet clay were prone to slip, causing variation in the apparent size and proportions of the footprints, and also affecting the gaps between the toes, characters that had taphonomic, but no taxonomic implications (Schmidt, 1928:422). A classic treatment of *Chirotherium* tracks was published by the paleontologist Wolfgang Soergel (1887–1946) in 1925. In his pursuit to unmask the nature of the *Chirotherium* tracks, Soergel adopted a commonsense approach, identifying the splayed-out digit as the fifth, rather than the first (thumb), thus removing the need for the assumption that the originator of the footprints walked with crossed legs. He proceeded to compare the fossil trackways with the morphology of the autopodium of the terrestrial Triassic reptiles known at his time, and found a group of thecodonts (i.e., archosaurs) then referred to as Pseudosuchia to come closest. A classic representative of psedosuchians was *Euparkeria*, which was too small, however, to match the size particularly of the larger footprints. Soergel consequently published a hypothetical reconstruction of *Chirotherium barthi* (1925, fig. 54). He depicted the reptile as an ungainly quadrupedal animal vaguely resembling a crocodile, with a long tail, a weakly digitigrade stance, and with hind limbs that are distinctly longer than the much shorter, much more lightly built fore limbs. The fifth digit in manus and pes he showed to be set off posteriorly, and diverging from the other four digits, as is indicated in the fossil footprints.

When describing the pes of the rauisuchid *Prestosuchus chiniquensis* from the Middle Triassic (Ladinian) of southern Brazil, Huene noted the aberrant morphology of metatarsal V, which necessitated a splayed position of the fifth toe relative to the four remaining ones: "this is then a foot close to the one *Chirotherium* must have had, with the fifth toe distinctly splayed out" (Huene, 1942: 179f). Recognizing *Ticinosuchus ferox* as a rauisuchid, it made lot of sense to reconstruct its locomotion to provide a basis from which to infer the type of track it would have left behind when walking across a slick, wet bed of clay. In spite of limb proportions not as markedly different as stipulated by Soergel, the close correspondence of the hypothetical footprints *Ticinosuchus* would have left behind to *Chirotherium* tracks became immediately obvious (Krebs, 1965:127, and fig. 66). Noting that the footprints left behind by *Ticinosuchus ferox* would fall into the size range of *Chirotherium barthi*, Krebs nevertheless refrained from getting bogged down any further in the confusing para-(ichno-) taxonomy of *Chirotherium* footprints. Haubold found *Ticinosuchus* to represent a poor model of *Chirotherium*, for both stratigraphic and morphological reasons (2006:26). Instead, he modeled *Chirotherium* on forms such as the archosauriform *Euparkeria* and the paracrocodyliform *Saurosuchus*. This perspective is reflected in the *Chirotherium* monument that today showcases the city center of Hildburghausen (fig. 10.2).

10.2. The *Chirotherium* monument in the market place in Hildburghausen, the town near which the *Chirotherium* tracks were discovered in 1834. The monument was opened in 2004. Casts of the tracks are mounted on the wall in the background. The animal is modeled on the archosauriform *Euparkeria* and the paracrocodylomorph *Saurosuchus* (Photo © H. Haubold, Halle/Saale).

Note

1. http://reader.digitale-sammlungen.de/de/fs1/object/display/ bsb10231937_00005.html (accessed October 4, 2018).

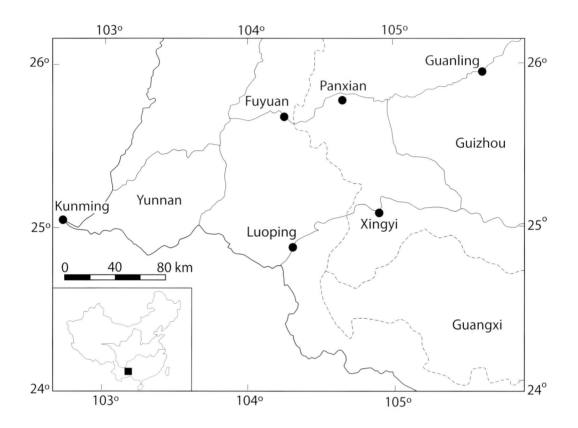

11.1. Map of Triassic marine fossil vertebrate localities in southwestern China (artwork Marlene Donnelly, the Field Museum, Chicago).

The Tethys Sea: Connections from East to West

11

The fossiliferous layers at Monte San Giorgio in the western Tethyan faunal province extend from the Anisian part of the lower and middle Grenzbitumenzone through the Ladinian part of the upper Grenzbitumenzone and on up through to the very top of the Meride Limestone, the latter the equivalent of the *Kalkschieferzone* of late Ladinian age (Furrer, 2003). Marine vertebrates of stratigraphically younger deposits up into the middle to upper Norian are known from other localities in the western Tethys, in particular sites in northern Italy (e.g., the Norian faunas from the Zorzino Limestone, Lombardia, and Preone-Tolmezzo, Friuli: Tintori and Felber, 2015, fig. 17). The latest record of Triassic marine reptiles in the western Tethyan faunal province is of the armored placodont *Psephoderma alpinum* from the Rhaetian of Lombardy, northern Italy (Pinna, 1990). By comparison, the record of Triassic marine fishes and reptiles comes in earlier in the eastern Tethyan faunal province, and also ends earlier, with several localities in southern and southwestern China (J.-L. Li, 2006) (fig. 11.1). This means that many of the major reptile clades that occur on both sides of the Tethys Sea tend to make an earlier appearance in the eastern than in the western Tethyan faunal province, and the same is true for fishes (Tintori and Felber, 2015). Whether or not this also reflects a phylogenetic pattern is still open to debate on a case-by-case basis, as no well-corroborated, and hence relatively stable global phylogenetic analysis of all the taxa involved is yet available. There is, however, one group of sauropterygians that sends a strong biogeographical signal connecting east and west: the pachypleurosaurs (see also chap. 6). Phylogenetic analysis showed *Odoiporosaurus* from Besano to be the sister taxon to the *Serpianosaurus–Neusticosaurus* clade from the Monte San Giorgio, with the *Anarosaurus–Dactylosaurus* clade from the Germanic Muschelkalk being the sister group of the pachypleurosaurs from the southern Alpine Triassic: ((*Anarosaurus, Dactylosaurus*) (*Odoiporosaurus* (*Serpianosaurus, Neusticosaurus*))) (Renesto, Binelli et al., 2014). Sister to this clade is a clade of Chinese pachypleurosaurs (J. Liu, Rieppel, et al., 2011). This taxic pattern suggests that pachypleurosaurs immigrated into the central European basin (Germanic basin) through the East Carpathian gate from the eastern Paleotethys, following the marine ingressions in latest Olenekian/earliest Anisian times (Renesto, Binelli et al., 2014:164; see also Rieppel and Hagdorn, 1997; Rieppel, 1999a). From there they dispersed in a southern direction into the intraplatform basins of the western Tethys in the area of today's Friuli (northeastern Italy) and Slovenia. *Odoiporosaurus* is the earliest evidence of pachypleurosaurs invading

Triassic Marine Reptiles from Southwestern China

the intraplatform basins of the southern Alps. During late Anisian and Ladinian times, the intraplatform basins of the southern (Lombardian) Alps are thought to have constituted a "speciation center" for the *Serpianosaurus–Neusticosaurus* clade (Renesto, Binelli et al., 2014:164; and references therein).

As discussed below, the addition of some newly discovered sauropteryians to the phylogenetic analysis threatens to disrupt the monophyly of pachypleurosaurs as commonly understood, which would refute the paleobiogeographic scenario outlined above. But even so, some forms that do occur at the Monte San Giorgio and Besano localities are remarkably close to taxa known from southern and southwestern Chinese localities and hence again document very close paleobiogeographic relationships between the western and eastern Tethyan faunal provinces in a more general sense (Tintori and Felber, 2015: 69).

The Spathian Chaohu Fauna

The earliest record of Triassic marine reptiles from southern China comprises the late Olenekian (Spathian, late Lower Triassic) faunas from near Chaohu City in Anhui Province, from Yuan'an and Nanzhang Counties in Hubei Province, and from Wuming District in the Guangxi Autonomous Region (J.-L Li, 2006). Of these, the last one, the Wuming locality in Guangxi, is the least well known. So far, it has yielded only the putative pistosaurian sauropterygian *Kwangsisaurus orientalis*, a very incomplete specimen of uncertain relationships (C.-C. Young, 1959; Rieppel, 1999b; J.-L. Li, 2006). The Yuan'an and Nanzhang localities in Hubei Province likewise remain comparatively poorly known, although the fauna is famous for the remarkable diversity of highly derived, strictly endemic hupehsuchian taxa, such as *Nanchangosaurus*, *Hupehsuchus*, *Parahupehsuchus*, *Eohupehsuchus*, and *Eretmorhipis* (K.-M. Wang, 1959; C.-C. Young, 1972; Carroll and Dong, 1991; X.-H. Cheng et al., 2014; Chen, Motani et al., 2014a, 2014b, 2015). These remarkable creatures comprise forms characterized by edentulous jaws, limbs transformed into paddles or flippers, pachyostotic ribs that in some species contact each other along their length so as to encase the trunk in a carapace-like structure, and osteoderms capping the bifurcated neural spines along the dorsal midline of the body. *Hupehsuchus* may have been the only lunge feeder known among Triassic marine reptiles (Motani, Chen, et al., 2015; see also Collin and Janis, 1997). From the same localities, incomplete (non-diagnostic) remains of a pachypleurosaur (cf. *Keichousaurus yunnanensis*) have been reported, as well as the enigmatic sauropterygian *Hanosaurus hupehensis* (C.-C. Young, 1965, 1972; Rieppel, 1998c).

Most important among these late Lower Triassic (Spathian) marine reptile localities is the Majiashan quarry in the Juchao District near the city of Chaohu, Hubei Province. This is located near the type locality of the early small ichthyopterygian *Chaohusaurus geishanensis*, of which it yielded additional specimens (C.-C. Young and Dong, 1972; see also McGowan and Motani, 2003). A second species in the same genus and from the same locality is *Chaohusaurus chaoxianensis* (Motani, Jiang, Tintori et al., 2015a). *Chaohusaurus zhangjiawanensis*, a third species in its genus, was collected near the Zhangjiawan village of Yangping, northern Yuanan County, in western Hubei Province (Chen, Sander et al., 2013). The Majiashan quarry near Chaohu has yielded other, even more basal ichthyopterygians briefly mentioned in chapter 4. Most remarkable is the small, short-snouted ichthyosauriform *Cartorhynchus lenticarpus* with unusually large flippers yet a poorly ossified wrist, characters that indicate possible amphibious habits (Motani, Jiang et al., 2015b). In that sense, the Majiashan quarry proved to be of particular importance in the elucidation of the rapid earliest diversification and subsequent selective demise of ichthyopterygians in the Early Triassic (Jiang, Motani, Huang et al., 2016; see also Stone, 2010). Possible fragments of utatsusaur, mixosaurid, and shastasaurid ichthyosaurs from the Majiashan quarry still require more scrutiny and/or more complete specimens before their identity can be confirmed. Other than ichthyopterygians, the Majiashan quarry has also yielded the earliest sauropterygian as yet known from the eastern Tethyan faunal province, *Majiashanosaurus discocoracoidis* (Jiang, Motani, Tintori et al., 2014).

Of Anisian age are again two historic finds that come from otherwise poorly known localities. From a locality near Qingzhen in Guizhou Province comes the putative pistosaur sauropterygian *Chinchenia suni*, incomplete and disarticulated material of uncertain affinities (C.-C. Young, 1965; Rieppel, 1999b). Equally incomplete are postcranial remains from Sanchiao near Guiyang, Guizhou Province, which have been described under the name of *Sanchiaosaurus dengi* (C.-C. Young, 1965; see also Rieppel, 1999b; J.-L. Li 2006). Significantly more numerous and much better preserved marine reptiles of Anisian (Pelsonian) age have more recently been collected in Panxian County, Guizhou Province, and Luoping County, Yunnan Province (Zhang, Zhou et al., 2008; Hu et al., 2011). It is with the discovery of this fauna that not only the unexpectedly rapid diversification of the marine reptiles in the eastern Tethyan faunal province became apparent but also their close relationship with the fauna of the western Tethyan faunal province. Many of the taxa recorded from the Panxian–Luoping fauna had previously not been known east of the Middle Triassic (Anisian) fauna from Makhtesh Ramon, Negev, Israel (Tintori and Felber, 2015:69; Rieppel, 1998d, 2001c; Rieppel, Mazin et al., 1999). On the other hand, the Panxian–Luoping fauna also comprises highly endemic taxa, among them some of the most bizarre Triassic marine reptiles known.

Although coming from two main localities roughly 100 km apart, the two fossil assemblages can be considered to represent the same fauna. Fossil collections have come from near the village of Yangjuan in the Xinmin District of Panxian County; extending eastward to Xuepu in the Xuepu District of Puan County, Guizhou Province; and on to Dawazi

The Anisian Panxian–Luoping Fauna

in the Luoxiong District of Luoping County, Yunnan Province. The ichthyosaur *Mixosaurus panxianensis* is considered the "index taxon" for this fauna (Jiang, Schmitz, Hao et al., 2006; see also Hao et al., 2006; G.-B. Liu and Yin, 2008; Motani, Jiang, Tintori et al., 2008; Jiang, Motani, Hao et al., 2009; Chen and Cheng, 2009, 2010). Of middle Pelsonian (Anisian) age, the diverse Panxian–Luoping fauna ranks amongst the oldest Middle Triassic marine reptile faunas worldwide (Sun, Sun et al., 2006; Sun, Hao et al., 2009; Zhang, Zhou et al., 2009; Jiang, Motani, Hao et al., 2009:452).

Other than *Mixosaurus panxianensis* the Panxian–Luoping fauna comprises the mixosauroid *Phalarodon* cf. *P. fraasi* (Merriam, 1910; see also Jiang, Hao et al., 2003; Jiang, Schmitz, Motani et al., 2007; *Phalarodon atavus*: Liu, Motani et al., 2013), and the cymbospondylid ichthyosaur *Xinminosaurus catactes* (Jiang, Motani, Schmitz et al., 2008). Sauropterygians represented in the Panxian –Luoping fauna encompass the whole taxonomic diversity, with the placodont *Placodus inexpectatus* (Jiang, Motani, Hao et al., 2008); the pachypleurosaur *Dianopachysaurus dingi* (J. Liu, Rieppel et al., 2011), *Dianmeisaurus gracilis* (Shang and Li, 2015; Shang, Li and Wu, 2017), and *Wumengosaurus delicatomandibularis*, the latter two of disputed pachypleurosaur affinities (Jiang, Rieppel, Motani et al., 2008; X.-C. Wu, Cheng, Li et al., 2011; Neenan, Li et al., 2015, confirmed a pachypleurosaur relationship of *Wumengosaurus*), as well as possible remains of *Keichousaurus* (Jiang, Motani, Hao et al., 2009); the nothosaur *Nothosaurus yangjuanensis* (Jiang, Maisch, Hao et al., 2005, 2006; Shang, 2006); and the lariosaur *Lariosaurus hongguoensis* (Jiang, Maisch, Sun et al., 2006).

Of particular interest is the occurrence of a "gigantic" nothosaur in the Panxian–Luoping fauna, *Nothosaurus zhangi* (J. Liu, Hu et al., 2014). The species compares in size to *Nothosaurus giganteus* from the Germanic Muschelkalk and must have occupied the niche of an apex predator in its biota (Rieppel, 2000a; Rieppel and Wild 1996). The simultaneous appearance of macropredators in the early Middle Triassic of the eastern and western Tethyan faunal province, as well as in the in the eastern Pacific faunal province (the ichthyosaur *Thalattoarchon* from the Anisian of northwestern Nevada), associated with a complex and diverse contemporary fauna, puts a temporal marker on the recovery of the global marine biota after the end-Permian mass extinction (Fröbisch et al., 2013; J. Liu, Hu et al., 2014).

Other taxa so far known exclusively from the Panxian–Luoping fauna defy easy classification within the framework of the major sauropterygian clades that have previously been recognized (Rieppel, 2000a). *Diandongosaurus acutidentatus* was described as a small-sized eosauropterygian from Luoping County that combines nothosaur and pachypleurosaur characters (Shang et al., 2011; X.-Q. Liu et al., 2015). The snout is not constricted, and the upper temporal fossa is smaller than the orbit, features that also characterize pachypleurosaurs, whereas the dentition, comprising enlarged procumbent teeth in the premaxilla and anterior part of the dentary, and paired fangs in the preorbital part of the maxilla, is similar to that of nothosaurs. A recent re-appraisal of the anatomy of *Diandongosaurus acutidentatus*, followed by a phylogenetic analysis, showed the taxon to group with *Keichousaurus* and *Dianopachysaurus* as a sister clade to nothosauroids (plus *Hanosaurus*), rather than with the European pachypleurosaurs (Sato, Cheng et al., 2014b). In the same analysis, *Wumengosaurus* is likewise pulled away from European pachypleurosaurs. On the other hand, Neenan, Li et al. found *Wumengosaurus* to represent a basal pachypleurosaur, whereas *Diandongosaurus* nested as the basal-most eusauropterygian (2015). The overall result of the inclusion of *Diandongosaurus* in a comprehensive phylogenetic analysis thus indicates the possibility that the pachypleurosaurs as traditionally understood might have to be recognized as a strictly European radiation (the *Anarosaurus–Dactylosaurus* clade in the Germanic basin, and the *Odoiporosaurus-Serpianosaurus–Neusticosaurus* clade in the Alpine Triassic), whereas all those taxa from the eastern Tethyan faunal province that have at one time or another been considered to represent pachypleurosaurs may no longer be considered to be part of that radiation (Y.-N. Cheng, Wu, Sato et al., 2012).

Dawazisaurus brevis from Luoping County is yet another small eosauropterygian from the Panxian–Luoping assemblage with a total length of less than 50 cm (Y.-N. Cheng, Wu, Sato et al., 2016). With only 16 dorsal vertebrae it has a remarkably short trunk, but in its overall habitus it again resembles pachypleurosaurs. The carpus and tarsus are, however, remarkably well ossified, while the phylogenetic analysis placed the taxon in a paraphyletic (unresolved) assemblage of "pachypleurosaur-like" sauropterygians that includes both European and Chinese taxa (Y.-N. Cheng, Wu, Sato et al., 2016, fig. 6a). It is probably fair to say that with the many new forms of marine reptiles discovered in the Middle Triassic of southern and southwestern China, the analysis of sauropterygian interrelationships is currently in a state of transition, as the cladograms that are currently being proposed all remain rather unstable.

Protorosaurs are represented in the Panxian–Luoping biota by the long-necked *Dinocephalosaurus orientalis*, a form endemic to the Panxian–Luoping fauna but strikingly convergent on *Tanystropheus* (C. Li, 2003; C. Li, Rieppel, and LaBarbera, 2004; Rieppel, C. Li, and Fraser, 2008). With a disproportionally elongate neck, and a relatively long tail, *Dinocephalosaurus* closely resembles *Tanystropheus* in its external appearance, and as in the latter taxon the much-elongated cervical ribs bundle up on either side along the ventrolateral aspect of the cervical vertebral column, bridging multiple intervertebral joints. Yet the cervical vertebral column is made up of 33 elements in *Dinocephalosaurus*, as opposed to the 13 cervicals in *Tanystropheus longobardicus*, and unlike in the latter genus, the cervical vertebrae are not hollow in *Dinocephalosaurus*, an indication of convergent evolution.

The saurosphargid clade is represented in the Panxian–Luoping biota by two genera, the more generalized *Largocephalosaurus* and the highly derived *Sinosaurosphargis*. Two species have been described in the genus *Largocephalosaurus*, although their distinctiveness, as also the distinctiveness of the genus *Largocephalosaurus* from the European genus *Saurosphargis*, remain open to discussion (Jiang, Rieppel, Motani et al., 2011). The description of *Largocephalosaurus polycarpon* was originally mainly based on an incomplete skull, given that the postcranium was incompletely prepared at the time of its description; the taxon was initially referred to the Eosauropterygia (L. Cheng, Chen, Zeng et al., 2012). *Largocephalosaurus qianensis* is based on three specimens, comprising well-preserved skulls and postcranial skeletons. The characteristic leaf-shaped teeth immediately reveal the saurosphargid nature of the genus (C. Li, Jiang, Cheng et al., 2014). The postcranial skeleton of *Largocephalosaurus* is very closely comparable to that of *Saurosphargis volzi* from the lower Muschelkalk (early Anisian) of Upper Silesia (Huene, 1936). Such similarity might seem to indicate placodont affinities, but the skull of *Largocephalosaurus* refutes earlier hypotheses that saurosphargids may share sauropterygian, or even more specifically placodont relationships (for a discussion see Rieppel, 2000a).

The turtle-like *Sinosaurosphargis yunguiensis* is a highly derived member of the saurosphargid clade again endemic to the Panxian–Luoping fauna. The skull of *Sinosaurosphargis* is that of a diapsid reptile without a lower temporal arch, and with an upper temporal fossa that is secondarily closed due to an expansion of neighboring elements. The body shape of *Sinosaurosphargis* is discoidal, the dorsal ribs broadened and in mutual contact along their length. The whole back of the animal is covered by a closed osteoderm layer that extends onto the neck, the proximal limb segments, and may have done so across the proximal tail (C. Li, Rieppel, Wu et al., 2011). The body armor in *Sinosaurosphargis* is thus of a distinctly different structure compared to the turtle carapace — again a distinctive sign of convergence. Nevertheless, given the discoidal body shape and the dorsal armor covering the trunk involving the dorsal ribs, the possibility has been raised that similar developmental mechanisms might have been at work in *Sinosaurosphargis* as are known to shape the turtle shell (Hirasawa et al., 2013)

From a locality near Dawazi, Luoping County, comes the enigmatic *Atopodentatus unicus*, an algae filter-feeding herbivore with a relatively small skull, the jaws expanded laterally into a hammerhead-like structure (L. Cheng, Chen, Shang et al., 2014; C. Li, Rieppel, Cheng et al., 2016). This unique fossil represents the earliest occurrence of a herbivore reptile in the marine biota, again testifying to the relatively rapid reconstitution of the marine vertebrate biodiversity after the end-Permian extinction by Anisian times.

Another endemic taxon of the Panxian–Luoping fauna is the archosaur *Qianosuchus mixtus*, a taxon that sports a mixture of terrestrial and aquatic adaptations (C. Li, Wu, Cheng et al., 2006). The limbs are gracile and slender, the hind limbs stronger and longer than the fore limbs, the erect gait suggesting versatile terrestrial locomotion. The deep tail, characterized by tall neural spines on the caudal vertebrae and elongate chevron bones, appears adapted to provide propulsion in an aquatic environment. *Qianosuchus* is among the first archosaurs to show any adaptations to an aquatic—in this case marine—lifestyle, preying on suitably sized individuals of sauropterygians, protorosaurs, and perhaps even mixosaurid ichthyosaurians in near-shore shallow water (C. Li, 2010).

A stratigraphically younger fossil assemblage is found at localities that again extend from east to west, that is, from Dingxiao and the Wusha District near Xingyi City, Guizhou Province, to Fuyuan and Luoping Counties of Yunnan Province. The fossils that constitute this Xingyi fauna originate from the Zhuganpo Member of the Falang Formation, and are of late Ladinian (possibly earliest Carnian) age (late Middle to early Upper Triassic) age. All fossiliferous localities of that assemblage yield the pachypleurosaur *Keichousaurus hui*, the first Triassic marine reptile ever to have been described from China and found in the hundreds in deposits near Dingxiao and Xingyi in Guizhou Province (C.-C. Young, 1958; Xue et al., 2015; pachypleurosaur affinities of *Keichousaurus* have been challenged by Holmes et al., 2008).

The only ichthyosaur so far described from the Xingyi fauna is *Qianichthyosaurus xingyiensis*, a separate species in a genus that is otherwise also known from the Guanling biota of Carnian age (Yang et al., 2013).

Placodonts are represented in the Xingyi fauna by the cyamodontoid *Glyphoderma kangi* from the Fuyuan locality (Yunnan Province), a possible subjective junior synonym of *Psephochelys polyosteoderma*, from the Carnian of Xinpu near Guanling, Guizhou Province (C. Li and Rieppel, 2002; Zhao, Li et al., 2008). More detailed comparisons and probably additional, new material are required, however, in order to confirm or reject such synonymy. In the most recent analysis of placodont interrelationships, *Glyphoderma* was found to be the sister taxon to *Placochelys* from the Upper Triassic (Carnian) of Hungary (Neenan, Li et al., 2015).

The abundant fossil record of *Keichousaurus hui* from the Xingyi fauna allows extensive paleobiological investigations paralleled only by the numerous specimens of *Neusticosaurus* recovered at Monte San Giorgio (Sander, 1989a; Xue et al., 2015). Sexual dimorphism is distinct in both *Keichousaurus* and *Neusticosaurus*, as is also the case in other pachypleurosaurs such as *Serpianosaurus* and *Dactylosaurus* (Rieppel, 1989b; Sander, 1989a ; Rieppel and Lin, 1995; Y.-N. Cheng, Holmes et al., 2009). With gravid females now known for *Keichousaurus*, it is possible to determine the sex of individual specimens (N.-Y. Cheng, Wu, and Ji, 2004). *Nothosaurus youngi* and *Lariosaurus xingyiensis* are both known from the Xingyi fauna (J.-L. Li, Liu et al., 2002; Rieppel, J.-L. Li et al., 2003; J.-L. Li and Rieppel, 2004; Ji, Jiang, Rieppel et al., 2014). It was, indeed, the description of that Chinese material that first signaled a possible breakdown of the genera *Nothosaurus* and *Lariosaurus* as previously conceived (Rieppel, 2000a; see also Klein, Voeten et al., 2016).

The Ladinian Xingyi Fauna

Qianxisaurus chajiangensis is a relatively small sauropterygian of approximately 80 cm total length that bears an overall resemblance to a large pachypleurosaur. A broad-based phylogenetic analysis did not support its inclusion in a monophyletic Pachypleurosauroidea that also comprises the European forms, however (Y.-N. Cheng, Wu, Sato et al., 2012).

Pistosaurs are represented in the Xingyi fauna by two monotypic genera. The first one described was *Yunguisaurus liae*, a typical pistosaur with a distinctly elongate neck (49 to 51 cervical vertebrae) and remarkable hyperphalangy in the fore and hind limbs (N.-Y. Cheng, Sato et al., 2006; Sato, Cheng et al., 2010, 2014a). The second pistosauroid from the Xingyi fauna is *Wangosaurus brevirostris*, a taxon that is a bit more difficult to classify, as it could be loosely described as a nothosaur skull sitting on a pistosaur neck that comprises 33 cervical vertebrae (Ma et al., 2015). The phalangeal formula of the manus remains unknown, whereas that of the pes is 3?-3-4-5-5. A third possible pistosauroid from the Xingyi fauna is *Dingxiaosaurus luyinensis*, described by G.-B. Liu et al. (2002). The taxon is based on two incomplete limbs with plesiosaur-like phalanges, along with some fragmentary vertebrae and ribs, and was declared a *nomen dubium* by N.-Y. Cheng, Sato et al. (2006).

Thalattosaurs are represented in the Xingyi fauna by three species in two genera. The first thalattosaur to be described from the Wusha locality was *Anshunsaurus wushaensis* on the basis of a nearly complete specimen of 2.6 m total length, but it is known from a juvenile specimen as well (Rieppel, Liu and Li, 2006; J. Liu, 2007). It differs from the geologically younger *Anshunsaurus huangguoshuensis* from the Carnian Guanling biota (see below), among other features, by a relatively smaller skull with an incomplete lower temporal arch. A second species of the same genus, also a representative of the Xingyi fauna, was described under the name of *Anshunsaurus huangnihensis* (L. Cheng, Chen, and Wang, 2007). The proposed synonymy of *A. huangnihensis* with *A. wushaensis* was rejected largely on the basis of characters of cranial morphology, a claim that warrants further critical evaluation, however (Zhao, Wang et al., 2008; L. Cheng, Chen, Zhang et al., 2011). The thalattosaur genus *Xinpusaurus* is again known from different species that occur in the Ladinian Xingyi fauna and in the Carnian Guanling biota, respectively. *Xinpusaurus xingyiensis* is a large representative of its genus collected at a locality near Wusha, Xingyi, with a preserved total length of 2.1 meters (Z.-G. Li et al., 2016). It resembles the even larger *Concavispina biseridens* from the Carnian Guanling biota (see below) in the concave dorsal margin of the neural spines but differs in proportional and morphological details as well as—and in particular—in the dentition (J. Liu, Zhao et al., 2013). Of the thalattosaurs represented in the Xingyi fauna, *Anshunsaurus*—unlike *Xinpusaurus*—resembles the western Tethyan genus *Askeptosaurus* rather closely both in overall size and in skeletal morphology (L. Cheng, Chen, Zhang et al., 2011).

A clade that is prominently represented in the Xingyi fauna is the protorosaurs, with two genera also known from the western Tethyan faunal province, that is, *Macrocnemus* and *Tanystropheus*. *Macrocnemus fuyua-nensis* from Fuyuan County, Yunnan Province, is slightly larger than *Macrocnemus bassanii* from the Middle Triassic of the southern Alps and differs from the latter only in limb proportions and, perhaps, the number of dorsal vertebrae (C. Li, Zhao et al., 2007; Jiang, Rieppel, Fraser et al., 2011). Interestingly, a specimen of *Macrocnemus* from the Grenzbitumen-zone (late Anisian) of Monte San Giorgio has recently been described as aff. *M. fuyuanensis* (Jaquier et al., 2017), for its close similarity with the Chinese form. Finally, the genus *Tanystropheus* was first reported from the Xingyi fauna on the basis of a juvenile specimen not identified to the species level (C. Li, 2007). A nearly complete skeleton of a mature individual from the Wusha District, Xingyi County, lacking the skull, however, revealed no diagnostic species-specific characteristics and for that reason was referred to *Tanystropheus* cf. *T. longobardicus* (Rieppel, Jiang et al., 2010).

Collected in Fuyuan County, Yunnan Province, and of Ladinian age (Xingyi fauna; Zhuganpo Member of the Falang Formation) comes another semi-aquatic stem-archosaur, *Litorosuchus somnii* (C. Li, Wu, Zhao et al., 2016). The nares in the stream-lined skull open dorsally, the elongate rostrum is furnished with tall and pointed teeth, tall neural spines and elongate chevrons result in a lateral compression of the long tail, and the hind limb morphology suggests webbed feet—features that collectively indicate an aquatic predatory lifestyle. The Middle Triassic of southwestern China thus provides evidence for important ecological diversification during early stages of archosauriform evolution that pre-dates the diversification of the Archosauria proper.

The Carnian Guanling Biota

The Guanling biota of Triassic (Carnian) marine invertebrates, fishes, and reptiles from the surroundings of Guanling (Guizhou Province, southwestern China) was the first one to be discovered and to attract international attention (Yin et al., 2000; Jiang, Motani, Li et al., 2005; X.-F. Wang et al., 2008). The deposits form part (the lower member) of the Xiaowa Formation (previously also referred to as Wayao Member of the Falang Formation) of Carnian age and is best known for their pelagic crinoids and marine reptiles, in particular ichthyosaurs, with species growing up to 10 meters in total length. Geographically, the fossiliferous localities are located in the Xinpu District (near Xiaowa, Dawa, Maowa, and Bamaolin villages) of Guanling County, extending westward into Qinglong County, Guizhou Province. The taxonomy of Triassic marine reptiles of the Guanling biota is somewhat impaired by its problematic earliest systematic treatment offered by Yin et al., but thanks to the abundance of the fossils that have since been collected, a certain consensus has emerged over the last few years (2000).

In contrast to the Ladinian Xingyi fauna, the Carnian Guanling biota is dominated by a notable taxic diversity of ichthyosaurs that occur in large numbers. In their early treatment of the marine reptiles from the Guanling biota, Yin et al. recognized five genera and species of ichthyosaurs: *Mixosaurus guanlingensis*, *Guizhouichthyosaurus tangae*, *Typicusichthyosaurus tsaihuae*, *Guanlingsaurus liangae*, and *Qianichthyosaurus zhoui* (they also erroneously referred the thalattosaur genus *Xinpusaurus* to ichthyosaurs) (2000). Of these, *Mixosaurus guanlingensis* was recognized as misidentified and declared a possible synonym of the other relatively small-sized ichthyosaur that occurs in the Guanling biota, *Qianichthyosaurus zhoui* by Jiang, Motani, Li et al. (2005; see also Maisch, Jiang et al., 2008; Maisch, Pan et al., 2010). *Typicusichthyosaurus tsaihuae* is treated as a *species inquirenda* by McGowan and Motani, the status of which still remains to be clarified, especially in view of preparational artifacts manifest in the holotype (McGowan and Motani, 2003; Maisch, Pan et al., 2010). A general consensus has emerged, however, to treat the remaining three genera and species as valid taxa (Jiang, Motani, Li et al., 2005; X.-F. Wang et al., 2008; Ji, Jiang, Motani et al., 2016). Of those, by virtue of its relatively small size reaching up to two meters total length, *Qianichthyosaurus zhoui* is the most frequently found ichthyosaur in the Guanling biota, known from juvenile specimens as well (C. Li, 1999; Nicholls, Wei et al., 2002; Maisch, Jiang et al., 2008).

Guizhouichthyosaurus tangae, of characteristic slender build, ranks among the larger ichthyosaurs represented in the Guanling biota (Maisch, Pan et al., 2006). Other large ichthyosaur specimens from the Guanling biota have been assigned to *Cymbospondylus asiaticus* and *Panjiangsaurus epicharis*, both of which are considered junior synonyms of *Guizhouichthyosaurus tangae* (C. Li and You, 2002; L. Cheng and Chen, 2003). The new combination of genus name and specific epithet *Shastasaurus tangae* has likewise been rejected (Jiang, Motani, Li et al., 2005; Shang and Li, 2009; Ji, Jiang, Motani et al., 2016). *Guizhouichthyosaurus tangae* is known for its sexual dimorphism, stomach content (small fishes, bivalves, and brachiopods), and the uptake of gastroliths (L. Cheng, Wings et al., 2006; L. Cheng and Chen, 2007; X.-F. Wang et al., 2008; Shang and Li, 2013). Another large ichthyosaur recognized in the Guanling biota is *Guanlingsaurus liangae*, interpreted as a short-snouted, toothless suction-feeding species in the genus *Shastasaurus* by Sander, Chen et al. (2011). The validity of the genus *Guanlingsaurus* was reestablished, however, on the basis of new material described by J. Cheng et al., who also questioned the ability of suction-feeding in view of the delicate structure of the hyoid apparatus of *Guanlingsaurus liangae* (2013).

Finally, a new genus and species not previously recognized among the ichthyosaurs from the Guanling biota was introduced by Chen, Cheng et al. under the name of *Callawayia wolonggangensis* (2007). Maisch referred the species *wolonggangensis* to the genus *Guizhouichthyosaurus*, a proposal that was not followed by Ji, Jiang, Motani et al. (Maisch, 2010:163; Ji, Jiang, Motani et al., 2016:e1025956-10). Referral of the Chinese species to the genus *Callawayia*, a genus otherwise known from British Columbia, continues to remain questionable, however (Maisch and Matzke, 2000:69; Ji, Jiang, Motani et al., 2016:e1025956-10).

The Guanling biota yielded the first placodont ever described from the eastern Tethys, the cyamodontoid *Sinocyamodus xinpuensis* (C. Li, 2000). The holotype is an exquisitely preserved specimen exposed in dorsal view, showing a carapace composed of osteoderms that do not cover the pectoral and pelvic girdle. The second sauropterygian known from the Guanling biota is again a cyamodontoid placodont, *Psephochelys polyosteoderma* (C. Li and Rieppel, 2002). The skull is well preserved, but the osteoderms in the central part of the carapace are poorly delineated. As indicated by the genus name, the taxon combines cranial characteristics of the western Tethyan cyamodontoid genera *Placochelys* and *Psephoderma* (Rieppel, 2000a). In the most recent phylogenetic analysis of placodont interrelationships, *Sinocyamodus* was found to be the sister taxon to *Cyamodus*, while *Psephochelys* came out as sister taxon to *Psephoderma* (Neenan, Li et al., 2015).

It was again a fossil recovered from the Guanling biota on the basis of which the first thalattosaur from the eastern Tethys was described, although originally erroneously identified as a sauropterygian: the askeposauroid *Anshunsaurus huangguoshuensis* (J. Liu, 1999; see also Rieppel, Liu and Bucher, 2000; J. Liu and Rieppel, 2005). Although initially referred to the Ichthyosauria, the thalattosaur *Xinpusaurus suni* was already named in the initial study of the reptiles from the Guanling biota on the basis of four poorly prepared specimens (Yin et al., 2000). The skull of the holotype along with an additional specimen were subsequently re-described by Luo and Yu (2002). An isolated skull referred to the same genus and species was described by J. Liu and Rieppel, as well as additional material referred to the same taxon (J. Liu, 2001; J. Liu and Rieppel; 2001; Rieppel and Liu, 2006). L. Cheng described a second species of *Xinpusaurus* from the Guanling biota, *Xinpusaurus bamaolinensis*, purportedly differing from the genotypical species by the extensive overbite of the premaxilla over the dentary in the slender rostrum (2003). *Xinpusaurus kohi* is a third species in the genus, again from the Guanling biota (Jiang, Maisch, and Sun, 2004). Taking individual variation as well as taphonomic implications into account, a case has been made to consider the latter two species as junior synonyms of the genotypical species, *Xinpusaurus suni* (J. Liu, 2013). Maisch subsequently defended the validity of *Xinpusaurus kohi*, a conclusion that was upheld by Z.-G. Li et al., who continued to treat *Xinpusaurus bamaolinensis* as a *species inquirenda*, however (Maisch, 2014; Z.-G. Li et al., 2016).

Concavispina biseridens is a thalattosaur from the Guanling biota, which in its overall morphology rather closely resembles *Xinpusaurus* but differs from the latter mostly in its larger overall size, a relatively larger skull, the concave dorsal margin of the neural spines, and a biserial arrangement of blunt teeth in the anterior part of the maxilla (J. Liu, Zhao et al., 2013). With a total preserved length of 364 cm, *Concavispina* is one of the largest thalattosaurs on record, exceeded in length only by *Miodentosaurus brevis* (Y.-N. Cheng, Wu and Sato, 2007; J. Liu, Zhao et al., 2013:347; see also X.-C. Wu, Cheng, Sato et al., 2009). *Miodentosaurus brevis* is yet another thalattosaur from the Guanling biota that reached up to 440 cm in total length (Zhao, Sato et al., 2010). It is notable not only for its size but also for its straight and relatively very short snout, and the reduced dentition (six teeth in the anterior part of the premaxilla and dentary; maxilla edentulous; palatal dentition absent). The Guanling biota yielded the most diverse assemblage of coexisting thalattosaur species known from anywhere in the world. It might even be that thalattosaur diversity in the Guanling biota is as yet incompletely understood, since some of the specimens and taxa originally named by Yin et al. might be thalattosaurs (Yin et al., 2000; Y.-N. Cheng, Wu, and Sato, 2007:259; X.-C. Wu, Cheng, Sato et al., 2009:2). The most recent phylogenetic analyses of thalattosaur interrelationships place *Anshunsaurus* and *Miodentosaurus* together with *Askeptosaurus* (and *Endennasaurus*) in the Askeptosauroidea, the sister taxon of the Thalattosauroidea, which includes all other thalattosaurs, *Xinpusaurus* and *Concavispina* among them (X.-C. Wu, Cheng, Sato et al., 2009; J. Liu, Zhao, Li et al., 2013). Within the Askeptosauroidea, *Askeptosaurus* is variably found to be more closely related to *Miodentosaurus*, or to *Anshunsaurus*, respectively (X.-C. Wu, Cheng, Sato et al., 2009; L. Cheng, Chen, Zhang et al., 2011; J. Liu, Zhao, Li et al., 2013). *Xinpusaurus* and *Concavispina* potentially come out as sister taxa within the Thalattosauroidea (sensu Nicholls, 1999). A weak signal indicates the possibility of a relationship of the *Xinpusaurus* clade with *Nectosaurus* from the Late Triassic (Carnian) of California (Müller, 2005, 2007; X.-C. Wu, Cheng, Sato et al., 2009).

The last and arguably most famous faunal elements of the Guanling biota are the ancestral turtles *Odontochelys semitestacea* and *Eorhynchochelys sinensis* (C. Li, Wu, Rieppel et al., 2008; C. Li, Fraser et al., 2018). Together, these stem turtles provide important insights into the early evolution of turtles, in particular with respect to the evolution of the turtle shell and beak. Many fossils of Triassic marine reptiles from southwestern China, most notably protorosaur material from the Xingyi fauna, remain as yet unprepared and/or undescribed. The present review of Triassic marine reptiles from southern and southwestern China can thus be no more than a progress report.

The Paleobiogeography of Triassic Marine Reptiles

In the discussion of ichthyosaur diversity in the Middle Triassic of Monte San Giorgio, the distinction of the splitters' versus lumpers' approach to alpha taxonomy was introduced (Simpson, 1961). The account of the Middle Triassic marine reptiles from southern and southwestern China certainly reflects a splitter's approach. It is, for example, not clear that the saurosphargid *Largocephalosaurus* from the Panxian-Luoping fauna is generically distinct from *Saurosphargis*, the latter known from the Germanic Muschelkalk. It is also conceivable that the thalattosaur *Concavispina* from the Guanling biota is congeneric with *Xinpusaurus* from equivalent deposits. It is noteworthy that the taxonomic approach to the Triassic marine reptiles in southern and southwestern China rarely takes into account ecological considerations such as issues relating to trophic specialization and niche partitioning. There is also an apparent stratification of the reptile diversity in the marine Triassic of southern and southwestern China, as documented by the Chaohu fauna (Lower Triassic, Olenekian, Spathian), followed by the Panxian-Luoping Fauna (Middle Triassic, Anisian, Pelsonian), and on to the Xingyi fauna (Middle Triassic, Ladinian), and the Guanling biota (Late Triassic, lower Carnian) (J.-L. Li, 2006). Many endemic forms characterizing these respective faunas lend credence to such stratification, but the possibility of a taxonomic bias influenced by stratigraphic considerations nevertheless remains to be investigated.

Given the wealth of material collected by different parties representing different institutions, the alpha taxonomy of the Triassic marine reptiles from southern and southwestern China is likely to remain fluid for some time to come. Global phylogenetic analyses including the newly discovered Chinese material likewise remain unstable and probably will continue to remain so at least for the near future. Nevertheless, the current state of knowledge does allow some paleobiogeographical conclusions at least at a very general level. Given their superior dispersal capabilities, due to their pelagic adaptations, the paleobiogeographical signal for ichthyosaurs—especially the larger species—may be expected to be weaker than for other clades of Triassic marine reptiles. *Mixosaurus panxianensis* shares western Tethyan relationships, as the most recent phylogenetic analysis found it to group in an unresolved trichotomy with *M. cornalianus*, and *M. kuhnschnyderi* (Ji, Jiang, Motani et al., 2016). *Guizhouichthyosaurus*, *Guanlingsaurus*, and "*C.*" *wolonggangensis* represent the shastasaurids in the eastern Tethys, as they group with *Besanosaurus*, *Shastasaurus*, and *Shonisaurus* from the western Tethyan and eastern Pacific faunal provinces, respectively (Ji, Jiang, Motani et al., 2016). *Qianichthyosaurus* is a toretocnemid, as it groups with *Toretocnemus* from the Upper Triassic (upper Carnian) of Shasta County, California (Ji, Jiang, Motani et al., 2016). Such relationships of *Qianichthyosaurus* had earlier already been recognized as a distinct signal for the trans-Pacific distribution of ichthyosaur taxa in the Triassic (Nicholls, Wei, and Manabe, 2002). The only non-ichthyosaur Triassic marine reptile from southwestern China for which a similar—albeit weaker— trans-Pacific signal obtains is the thalattosaur *Xinpusaurus*, as it is possibly related to the Late Triassic genus *Nectosaurus* from northern California.

Given that the non-ichthyosaur Triassic marine reptiles occur in three principal faunal provinces—the eastern Tethys (southern China), western Tethys (Europe and circum-Mediterranean localities), and the eastern Pacific (western North America)—a three-area statement can be generated which, for non-ichthyosaurian clades of Triassic marine reptiles, clearly indicates that the eastern and western Tethyan faunal provinces are much more closely related to one another than either of those two is to the eastern Pacific faunal province (Rieppel, 1999a). Many clades common to the two Tethyan faunal provinces, such as

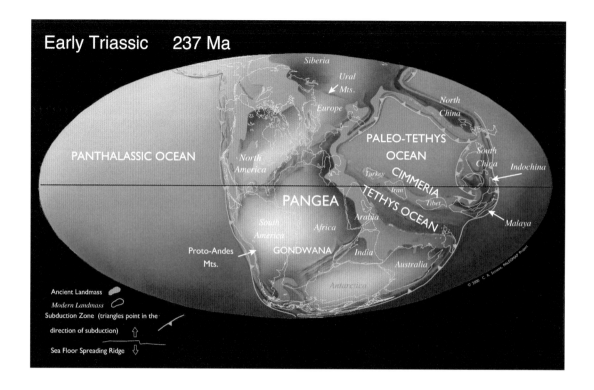

11.2. Paleogeographic reconstruction of the Early Triassic continents. Reproduced with permission; copyright by Christopher R. Scotese, Director, PALEOMAP Project.

placodonts, pachypleurosaurs, nothosaurs, and saurosphargids have never been reliably identified in the eastern Pacific faunal province. The only non-ichthyosaur Triassic marine reptiles that have been reported from the eastern Pacific faunal province are the pistosaurian sauropterygians *Augustasaurus* and *Corosaurus*, and thalattosaurs (Storrs, 1991a, 1991b; Nicholls and Brinkman, 1993b; Sander, Rieppel, and Bucher, 1997; Nicholls, 1999). In contrast, the eastern Tethyan occurrences of the sauropterygian *Placodus*, and the protorosaurs *Macrocnemus*, and *Tanystropheus* document a particularly close affinity with western Tethyan congeneric forms, as morphological differences are either minimal, or even absent, as in the postcranium of *Tanystropheus* from the Xingyi fauna referred to cf. *T. longobardicus* (Rieppel, Jiang, Fraser et al., 2010). The Chinese thalattosaur *Anshunsaurus* was likewise found to be very closely related to *Askeptosaurus* from the western Tethys (L. Cheng, Chen, and Zhang et al., 2011).

Triassic marine reptiles other than ichthyosaurs were generally shallow-water inhabitants, living in near-shore habitats, intraplatform basins, or shallow epicontinental seas. Their limited capabilities of dispersal are indicated by a high degree of endemism not just at the species level, but also with respect to some of the more highly derived forms from southwestern China, such as the protorosaur *Dinocephalosaurus*, or the saurosphargid *Sinosaurosphargis* from the Anisian Panxian–Luoping biota. Nevertheless, the close affinities manifest between the eastern and western Tethyan faunal provinces can well be explained because there existed, in the Lower to Middle Triassic, a "shallow water archipelago" with a "water depth no more than ten meters," spanning the distance between the eastern and western Tethys at equatorial latitude (fig. 11.2). This shallow water archipelago, called Cimmeria, involved a string of small Gondwanan microcontinents or terranes rifting northward, where "some of the land may have emerged above water." The water current predominant along this shallow water archipelago ran from east to west (Christopher Scotese, personal communication, August 10, 2016), which accords well with the dispersalist scenario for pachypeurosaurs outlined in the first section of this chapter—assuming the phylogenetic relationships underlying that scenario stand the test of time. It is conceivable that this shallow water archipelago served as a route for dispersal between the eastern and western Tethys during the Triassic for other clades of marine reptiles as well. Whether any one of those clades represented in both faunal provinces more specifically had an eastern, or western origin remains subject to debate, the discovery of new fossils, and ongoing phylogenetic analysis. The current literature suggests that a faunal exchange between the eastern and western Tethys occurred in both directions: pachypleurosaurs appear to have dispersed from east to west, whereas a western Tethyan origin has been suggested for placodonts (Rieppel and Hagdorn, 1997; Rieppel, 1999a; Neenan, Klein, and Scheyer, 2013).

Epilogue—In the Shadow of the Chinese Dragon

For decades, the Monte San Giorgio was the undisputed epicenter for the study of Triassic marine reptiles. A *Konservat-Lagerstätte* of Middle Triassic marine biota comparable to the Besano–Monte San Giorgio localities in terms of richness and diversity of fossil vertebrates was not known to exist anywhere else in the world. The study of fossil fishes and reptiles from the bituminous black shales of Besano (Besano Formation) goes back to the very beginning of vertebrate paleontology in Italy. Excavation activities at Monte San Giorgio by Bernhard Peyer, and especially the field campaign conducted by Emil Kuhn-Schnyder in the Grenzbitumenzone at Point 902, resulted in an incredibly rich and diverse collection of Middle Triassic marine fishes and reptiles. As a result, the Paleontological Institute and Museum of the University of Zurich has become the most important repository of fossil material from Monte San Giorgio. Fossils collected at the Besano locality after World War II are kept at the Museo Civico di Storia Naturale di Milano (earlier finds were destroyed during the Second World War). Specimens that have more recently been collected, and continue to be collected at Monte San Giorgio, go to the Cantonal Museum of Natural History in Lugano. Other than the odd specimens here or there—either commercially traded, or gifted—the bulk of the Besano–Monte San Giorgio fossil material is thus rather easily accessible for study and comparison. Historically, this material has played an immeasurably important role in solving many a paleontological riddle, as discussed in the preceding chapters. With regard to the taxonomy of Triassic hybodontiform sharks, the Besano–Monte San Giorgio sediments yielded illuminating material that preserves fin spines, scales, jaws, and teeth in association. The riddle concerning the fin spines of *Nemacanthus tuberculatus* found its solution in a complete and well-preserved specimen of the enigmatic shark *Acronemus*. A beautiful, completely preserved and articulated specimen of the actinopterygian *Birgeria stensioei* clarified the skeletal anatomy of this widespread Triassic chondrostean. A large number of beautifully preserved specimens allows the investigation of the anatomy, taxonomic diversity, and paleobiology in the actinopterygian genus *Saurichthys*. The articulated specimen of the Middle Triassic actinistian *Ticinepomys* reveals relationships with the living coelacanth. As for reptiles, *Paraplacodus* represents an important early stage in the evolution of placodonts, especially with regard to the differentiation of the crushing dentition. The postcranial morphology of cyamodontoid (armored) placodonts was clarified with the collection of an articulated specimen of *Cyamodus hildegardis*. A previously unsurpassed record of pachypleurosaurs from successive stratigraphic horizons allowed detailed investigations of the anatomy, evolution, and life history in the species included in the clade. The skeletal morphology of complete and articulated specimens of nothosaurs in the genera *Nothosaurus* and *Lariosaurus* (*Ceresiosaurus*) facilitated the interpretation of disarticulated material from other Middle Triassic localities, most important, the Germanic Muschelkalk. It was again a complete and articulated specimen from Monte San Giorgio that allowed recognition of the putatively earliest pterosaur *Tribelesodon* as having been based on a misinterpretation of the skeletal anatomy of a protorosaur with extreme neck elongation, that is, *Tanystropheus*. Several articulated specimens of *Macrocnemus* from Besano–Monte San Giorgio allowed not only functional anatomical and paleobiological analysis but also provided the starting point for investigation of progressive neck elongation in protorosaurs in general. The Monte San Giorgio yielded the first thalattosaurs to be known from complete and articulated skeletons, such as *Askeptosaurus* and *Clarazia*, and also revealed such extreme trophic specialization among thalattosaurs as seen in *Hescheleria*. And finally, the rauisuchid archosaur *Ticinosuchus* provided definitive clues in the interpretation of the much debated and initially badly misinterpreted *Chirotherium* tracks.

Not all of these important insights were gained in a straightforward manner, as sketched in the preceding chapters. Many of these discoveries resulted from research that took a roundabout path, as is often the case when no extant representatives are available that could provide a model for the interpretation of fossils. It is interesting to follow research that was influenced by a prevailing interest in the origin of lepidosaurs (rhynchocephalians and squamates). Adopting the tuatara (*Sphenodon*) and lizards as extant models led to repeated, and often times heated, yet ultimately badly misguided discussions in the interpretation of the protorosaurs *Macrocnemus* and *Tanystropheus*, and of the thalattosaur *Askeptosaurus* as being close to the ancestry of squamates. And yet, in spite of the outstanding historical importance of the research on the Middle Triassic marine fishes and reptiles from Besano and Monte San Giorgio, interest in Triassic, indeed Mesozoic marine reptiles was slow to gain pace at an international level. Robert Carroll's *Vertebrate Paleontology and Evolution* became the standard textbook in vertebrate paleontology when it was published in 1988. The photograph of an iconic Holzmaden ichthyosaur fossilized in the process of giving birth to offspring adorns its dust cover and front page. All the same: fossil reptiles in general are covered in a total of some 146 pages, of which only 18 pages—a mere 12.3%—are dedicated to Mesozoic marine reptiles. The year 1997 saw the publication of the proceedings of a symposium on *Ancient Marine Reptiles*, which was the first book "devoted solely to aquatic reptiles" since Williston's *Water Reptiles of the Past and Present*, the latter published in 1914 (Nicholls and Callaway, 1997:xvii). The book lists 28 contributors, who collectively represented the vast majority of workers at the time actively involved in research on Mesozoic marine reptiles of all kinds.

This was to change rapidly, however, following the initial discovery of the early Late Triassic Guanling biota in Guizhou Province, southwestern China, in the late 1990s.

Significant parts of the Zurich Monte San Giorgio collection remain unprepared and undescribed to the present day. Yet it is unclear how much more taxic diversity would be revealed through continuing excavations at Monte San Giorgio and/or Besano. Recent taxonomic treatments of Middle Triassic vertebrates from Besano and Monte San Giorgio, involving actinopterygian fishes (*Saurichthys*) and ichthyosaurs (mixosaurs), reveal a tendency toward a very fine-grained alpha taxonomy. In contrast, the discovery and excavation of Triassic marine vertebrate faunas in southern and southwestern China, at several localities ranging over a vast geographical area, and collectively spanning the time from the late Lower Triassic (Spathian) to the early Upper Triassic (Carnian), has resulted in the collection and description of a host of new, previously unknown species of actinopterygian fishes and reptiles on the basis of a rich, and mostly superbly preserved fossil record. Some of the newly discovered taxa show a high degree of endemism, whereas others document the close relationships that this eastern Tethyan faunal province shares with the western Tethyan faunal province, with localities not just at Besano and Monte San Giorgio, but throughout Europe and the circum-Mediterranean territories.

The incredibly rich fossil record of Triassic marine reptiles from southern and southwestern China has motivated intensive research, attracting a large number of expert scientists, both Chinese and international, as well as an entire new cohort of Chinese graduate students. All this activity has quite generally generated new interest in Mesozoic marine reptiles at an international scale. The Chinese fossil record has offered stunning new insights into the taxonomic and ecological diversity of Triassic marine reptiles, and it has recalibrated our understanding of the recovery of the marine biota after the end-Permian mass extinction. Even so, it is unquestionable that the Middle Triassic marine biota from Besano–Monte San Giorgio will forever retain their historical significance and will necessarily continue to provide a benchmark for taxonomic comparison in systematic work on Middle Triassic marine fishes and reptiles. But it is equally unquestionable that the highest metabolic rate in research on Triassic marine fishes and reptiles conducted today is seen in southern and southwestern China (Tintori and Felber, 2015).

Literature Cited

L

Agassiz, L. 1833–1844. Recherches sur les Poissons Fossiles. Imprimerie de Petitpierre, Neuchâtel.

Albers, P. C. H. 2011. New *Nothosaurus* skulls from the Lower Muschelkalk of the western Lower Saxony Basin (Winterswijk, the Netherlands) shed new light on the status of *Nothosaurus winterswijkensis*. Netherlands Journal of Geosciences 90:15–22.

Aldinger, H. 1931. Über Reste von *Birgeria* (Pisces, Palaeoniscidae) aus der alpinen Trias. Neues Jahrbuch für Mineralogie, Geologie und Paläontologie, Abt. B, Beilage-Band 66:167–180.

———. 1937. Permische Ganoidfische aus Ostgrönland. Meddelelser om Gronland 102:1–392.

Alessandri, G. de 1910. Studii sui pesci Triasici della Lombardia. Memorie della Società Italiana di Scienze Naturali 7:1–147.

Andrews, S. M., B. G. Gardiner, R. S. Miles, and C. Patterson. 1967. Pisces; pp. 637–683 in W. B. Harland (ed.), The Fossil Record. Geological Society, London.

Araújo, R., and F. Correia. 2015. Soft-tissue anatomy of the plesiosaur pectoral girdle inferred from basal Eosauropterygia taxa and the extant phylogenetic bracket. Paleontologia Electronica 18.1.8A.

Araújo, R., and M. J. Polcyn. 2013. A biomechanical analysis of the skull and adductor chamber muscles in the Late Cretaceous plesiosaur *Libonectes*. Paleontologia Electronica 16.2.10A.

Argyriou, T., M. Clauss. E. E. Maxwell, H. Furrer, and M. Sánchez-Villagra. 2016. Exceptional preservtion reveals gastrointestinal anatomy and evolution in early actinopterygian fishes. Scientific Reports 6:18758; DOI: 10.1038/srep18758.

Arratia, G. 2013. Morphology, taxonomy, and phylogeny of Triassic pholidophorid fishes (Actinopterygii, Teleostei). Journal of Vertebrate Paleontology 33 (S1):1–138.

Arthaber, G. von. 1922. Über Entwicklung, Ausbildung und Absterben der Flugsaurier. Palaeontologische Zeitschrift 4:1–47.

———. 1924. Die Phylogenie der Nothosaurier. Acta Zoologica, Stockholm 5:439–516.

Balsamo-Crivelli, G. 1839. Descrizione di un nuovo rettile fossile della famiglia dei Paleosauri, e di due pesci fossili trovati nel calcareo nero, sopra Varenna sul lago di Como, dal nobile sig. Lodovico Trotti, con alcune riflessioni geologiche. Il Politecnico 1:421–431.

Bardet, N., J. Falconnet, V. Fischer, A. Houssaye, S. Jouve, X. Pereda Suberbiola, A. Pérez-García, J.-C. Rage, and P. Vincent. 2014. Mesozoic marine reptile paleobiogeography in response to drifting plates. Gondwana Research 26:869–887.

Baron, M. G., D. B. Norman, and P. M. Barrett. 2017. A new hypothesis of dinosaur relationships and early dinosaur evolution. Nature 543:501–506.

Bassani, F. 1886. Sui fossili e sull'età degli schisti bitumiosi triasici di Besano in Lombardia. Atti della Società Italiana di Scienze Naturali 29:15–72.

———. 1895 (1896). La Ittiofauna della Dolomia Principale di Giffoni (Provincia di Salerno). Palaeontographia Italica 1:169–210.

Baur, G. 1887a. Ueber den Ursprung der Extremitäten der Ichthyopterygier. Bericht über die XX. Versammlung des Oberrheinischen geologischen Vereins 1887:17–20.

———. 1887b. On the phylogenetic arrangement of the Sauropsida. Journal of Morphology 1:93–104.

———. 1889. *Palaeohatteria* Credner, and the Proganosauria. American Journal of Science 137 (3):310–313.

Beard, S. R., and H. Furrer. 2017. Land or water: using taphonomic models to determine the lifestyle of the Triassic protorosaur *Tanystropheus* (Diapsida, Archosauromorpha). Palaeobiodiversity and Palaeoenvironments; DOI: https://doi.org/10.1007/s12549-017-0299-7.

Bellotti, C. 1857. Desrizione di alcune nuove specie di pesci fossili di Perledo e di altri località Lombarde; pp. 419–438 in A. Stoppani (ed.), Studii Geologici e Paleontologici sulla Lombardia. C. Turati, Milan.

Beltan, L. 1977. La parturition d'un actinoptérigien de l'Eotrias du Nord-Ouest de Madagascar. Comptes Rendus de l'Académie des Sciences de Paris 284 (D):2223–2225.

———. 1980. Eotrias du Nord-Ouest de Madagascar: étude de quelques poissons don't un est en parturition. Annales de la Société de Géologie du Nord 99:453–464.

Bemis, W. E., E. K. Findeis, and L. Grande. 1997. An overview of Acipenseriformes. Environmental Biology of Fishes 48:25–71.

Benson, R. B., and P. S. Druckenmiller. 2014. Faunal turnover of marine tetrapods during the Jurassic-Cretaceous transition. Biological Reviews 89:1–23.

Benton, M. J. 1985. Classification and phylogeny of the diapsid reptiles. Zoological Journal of the Linnean Society 84:97–164.

———. 1986. More than one event in the Late Triassic mass extinction. Nature 321:857–861.

———. 1994. Late Triassic to Middle Jurassic extinctions among continental tetrapods: testing the pattern; pp. 366–397 in N. C. Fraser and H.-D. Sues (eds.), In the Shadow of the Dinosaurs: Early Mesozoic Tetrapods. Cambridge University Press, Cambridge.

———. 1999. *Scleromochlus* and the origin of dinosaurs and pterosaurs. Philosophical Transactions of the Royal Society of London B 354:1423-1446.

———. 2005. Vertebrate Paleontology. 3rd ed. Wiley-Blackwell, Malden, 455 pp.

Benton, M. J., and J. L. Allen. 1997. *Boreopricea* from the Lower Triassic of Russia, and the relationships of the prolacertiform reptiles. Palaeontology 40:931–953.

Benton, M. J., Q. Zhang, M. Hu, Z.-Q. Chen, W. Wen, J. Liu, J. Huang, C. Zhou, Y. Xie, J. Tong, and B. Choo. 2013. Exceptional vertebrate biotas from the Triassic of China, and the expansion of marine ecosystems after the Permo-Triassic mass extinction. Earth-Science Reviews 125:199–243.

Bernard, A., C. Lécuyer, P. Vincent, R. Amiot, N. Bardet, E. Buffetaut, G. Cuny, F. Fournel, F. Martineau, J.-M. Mazin, and A. Prieur. 2010. Regulation of body temperature by some Mesozoic marine repiles. Science 328:1379–1382.

Bernasconi, S. M. 1994. Geochemical and Microbial Controls on Dolomite Formation in Anoxic Environments: a Case Study from the Middle Triassic (Ticino, Switzerland). Contributions to Sedimentology, no. 19. Schweizerbart'sche Verlagsbuchhandlung, Stuttgart, 109 pp.

Bernasconi, S. M., and A. Riva. 1993. Organic geochemistry and depositional environment of a hydrocarbon source rock: the Middle Triassic Grenzbitumenzone Formation, southern Alps, Italy/Switzerland; pp. 179–190 in A. M. Spencer (ed.), Generation, Accumulation and Production of Europe's Hydrocarbons, Vol. 3, Special Publication of the European Association of Petroleum Geologists, no. 3. Springer, Berlin.

Bernhardi, R. 1834. Mittheilungen an den Geheimrath v. Leonhard gerichtet. Neues Jahrbuch für Mineralogie, Geognosie, Geologie und Petrefaktenkunde 1834:641–642.

Besmer, A. 1947. Die Triasfauna der Tessiner Kalkalpen. XVI. Beiträge zur Kenntnis des Ichthyosauriergebisses. Schweizerische Paläontologische Abhandlungen 65:1–21.

Bever, G. S., T. Lyson, D. J. Field, and B.-A. S. Bhullar. 2015. Evolutionary origin of the turtle skull. Nature 525:239–242.

Bickelmann, C., and P. M. Sander. 2008. A partial skeleton and isolated humeri of *Nothosaurus* (Reptilia: Eosauropterygia) from Winterswijk, the Netherlands. Journal of Vertebrate Paleontology 28:326–338.

Bizzarini, F., and G. Muscio. 1994. Un nuovo rettile (Repilia, Prolacertiformes) dal Norico di Preone (Udine, Italia Nordorientale). Nota preliminare. Gortania—Atti del Museo Friulano di Storia Naturale 16:67–76.

Blackburn, D. G. 1992. Convergent evolution of viviparity, matrotrophy, and specializations for fetal nutrition in reptiles and other vertebrates. American Zoologist 32:313–321.

Blackwelder, R. E. 1967. Taxonomy. A Text and Reference Book. John Wiley, New York, 698 pp.

Bonaparte, J. F., and H.-D. Sues. 2006. A new species of *Clevosaurus* (Lepidosauria: Rhynchocephalia) from the Upper Triassic of Rio Grande do Sul, Brazil. Palaeontology 49:917–923.

Böttcher, R. 1990. Neue Erkenntnisse über die Fortpflanzungsbiologie der Ichthyosaurier (Reptilia). Stuttgarter Beiträge zur Naturkunde. Serie B (Geologie und Paläontologie) 164:1–51.

Boulenger, G. A. 1898. On a nothosaurian reptile from the Trias of Lombardy, apparently referable to *Lariosaurus*. Transactions of the Zoological Society of London 14:1–10.

Bowden, A. J., G. R. Tresise, and W. Simkiss. 2010. *Chirotherium*, the Liverpool footprint hunters and their interpretation of the Middle Triassic environment. Geological Society, London, Special Publications 343:209–228.

Brack, P., and H. Rieber. 1986. Stratigraphy and ammonoids of the lower Buntsandstein Beds of the Brescian Prealps and Guidicarie and their significance for the Anisian/Ladinian boundary. Eclogae geologicae Helvetiae 79:181–225.

———. 1993. Towards a better definition of the Anisian/ Ladinian boundary: new biostratigraphic data and correlations of boundary sections from the Southern Alps. Eclogae geologicae Helvetiae 86:415–527.

Brack, P., H. Rieber, A. Nicora, and R. Mundil. 2005. The Global boundary Stratotype Section and Point (GSSP) of the Ladinian Stage (Middle Triassic) at Bagolino (Southern Alps, Northern Italy) and its implications for the Triassic time scale. Episodes 28:233–244.

Brazeau, M. D., and M. Friedmann. 2015. The origin and early phylogenetic history of jawed vertebrates. Nature 520:490–497.

Brinkmann, W. (ed.). 1994. Paläontologisches Museum der Universität Zürich. Führer durch die Ausstellung. Verlag des Paläontologischen Instituts und Museums der Universität Zürich, Zurich, 108 pp.

———. 1996. Ein Mixosaurier (Reptilia, Ichthyosauria) mit Embryonen aus der Grenzbitumenzone des Monte San Giorgio (Schweiz, Kanton Tessin). Eclogae geologicae Helvetiae 89:1321–1344.

———. 1997. Die Ichthyosaurier (Reptilia) aus der Mitteltrias des Monte San Giorgio (Tessin, Schweiz) und von Besano (Lombardei, Italien)—der aktuelle Forschungsstand. Vierteljahrsschrift der Naturforschenden Gesellschaft in Zürich 142:69–78.

———. 1998a. *Sangiorgiosaurus* n. g.—eine neue Mixosaurier-Gattung (Mixosauridae, Ichthyosauria) mit Quetschzähnen aus der Grenzbitumenzone (Mitteltrias) des Monte San Giorgio (Schweiz, Kanton Tessin). Neues Jahrbuch für Geologie und Paläontologie, Abhandlungen 207:125–144.

———. 1998b. Die Ichthyosaurier (Reptilia) aus der Grenzbitumenzone (Mitteltrias) des Monte San Giorgio (Tessin, Schweiz)—neue Ergebnisse. Vierteljahrsschrift der Naturforschenden Gesellschaft in Zürich 143:165–177.

———. 1999. *Ichthyosaurus cornalianus* Bassani, 1866 (currently *Mixosaurus cornalianus*, Reptilia, Ichthyosauria); proposed designation of a neotype. Bulletin of Zoological Nomenclature 56:247–249.

———. 2004. Mixosaurier (Reptilia, Ichthyosauria) mit Quetschzähnen aus der Grenzbitumenzone (Mitteltrias) des Monte San Giorgio (Schweiz, Kanton Tessin). Schweizerische Paläontologische Abhandlungen 124:1–84.

Brinkmann, W., and R. Mutter. 1999. Biologie der Fische und Saurier des Monte San Giorgio; pp. 115–145 in H. Rieber (ed.), Paläontologie in Zürich. Fossilien und ihre Erforschung in Geschichte und Gegenwart. Zoologisches Museum der Universität, Zurich.

Brochu, C. A. 2001. Progress and future directions in archosaur phylogenetics. Journal of Paleontology 75:1185–1201.

Broili, F. 1915. Beobachtungen an *Tanystropheus conspicuus* H. v. Meyer. Neues Jahrbuch für Mineralogie, Geologie und Paläntologie 1915 (II):51–62.

———. 1927. Ein Sauropterygier aus den Arlbergschichten. Sitzungsberichte der mathematisch-physikalischen Klasse der Bayerischen Akademie der Wissenschaften zu München 1927:205–228.

Broili, F., and E. Fischer. 1917. *Trachelosaurus fischeri* nov. gen. nov. sp. Ein neuer Saurier aus dem Buntsandstein von Bernburg. Jahrbuch der Königlich Preussischen geologischen Landesanstalt zu Berlin 37:359–414.

Bronn, H. G. 1835. Petrefaktenkunde. Jahrbuch für Mineralogie, Geognosie, Geologie und Petrefaktenkunde 1835:230–234.

Brough, J. 1939. The Triassic fishes of Besano, Lombardy. British Museum (Natural History), London, 117 pp.

Brown, W. H. 1890. Dates of publication of "Recherches sur les poissons fossiles," by L. Agassiz; pp. xxv–xxlx in A. S. Woodward and C. D. Sherborn (eds.), A Catalogue of British Fossil Vertebrata. Dulau, London.

Buffetaut, E. 1987. A Short History of Vertebrate Paleontology. Croom Helm, London, 223 pp.

Bürgin, T. 1990. Reproduction in Middle Triassic actinopterygians; complex fin structures and evidence of viviparity. Zoological Journal of the Linnean Society 100:379–391.

———. 1992. Basal ray-finned fishes (Osteichthyes; Actinopterygii) from the Middle Triassic of Monte San Giorgio (Canton Tessin, Switzerland). Schweizerische Paläontologische Abhandlungen 114:1–164.

———. 1996. Diversity in the feeding apparatus of perleidid fishes (Actinopterygii) from the Middle Triassic of Monte San Giorgio; pp. 555–565 in G. Arratia and G. Viohl (eds.), Mesozoic Fishes—Systematics and Paleoecology. Friedrich Pfeil, Munich.

———. 1999a. New actinopterygian fishes (Osteichtyes) from the lower Meride Limestone (lower Ladinian) of Aqua del Ghiffo (Monte San Giorgio, southern Switzerland). Rivista del Museo civico di Scienze Naturali "Enrico Caffi" Bergamo 20:57–62.

———. 1999b. Middle Triassic marine fish faunas from Switzerland; pp. 481–494 in G. Arratia and G. Viohl (eds.), Mesozoic Fishes—Systematics and Paleoecology. Friedrich Pfeil, Munich.

Bürgin, T., O. Rieppel, P. M. Sander, and K. Tschanz. 1989. The fossils of Monte San Giorgio. Scientific American 260 (6):74–81.

Burke, A. C. 2015. Origin of the turtle body plan; pp. 77–89 in K. P. Dial, N. Shubin, and E. L. Brainerd (eds.), Great Transformations in Vertebrate Evolution. University of Chicago Press, Chicago.

Callaway, J. M. 1997. A new look at Mixosaurus; pp. 45–59 in C. M. Callaway and E. L. Nicholls (eds.), Ancient Marine Reptiles. Academic Press, San Diego, California.

Camp, C. L. 1945. Prolacerta and the protorosaurian reptiles. American Journal of Science, 243:17–32; 84–101.

———. 1976. Vorläufige Mitteilung über grosse Ichthyosaurier aus der Oberen Trias von Nevada. Sitzungsberichte der Österreichischen Akademie der Wissenschaften, Mathematisch-Naturwissenschaftliche Klasse, Abteilung I 185:125–134.

Capetta, H. 1987. Chondrichthyes II. Mesozoic and Cenozoic Elasmobranchii. Handbook of Paleoichthyology, Vol. 3b. Gustav Fischer, Stuttgart.

Carroll, R. L., 1985a. A pleurosaur from the Lower Jurassic and the taxonomic position of the Sphenodontia. Palaeontographica A 189:1–28.

———. 1985b. Evolutionary constraints in aquatic diapsid reptiles. Special Papers in Palaeontology 33:145–155.

———. 1988. Vertebrate Paleontology and Evolution. W. H. Freeman, New York, 698 pp.

Carroll, R. L., and Z.-M. Dong. 1991. Hupehsuchus, an enigmatic aquatic reptile from the Triassic of China, and the problem of establishing relationships. Philosophical Transactions of the Royal Society of London B 331:131–153.

Carroll, R. L., and P. Gaskill. 1985. The nothosaur Pachypleurosaurus and the origin of plesiosaurs. Philosophical Transactions of the Royal Society of London B 309:343–393.

Casey, M., N. C. Fraser, and M. Kowalewski. 2007. Quantitative taphonomy of a Triassic reptile: Tanytrachelos ahynis from the Cow Branch Formation, Dan River Basin, Solite Quarry, Virginia. Palaios 22:598–611.

Cavin, L., H. Furrer, and C. Obrist. 2013. New coelacanth material from the Middle Triassic of eastern Switzerland, and comments on the taxic diversity of actinistians. Swiss Journal of Geosciences 106:161–177.

Chatterjee, S. 1986. Malerisaurus langstoni, a new diapsid reptile from the Triassic of Texas. Journal of Vertebrate Paleontology 6:297–312.

Chen, X.-H., and L. Cheng. 2009. The discovery of Mixosaurus (Reptilia, Ichthyopterygia) from the Middle Triassic of Luoping, Yunnan Province. Acta Geologica Sinica 83:1214–1220. (In Chinese with English summary)

———. 2010. A new species of Mixosaurus (Reptilia: Ichthyosauria) from the Middle Triassic of Pu'an, Guizhou, China. Acta Palaeontologica Sinica 49:251–260. (In Chinese with English summary)

Chen, X.-H., L. Cheng, and P. M. Sander. 2007. A new species of Callawayia (Reptilia: Ichthyosauria) from the Late Triassic in Guanling, Guizhou. Geology in China 34:974–982.

Chen, X.-H., R. Motani, L. Cheng, D.-Y. Jiang, and O. Rieppel. 2014a. A carapace-like bony "body tube" in an Early Triassic marine reptile and the onset of marine tetrapod predation. PLoS ONE 9:e94396; DOI: 10.1317/journal.pone.0094396.

———. 2014b. A small short-necked Hupehsuchian from the Lower Triassic of Hubei Province, China. PLoS One 9:e115244; DOI: 10.1371/journal.pone.0115244.

———. 2015. A new specimen of Carroll's mystery hupehsuchian from the Lower Triassic of China. PLoS ONE 10:e0126024; DOI: 10.1371/journal.pone.0126024.

Chen, X-H., P. M. Sander, L. Cheng, and X.-F. Wang. 2013. A new primitive ichthyosaur from Yuanan, South China. Acta Geologica Sinica 87:672–677.

Cheng, J., D.-Y. Jiang, R. Motani, W.-C. Hao, Z.-Y. Sun, and T. Cai. 2013. A new juvenile specimen of Guanlingsaurus (Ichthyosauria, Shastasauridae) from the Upper Trassic of southwestern China. Journal of Vertebrate Paleontology 33:340–348.

Cheng, L. 2003. A new species of Triassic Thalattosauria from Guanling, Guizhou. Geological Bulletin of China 22:274–277.

Cheng, L., and X.-H. Chen. 2003. A new species of large-sized and long-body ichthyosaur from the Late Triassic Guanling biota, Guizhou, China. Geological Bulletin of China 22:228–235.

———. 2007. Gut contents in the Triassic ichthyosaur Panjiangsaurus from the Guanling Biota in Guizhou. Geology in China 34:61–65.

Cheng, L., X.-H. Chen, Q.-H. Shang, and X.-C. Wu. 2014. A new marine reptile from the Triassic of China, with highly specialized feeding adaptation. Die Naturwissenschaften 101:251–259.

Cheng, L., X.-H. Chen, and C. Wang. 2007. A new species of Late Trissic Anshunsaurus (Reptilia: Thalattosauria) from Guizhou Province. Acta Geologica Sinica 81:1346–1351.

Cheng, L., X.-H. Chen, X. Zeng, and Y. Cai. 2012. A new eosauropterygian (Diapsida: Sauropterygia) from the Middle

Triassic of Luoping, Yunnan Province. Journal of Earth Science 23:33–40.

Cheng, L., X.-H. Chen, B. Zhang, and Y. Cai. 2011. New study of *Anshunsaurus huangnihensis* Cheng, 2007 (Reptilia: Thalattosauria): revealing its transitional position in Askeptosauridae. Acta Geologica Sinica 85:1231–1237.

Cheng, L., O. Wings, X.-H. Chen, and P. M. Sander. 2006. Gastroliths in the Triassic ichthyosaur *Panjiangsaurus* from China. Journal of Paleontology 80:583–588.

Cheng, Y.-N., R. Holmes, X.-C. Wu, and N. Alfonso. 2009. Sexual dimorphism and life history of *Keichousaurus hui* (Reptilia: Sauropterygia). Journal of Vertebrate Paleontology 29:401–408.

Cheng, Y.-N, T. Sato, X.-C. Wu, and C. Li. 2006. First complete pistosauroid from the Triassic of China. Journal of Vertebrate Paleontology 26:501–504.

Cheng, Y.-N., X.-C. Wu, and Q. Ji. 2004. Triassic marine reptile gave birth to live young. Nature 432:383–386.

Cheng, Y.-N., X.-C. Wu, and T. Sato. 2007. A new thalattosaurian (Reptilia: Diapsida) from the Upper Triassic of Guizhou, China. Vertebrata PalAsiatica 7:246–260.

Cheng, Y.-N., X.-C. Wu, T. Sato, and H.-Y. Shan. 2012. A new eosauropterygian (Diapsida, Sauropterygia) from the Triassic of China. Journal of Vertebrate Paleontology 32:1335–1349.

———. 2016. *Dawazisaurus brevis*, a new eosauropterygian from the Middle Triassic of Yunnan, China. Acta Geologica Sinica (English edition) 90:401–424.

Cheng, X.-H., R. Motani, L. Cheng, D.-Y. Jiang, and O. Rieppel. 2014. The enigmatic marine reptile *Nanchangosaurus* from the Lower Triassic of Hubei, China and the phylogenetic affinities of Hupehsuchia. PLoS ONE 9:e102361; DOI:10.1371/journal.pone.0102361.

Cloutier, R., and P. E. Ahlberg. 1996. Morphology, characters, and the interrelationships of basal sarcopterygians; pp. 445–479 in M. L. Stiassny, L. R. Parenti, and G. D. Johnson (eds.), Interrelationships of Fishes. Academic Press, San Diego, California.

Colbert, E. H. 1945a. *Prolacerta* and the protorosaurian reptiles. American Journal of Science 243:17–32, 84–101.

———. 1945b. The Dinosaur Book. The Ruling Reptiles and Their Relatives. American Museum of Natural History, New York, 156 pp.

———. 1955. Evolution of the Vertebrates. John Wiley, New York, 479 pp.

———. 1987. The Triassic reptile *Prolacerta* in Antarctica. American Museum Novitates 2882:1–19.

Collin, R., and C. M. Janis. 1997. Morphological constraints on tetrapod feeding mechanisms: why were there no suspension-feeding marine reptiles; pp. 451–466 in J. M. Callaway and E. L. Nicholls (eds.), Ancient Marine Reptiles. Academic Press, San Diego, California.

Conybeare, W. D. 1824. On the discovery of an almost perfect skeleton of the *Plesiosaurus*. Transactions of the Geological Society of London 5:559–594.

Cope, E. D. 1885. On the evolution of the Vertebrata, progressive and retrogressive. American Naturalist 19:234–247.

Cornalia, E. 1854. Notizie zoologiche sul *Pachypleura Edwardsii* COR. Nuovo sauro acrodonte degli strati triasici di Lombardia. Giornale del Reale Instituto Lombardo di Scienze, Lettre ed Arti 6:45–56.

Cox, B. 1975. The longest-necked lizard? Nature 254:654–655.

Cramp, R. L., E. A. Meyer, N. Sparks, and C. E. Franklin. 2008. Functional and morphological plasticity of crocodile (*Crocodylus porosus*) salt glands. Journal of Experimental Biology 211:1482–1489.

Curioni, G. 1847. Cenni sopra un nuovo saurio fossile dei monti di Perledo sul Lario e sul terreno che lo racchiude. Giornale del Reale Instituto Lombardo di Scienze, Lettre ed Arti 16:157–170.

———. 1863. Sui giacimenti metalliferi e bituminosi nei terreni Triasici di Besano. Memorie del Reale Instituto Lombardo di Scienze, Lettre ed Arti 9 (ser. 2, vol. 3): 241–268.

Dal Sasso, C., and G. Pinna. 1996. *Besanosaurus leptorhynchus* n. gen. n. sp., a new shastasaurid ichthyosaur from the Middle Triassic of Besano (Lombardy, N. Italy). Paleontologia Lombarda N.S. 4:3–23.

Dalla Vecchia, F. M. 1993. Reptile remains from the Middle-Upper Triassic of the Carnic and Julian Alps (Friuli-Venezia Giulia, northeastern Italy). Gortania–Atti del Museo Friulano di Storia Naturale 15:1–49–66.

———. 1996. Archosaurian trackways in the Upper Carnian of Dogna Valley (Udine, Friuli, NE Italy). Natura Nascosta 20:5–17.

Damme, J. v., and P. Aerts. 1997. Kinematics and functional morphology of aquatic feeding in Australian snake-necked turtles (Pleurodira: *Chelodina*). Journal of Morphology 133:113–125.

Dawson, W. R., G. A. Bartholomew, and A. F. Bennett. 1977. A reappraisal of the aquatic specializations of the Galapagos marine iguana (*Amblyrhynchus cristatus*). Evolution 31:891–897.

Day, E. C. H. 1864. On *Acrodus anningiae*, Agass.; with remarks upon the affinities of the genera *Acrodus* and *Hybodus*. Geological Magazine 1:57–64.

DeBraga, M., and R. L. Carroll. 1993. The origin of mosasaurs as a model of macroevolutionary patterns and processes. Evolutionary Biology 27:245–322.

Deecke, W. 1886. Über *Lariosaurus* und einige andere Saurier der Lombardischen Trias. Zeitschrift der deutschen Geologischen Gesellschaft 38:170–197.

———. 1889. Über Fische aus verschiedenen Horizonten der Trias. Palaeontographica 35:97–138.

———. 1926. Fossilium Catalogus I: Animalia. Pars 33: Pisces triadici. W. Junk, Berlin, 201 pp.

Deeming, D. C., L. B. Halstead, M. Manabe, and D. M. Unwin. 1993. An ichthyosaur embryo from the Lower Lias (Jurassic, Hettangian) of Somerset, England, with comments on the reproductive biology of ichthyosaurs. Modern Geology 18:423–442.

De la Beche, H.T., and W. D. Conybeare. 1821. Notice of the discovery of a new fossil animal, forming a link between the *Ichthyosaurus* and the crocodile, together with general remarks on the osteology of *Ichthyosaurus*. Transactions of the Geological Society of London 1 (2):381–389.

Dick, J. R. F. 1978. On the Carboniferous shark *Tristychius arcuatus* AGASSIZ from Scotland. Transactions of the Royal Society of Edinburgh 70:63–109.

Dilkes, D. W. 1998. The early Triassic rhynchosaur *Mesosuchus browni* and the interrelationships of basal archosauromorph reptiles. Transactions of the Royal Society of London B 353:501–541.

Drevermann, F. 1915. Über *Placodus*. Centralblatt für Mineralogie, Geologie und Palaeontologie 1915:402–405.

———. 1933. Die Placodontier. 3. Das Skelett von *Placodus gigas* Agassiz im Senckenberg-Museum. Abhandlungen der senckenbergischen naturforschenden Gesellschaft 38:319–364.

Dunson, W. A. 1976. Salt glands in reptiles; pp. 413–445 in C. Gans and W. R. Dawson (eds.), Biology of the Reptilia, Vol. 5, Physiology A. Academic Press, London.

Dutuit, J.-M. 1979. Un pseudosuchien du Trias continental marocain. Annales de Paléontologie (Vertébrés) 65:55-68.

Egerton, P. G. 1854. On some new genera and species of fossil fishes. Annals and Magazine of Natural History 13 (2):433–436.

Enault, S., S. Adnet, and M. Debiais-Thibaud. 2016. Skeletogenesis during the late embryonic development of the catshark *Scylliorhinus canicula* (Chondrichthyes; Neoselachii). MorphoMuseuM 1(4); DOI: 10.18563/m3.1.4.e2.

Etter, W. 1994. A new penaeid shrimp (*Antrimpos mirigiolensis* n. sp., Crustacea, Decapoda) from the Middle Triassic of the Monte San Giorgio (Ticino, Switzerland). Neues Jahrbuch für Geologie und Paläontologie, Monatshefte 1994:223–230.

———. 2002. Monte San Giorgio: remarkable Triassic marine vertebrates; pp. 221–242 in D. J. Bottjer, W. Etter, J. W. Hagdorn, and C. M. Tang (eds.), Exceptional Fossil Preservation. A Unique View on the Evolution of Marine Life. Columbia University Press, New York.

Evans, S. E. 1984. The classification of the Lepidosauria. Zoological Journal of the Linnean Society 82:887–100.

———. 1988. The early history and relationships of the Diapsida; pp. 221–260 in M. J. Benton (ed.), The Phylogeny and Classification of the Tetrapods, Vol. 1, Amphibians, Reptiles, Birds. Clarendon Press, Oxford.

———. 2003. At the feet of the dinosaurs: the early history and radiation of lizards. Biological Reviews 78:513–551.

Evans, S. E., and J. Klembara. 2005. A choristoderan reptile (Reptilia: Diapsida) from the Lower Miocene of northwest Bohemia (Czech Republic). Journal of Vertebrate Paleontology 25:171–184.

Ezcurra, M. D. 2016. The phylogenetic relationships of basal archosauromorphs, with an emphasis on the systematics of proterosuchian archosauriforms. PeerJ; DOI: 10.7717/peerj.1778.

Fabbri, M., F. M. Dalla Vecchia, and A. Cau. 2014. New information on *Bobosaurus forojuliensis* (Reptilia: Sauropterygia): implications for plesiosaurian evolution. Historical Biology 26:661–669.

Felber, M., H. Furrer, and A. Tintori. 2000. Geo-Guida del Monte San Giorgio (Ticino/Svizzera—Provincia di Varese/Italia). Carta escursionistica scientifico – didattica 1: 14 000. Geologia Insubrica, allegato al vol. 5–fasc. 1.

———. 2004. The Triassic of Monte San Giorgio in the World Heritage List of UNESCO: an opportunity for science, the local people, and tourism. Eclogae geologicae Helvetiae 97:1–2.

Fischer, H. 1963. Bernhard Peyer (1885–1963). Vierteljahrsschrift der Naturforschenden Gesellschaft in Zürich 108:467–469.

Fischer, V., N. Bardet, R. B. Benson, M. S. Arkhangelsky, and M. Friedmann, 2016. Extinction of fish-shaped marine reptiles associated with reduced evolutionary rates and global environmental volatility. Nature Communications 7:10825; DOI: 10.1038/ncomms10B25.

Fischer, W. 1959. Neue Funde von *Henodus chelyops* v. Huene im Tübinger Gipskeuper. Neues Jahrbuch für Geologie und Paläontologie, Monatshefte 1959:241–247.

Flügel, E. 1991. Triassic and Jurassic marine calcareous algae: a critical review; pp. 481–503 in R. Riding (ed.), Calcareous Algae and Stromatolites. Springer, Berlin.

———. 2004. Microfacies of Carbonate Rocks. Analysis, Interpretation, and Application. Springer, Berlin, 984 pp.

Forey, P. L. 1998. History of the Coelacanth Fishes. Chapman & Hall, London, 419 pp.

———. 2004. Systematics and paleontology; pp. 149–180 in D. M. Williams and P. L. Forey (eds.), Milestones in Systematics. CRC Press, Boca Raton, Florida.

Fraas, O. 1881. *Simosaurus pusillus* aus der Lettenkohle von Hoheneck. Jahreshefte des Vereins für vaterländische Naturkunde in Württemberg 37:319–324.

———. 1889. Kopfstacheln von *Hybodus* und *Acrodus*, sog. *Ceratodus heteromorphus* Ag. Jahreshefte des Vereins für vaterländische Naturkunde in Württemberg 45:233–240.

———. 1891. Die Ichthyosaurier der süddeutschen Trias- und Jura-Ablagerungen. Laupp, Tübingen, 81 pp.

———. 1896. Die schwäbischen Trias-Saurier nach dem Material der Kgl. Naturalien-Sammlung in Stuttgart zusammengestellt. Mitteilungen aus dem Naturalien-Kabinett Stuttgart 5:1–88.

Franz-Odendaal, T. A. 2006. Intramembranous ossification of scleral ossicles in *Chelydra serpentina*. Zoology 109:75–81.

Franz-Odendaal, T. A., and M. K. Vickaryous. 2006. Skeletal elements in the vertebrate eye and adnexa. Morphological and developmental perspectives. Developmental Dynamics 235:1244–1255.

Fraser, N. C., D. A. Grimaldi, P. E. Olsen, and B. Axsmith. 1996. A Triassic Lagerstätte from eastern North America. Nature 380:615–619.

Fraser, N. C., S. Nosotti, and O. Rieppel. 2004. A re-evaluation of two species of *Tanystropheus* (Diapsida, Protorosauria) from Monte San Giorgio, Southern Alps. Journal of Vertebrate Paleontology 24:60A.

Fraser, N. C., and O. Rieppel. 2006. A new protorosaur (Diapsida) from the upper Buntsandstein of the Black Forest, Germany. Journal of Vertebrate Paleontology 26:866–871.

Frauenfelder, A. 1916. Beiträge zur Geologie der Tessiner Kalkalpen. Eclogae geologicae Helvetiae 14:247–371.

Freyberg, B.v. 1972. Die erste erdgeschichtliche Erforschungsphase Mittelfrankens (1840–1847). Eine Briefsammlung zur Geschichte der Geologie. Erlanger geologische Abhandlungen 92:1–33.

Fricke, H., and K. Hissmann. 1992. Locomotion, fin coordination and body form of the living coelacanth *Latimeria chalumnae*. Environmental Biology of Fishes 34:329–356.

Fricke, H., O. Reinicke, H. Hofer, and W. Nachtigall. 1987. Locomotion of the coelacanth *Latimeria chalumnae* in its natural environment. Nature 329:331–333.

Fröbisch, N. B., J. Fröbisch, P. M. Sander, L. Schmitz, and O. Rieppel. 2006. A new species of *Cymbospondylus* (Diapsida, Ichthyosauria) from the Middle Triassic of Nevada and a re-evaluation of the skull osteology of the genus. Zoological Journal of the Linnean Society 147:515–538.

———. 2013. Macropredatory ichthyosaur from the Middle Triassic and the origin of modern trophic networks. Proceedings of the National Academy of Sciences 110:1393–1397.

Fürbringer, M. 1900. Beitrag zur Systematik und Genealogie der Reptilien. Gustav Fischer, Jena, 91 pp.

Furrer, H. 1995. The Kalkschieferzone (Upper Meride Limestone; Ladinian) near Meride (Canton Ticino, Southern Switzerland) and the evolution of a Middle Triassic intraplatform basin. Eclogae geologicae Helvetiae 88:827–852.

———. 1999. Aktuelle Grabungen in den unteren Meride-Kalken bei Acqua del Ghiffo; pp. 87–103 in H. Rieber (ed.), Paläontologie in Zürich. Fossilien und ihre Erforschung in Geschichte und Gegenwart. Zoologisches Museum der Universität, Zurich.

———. 2003. Der Monte San Giorgio im Südtessin—vom Berg der Saurier zur Fossil-Lagerstätte internationaler Bedeutung. Neujahrsblatt der Naturforschenden Gesellschaft in Zürich, 206. Stück. Koprint AG, Alpnach Dorf, 64 pp.

Gaffney, E. S., and M. C. McKenna. 1979. A Late Permian captorhinid from Rhodesia. American Museum Novitates 2688:1–15.

Gao, K., S. Evans, J. Qiang, M. Norell, and J. S. An. 2000. Exceptional fossil material of a semi-aquatic reptile from China: the resolution of an enigma. Journal of Vertebrate Paleontology 20:417–421.

Gardiner, B. G. 1993. Osteichthyes: basal actinopterygians; pp. 611–619 in M. J. Benton (ed.), The Fossil Record 2. Chapman & Hall, London.

Gardiner, B. G., and B. Schaeffer. 1989. Interrelationships of lower actinopterygian fishes. Zoological Journal of the Linnean Society 97:135–187.

Gauthier, J. A. 1984. A cladistc analysis of the higher systematic categories of the Diapsida. PhD Dissertation, University of California, Berkeley.

Godfrey, S. J. 1984. Plesiosaur subaqueous locomotion: a reappraisal. Neues Jahrbuch für Geologie und Paläontologie, Monatshefte 1984:661–672.

Gottmann-Quesada, A., and P. M. Sander. 2009. A redescription of the early archosauromorph Protorosaurus speneri Meyer, 1832, and its phylogenetic relationships. Palaeontographica A 287:123–220.

Gower, D. J. 2000. Rauisuchian archosaurs (Reptilia, Diapsida): an overview. Neues Jahrbuch für Geologie und Paläontologie, Abhandlungen 218:447–488.

Gower, D. J., and E. Weber. 1998. The braincase of Euparkeria, and the evolutionary relaionships of birds and crocodilians. Biological Reviews 73:367–411.

Grande, L. 2010. An empirical synthetic pattern study of gars (Lepisosteiformes) and closely related species, based mostly on skeletal anatomy. The resurrection of Holostei. Copeia 2010 (supplement):1–871.

Grande, L., and W. E. Bemis. 1991. Osteology and phylogenetic relationships of fossil and Recent paddlefishes (Polyodontidae) with comments on the interrelationships of Acipenseriformes. Journal of Vertebrate Paleontology 11, Supplement to Number 1:1–121.

Griffith, J. 1957. Studies on the Saurichthyidae, a family of Mesozoic fishes. PhD Dissertation, University of London.

———. 1959. On the anatomy of two saurichthyid fishes, Saurichthys striolatus (Bronn) and S. curioni (Bellotti). Proceedings of the Zoological Society of London 132:587–606.

Grigg, G., and D. Kirshner. 2015. Biology and Evolution of Crocodylians. Cornell University Press, Ithaca, 649 pp.

Guttormsen, S. E. 1937. Die Triasfauna der Tessiner Kalkalpen. XIII. Beiträge zur Kenntniss des Ganoidengebisses, insbesondere des Gebisses von Colobodus. Abhandlungen der schweizerischen Paläontologischen Gesellschaft 60:1–41.

Haeckel, E. 1866. Generelle Morphologie der Organismen. Zweiter Band: Allgemeine Entwickelungsgeschichte der Organismen. Georg Reimer, Berlin.

Haeckel, E., K. Hescheler, and H. Eisig. 1916. Aus dem Leben und Wirken von Arnold Lang. Dem Andenken des Freundes und Lehrers gewidmet. Gustav Fischer, Jena.

Hänni, K. 2004. Die Gattung Ceresiosaurus. Ceresiosaurus calcagnii PEYER und Ceresiosaurus lanzi n.sp. (Lariosauridae, Sauropterygia). Vdf Hochschulverlag AG, ETH Zürich, Zurich, 147 pp.

Hao, W.-C., Y.-L. Sun, D.-Y. Jiang, and Z.-Y. Sun. 2006. Advance in studies of the Panxian Fauna. Acta Scientiarum Naturalium Universitatis Pekinensis 42:817–823.

Harland, W. B. 1997. The geology of Svalbard. Memoirs of the Geological Society of London 17:1–521.

Haubold, H. 2006. Die Saurierfährten Chirotherum barthii Kaup, 1835—das Typusmaterial aus dem Buntsandstein bei Hildburghausen/Thüringen und das "Chirotherium-Monument." Veröffentlichungen des Naturhistorischen Museum Schleusingen 21:3–31.

Hauff, B., and R. B. Hauff. 1981. Das Holzmadenbuch. Repro-Druck, Fellbach, 136 pp.

Heatwole, H. 1978. Adaptations of marine snakes: unusual physiological strategies have enabled some snakes to live in saltwater environment. American Scientist 66:594–604.

Heckel, J. J. 1849. Untersuchung der fossilen Fische des Österreichischen Kaiser-Staates. Neues Jahrbuch für Mineralogie, Geognosie, Geologie, und Petrefakten-Kunde 1849:499–500.

Heckert, A. B., S. G. Lucas, L. F. Rinehart, J. A. Spielmann, A. P. Hunt, and R. Kahle. 2006. Revision of the archosauomorph reptile Trilophosaurus, with a description of Trilophosaurus jacobsi, from the Upper Triassic Chinle Group, West Texas, USA. Palaeontology 49:621–640.

Hilton, R. P. 2003. Dinosaurs and other Mesozoic reptiles of California. University of California Press, Berkeley, 318 pp.

Hirasawa, T., H. Nagashima, and S. Kuratani. 2013. The endoskeletal origin of the turtle carapace. Nature Communications 4:2107; DOI: 10.1038/ncomms3107.

Hoffstetter, R. 1955. Rhynchocephalia; pp. 556–576 in J. Piveteau (ed.), Traité de Paléontologie, Vol. 5. Masson, Paris.

Holmes, R., Y.-N. Cheng, and X.-C. Wu. 2008. New information on the skull of Keichosaurus hui (Reptilia, Sauropterygia) with comments on sauropterygian interrelationships. Journal of Vertebrate Paleontology 28:76–84.

Hu, S., Q. Zhang, Z. Chen, C. L. T. Zhou, T. Lü, T. Xie, W. Wen, J. Huang, and M. J. Benton. 2011. The Luoping biota: exceptional preservation, and new evidence on the Triassic recovery from end-Permian mass extinction. Proceedings of the Royal Society of London B 278:2274–2282.

Huene, F. v. 1907–1908. Die Dinosauier der europäischen Triasformation. Geologische und Palaeontologische Abhandlungen Suppl. 1:1–419.

———. 1931. Über Tanystropheus und verwandte Formen. Neues Jahrbuch für Mineralogie, Geologie und Paläontologie, Beilageband B 67:65–86.

———. 1936. *Henodus chelyops*, ein neuer Placodontier. Palaeontographica A 36:99–148.

———. 1942. Die fossilen Reptilien des südamerikanischen Gondwanalandes. Ergebnisse der Sauriergrabungen in Südbrasilien 1928/29. C. H. Beck, München, 332 pp.

———. 1948. Short review of the lower tetrapods; pp. 65–106 in A. J. Du Toit (ed.), Royal Society of South Africa Special Publication, Robert Broom Commemorative Volume. Royal Society of South Africa, Cape Town.

Hugi, J., and T. M. Scheyer. 2012. Ossification sequences and associated ontogenetic changes in the bone histology of pachypleurosaurids from Monte San Giorgio (Switzerland/Italy). Journal of Vertebrate Paleontology 32:315–327.

Hugi, J., T. M. Scheyer, P. M. Sander, N. Klein, and M. R. Sánchez-Villagra. 2011. Long bone microstructure gives new insights into the life of pachypleurosaurids from the Middle Triassic of Monte San Giorgio, Switzerland/Italy. Comptes Rendus Palevol 10:423–426.

Hulke, J. W. 1873. Memorandum on some fossil vertebrate remains collected by the Swedish expedition to Spitzbergen in 1864 and 1868. Bihang till Kungliga Svenska Vetenskapsakademiens Handlinger I, Afdelning IV 9:1–11.

Hull, D. L. 1988. Science as a Process: An Evolutionary Account of the Social and Conceptual Development of Science. University of Chicago Press, Chicago, 586 pp.

Hutchinson, M. N., A. Skinner, and M. S. Y. Lee. 2012. *Tikiguania* and the antiquity of squamate reptiles (lizards and snakes). Biology Letters 8:665–669.

Huxley, J. S. 1958. Evolutionary processes and taxonomy with special reference to grades. Uppsala Universitets Arsskrift 6:21–39.

Huxley, T. H. 1872. A Manual of the Anatomy of the Vertebrated Animals. D. Appleton, New York, 431 pp.

ICZN, 2005. Opinion 2120 (Case 3230). *Colobodus* Agassiz 1844 (Osteichthyes, Perleidiformes): existing usage conserved by the designation of *C. bassanii* de Alessandri, 1910 as the type species. Bulletin of Zoological Nomenclature 62:51–52.

Jaekel, O., 1889. Selachier aus dem oberen Muschelkalk Lothringens. Abhandlungen zur geologischen Specialkarte von Elsass-Lothringen, Band II, Heft IV. Strassburger Druckerei und Verlagsanstalt, Strassburg, pp. 275–332.

Jäger, G. F. 1852. Ueber die Fortpflanzungsweise des Ichthyosaurus. Gelehrte Anzeigen, Bulletin der Königlich Bayerischen Akademie der Wissenschaften 1851:33–36.

Jalil, N.-E., and K. Peyer. 2007. A new rauisuchian (Archosauria, Suchia) from the Upper Triassic of the Argana Basin, Morocco. Palaeontology 50:417–430.

Jaquier, V. P., N. C. Fraser, H. Furrer, and T. M. Scheyer. 2017. Osteology of a new specimen of *Macrocnemus* aff. *M. fuyuanensis* (Archosauromorpha, Protorosauria) from the Middle Triassic of Europe: potential implications for species recognition and paleogeography of tanystropheid protorosaurs. Frontiers in Earth Science 5:91; DOI: 10.3389/feart.2017.00091.

Jaquier, V. P., and T. M. Scheyer. 2017. Bone histology of the Middle Triassic long–necked reptiles *Tanystropheus* and *Macrocnemus* (Archosauromorpha, Protorosauria). Journal of Vertebrate Paleontology 37:2, e1296456; DOI: 10.1080/02724634.2017.1296456.

Ji, C., D.-Y. Jiang, R. Motani, O. Rieppel, W.-C. Hao, and Z.-Y. Sun. 2016. Phylogeny of the Ichthyopterygia incorporating recent discoveries from South China. Journal of Vertebrate Paleontology 36:1, e1025956; DOI: 10.1080/02724634.2015.1025956.

Ji, C., D.-Y. Jiang, O. Rieppel, R. Motani, A. Tintori, and Z.-Y. Sun. 2014. A new specimen of *Nothosaurus youngi* from the Middle Triassic of Guizhou, China. Journal of Vertebrate Paleontology 34:465–470.

Jiang, D.-Y., W.-C. Hao Y.-L. Sun, M. W. Maisch, and A. T. Matzke. 2003. The mixosaurid ichthyosaur *Phalarodon* from the Middle Triassic of China. Neues Jahrbuch für Geologie und Paläontologie, Monatshefte 11:656–666.

Jiang, D.-Y., M. W. Maisch, W.-C. Hao, H. U. Pfretzschner, Y.-L. Sun, and Z.-Y. Sun. 2005. *Nothosaurus* sp. (Reptilia, Sauropterygia, Nothosauridae) from the Anisian (Middle Triassic) of Guizhou, southwestern China. Neues Jahrbuch für Geologie und Paläontologie, Monatshefte 2005:565–576.

Jiang, D.-Y., M. W. Maisch, W.-C. Hao, Y.-L. Sun, and Z.-Y. Sun. 2006. *Nothosaurus yangjuanensis* n. sp. (Reptilia, Sauropterygia, Nothosauridae) from the middle Anisian (Middle Triassic) of Guizhou, southwestern China. Neues Jahrbuch für Geologie und Paläontologie, Monatshefte 2006: 257–276.

Jiang, D.-Y., M. W. Maisch, and Y.-L. Sun. 2004. A new species of *Xipusaurus* (Thalattosauria) from the Upper Triassic of China. Journal of Vertebrate Paleontology 24:80–88.

Jiang, D.-Y., M. W. Maisch, Z.-Y. Sun, Y.-L. Sun, and W.-C. Hao. 2006. A new species of *Lariosaurus* (Reptilia, Sauropterygia)) from the Middle Anisian (Middle Triassic) of southwestern China. Neues Jahrbuch für Geologie und Paläontologie, Abhandlungen 242:19–42.

Jiang, D.-Y., R. Motani, W.-C. Hao, O. Rieppel, Y.-L. Sun,L. Schmitz, and Z.-Y. Sun. 2008. First record of Placodontoidea (Reptilia, Sauropterygia, Placodontia) from the eastern Tethys. Journal of Vertebrate Paleontology 28:904–908.

Jiang, D.-Y., R. Motani, C. Li, W-C. Hao, Y.-L. Sun, Z.-Y. Sun, and L. Schmitz. 2005. Guanling biota: a marker of Triassic biotic recovery from the end-Permian extinction in the ancient Guizhou Sea. Acta Geologica Sinica 79:729–738.

Jiang, D.-Y., R. Motani, W.-C. Hao, O. Rieppel, Y.-L. Sun, A. Tintori, Z.-Y. Sun, and L. Schmitz. 2009. Biodiversity and sequence of the Middle Triassic Panxian Marine Reptile Fauna, Guizhou Province, China. Acta Geologica Sinica 83:451–459.

Jiang, D.-Y., R. Motani, J.-D. Huang, A. Tintori, Y.-C. Hu, O. Rieppel, N. C. Fraser, C. Ji, N. P. Kelley, W.-L. Fu, and R. Zhang. 2016. A large aberrant stem ichthyosauriform indicating early rise and demise of ichthyosauromorphs in the wake of the end-Permian extinction. Scientific Reports 6:26232; DOI: 10.1038/srep26232.

Jiang, D.-Y., R. Motani, L. Schmitz, O. Rieppel, W.-C. Hao, Y.-L. Sun, and Z.-Y. Sun. 2008. New primitive ichthyosaurian (Reptilia, Diapsida) from the Middle Triassic of Panxian (Guizhou, southwestern China) and its position in the Triassic Biotic Recovery. Progress in Natural Science 18:1315–1319.

Jiang, D.-Y, R. Motani, A. Tintori, O. Rieppel, G.-B. Chen, J.-D. Huang, R. Zhang, Z.-Y. Sun, and C. Ji. 2014. The Early Triassic eosauropterygian *Majiashanosaurus discocoracoidis*, gen. et sp. nov. (Reptilia, Sauropterygia), from Chaohu, Anhui Province, People's Republic of China. Journal of Vertebrate Paleontology 34:1044–1052.

Jiang, D.-Y., O. Rieppel, N. C. Fraser, R. Motani, W.-C. Hao, A. Tintori, Y.-L. Sun, and Z.-Y. Sun. 2011. New information on the protorosaurian reptile *Macrocnemus fuyuanensis* Li et al., 2007, from the Middle/Upper Triassic of Yunnan, China. Journal of Vertebrate Paleontology 31:1230–1237.

Jiang, D.-Y., O. Rieppel, R. Motani, W.-C. Hao, Y.-L. Sun, L. Schmitz, and Z.-Y. Sun. 2008. A new Middle Triassic Eosauropterygian (Reptilia, Sauropterygia) from southwestern China. Journal of Vertebrate Paleontology 28:1055–1062.

Jiang, D.-Y., O. Rieppel, R. Motani, W.-C. Hao, and A. Tintori. 2011. Marine reptile *Saurosphargis* from Anisian (Middle Triassic) of Panxian, Guizhou, southwestern China. Journal of Vertebrate Paleontology 31 (Supplement 2):132.

Jiang, D.-Y., L. Schmitz, W.-C. Hao, and Y.-L.Sun. 2006. A new mixosaurid ichthyosaur from the Middle Triassic of China. Journal of Vertebrate Paleontology 26:60–69.

Jiang, D.-Y., L. Schmitz, R. Motani, W.-C. Hao, and Y.-L. Sun. 2007. The mixosaurid ichthyosaur *Phalarodon* cf. *P. fraasi* from the Middle Triassic of Guizhou Province, China. Journal of Paleontology 81:602–605.

Jones, M. E. H., C. L. Anderson, C. A. Hipsley, J. Müller, E. S. Evans, and R. R. Schoch. 2013. Integration of molecules and new fossils supports a Triassic origin for Lepidosauria (lizards, snakes, and tuatara). BMC Evolutionary Biology 13:208. http://www.biomedcentral.com/1471-2148/13/208.

Jurcsak, T. 1978. Rezultate noi in studiul saurienilor fosili de la Alesd. *Nymphaea* 6:15–60.

Kaup, J. J. 1835. Mittheilungen, an Professor Bronn gerichet. Jahrbuch für Mineralogie, Geognosie, Geologie und Petrefaktenkunde 1835:327–328.

Kelly, N. P., R. Motani, P. Embree, and M. J. Orchard. 2016. A new Lower Triassic ichthyopterygian assemblage from Fossil Hill, Nevada. PeerJ 4:e1626; DOI: 10.7717/peerj.1626.

Klein, N. 2009. Skull morphology of *Anarosaurus heterodontus* (Reptilia: Sauropterygia: Pachypleurosauria) from the lower Muschelkalk of the Germanic Basin (Winterswijk, the Netherlands). Journal of Vertebrate Paleontology 29:665–676.

Klein, N., and P. C. H. Albers. 2009. A new species of the sauropsid reptile *Nothosaurus* from the Lower Muschelkalk of the Western Germanic Basin, Winterswijk, the Netherlands. Acta Palaeontologica Polonica 54:589–598.

Klein, N., and O. J. Sichelschmidt. 2014. Remarkable dorsal ribs with distinct uncinate processes from the early Anisian of the Germanic Basin (Winterswijk, the Netherlands). Neues Jahrbuch für Geologie und Paläontologie 271:307–314.

Klein, M., D. F. A. E. Voeten, A. Haarhuis, and R. Bleeker. 2016. The earliest record of the genus *Lariosaurus* from the early Middle Anisian (Middle Triassic) of the Germanic Basin. Journal of Vertebrate Paleontology 36:4, e1163712; DOI: 10.1080/02724634.2016.1163712.

Kogan, I., S. Pacholak, M. Licht, J. W. Schneider, C. Brücker, and S. Brandt. 2015. The invisible fish: hydrodynamic constraints for predator-prey interaction in fossil fish *Saurichthys* compared to recent actinopterygians. Biology Open 4:1715–1726.

Kogan, I., and C. Romano. 2016. Redescription of *Saurichthys madagascariensis* Piveteau, 1945 (Actinopterygii, Early Triassic), with implications for the early saurichthyid morphotype. Journal of Vertebrate Paleontology 36:4,

e1151886; DOI: http://dx.doi.org/10.1080/02724634.2016.1151886.

Koken, E. 1893. Beiträge zur Kenntnis der Gattung *Nothosaurus*. Zeitschrift der deutschen Geologischen Gesellschaft 45:337–377.

Krebs, B. 1965. Die Triasfauna der Tessiner Kalkalpen. XIX. *Ticinosuchus ferox* nov. gen. nov. sp. Schweizerische Paläontologische Abhandlungen 81:1–140.

Kuhn, O. 1934. Fossilium Catalogus. I: Animalia. Pars 69: Sauropterygia. W. Junk, 's-Gravenhage, 127 pp.

Kuhn-Schnyder, E. 1942. Über einen weiteren Fund von *Paraplacodus broilii* Peyer aus der Trias des Monte San Giorgio. Eclogae geologicae Helvetiae 35:174–183.

———. 1945. Über *Acrodus*-Funde aus dem Grenzbitumenhorizont der anisischen Stufe der Trias des Monte San Giorgio (Kt. Tessin). Eclogae geologicae Helvetiae 38:662–673.

———. 1946. Über einen Fund von *Birgeria* aus der Trias des Monte San Giorgio (Kt. Tessin). Eclogae geologicae Helvetiae 39:363–364.

———. 1947. Der Schädel von *Tanystropheus*. Eclogae geologicae Helvetiae 40:390.

———.1952. Die Trisfauna der Tessiner Kalkalpen. XVII. *Askeptosaurus italicus* Nopcsa. Schweizerische Palaeontologische Abhandlungen 69:1–73.

———. 1954a. Über die Herkunft der Eidechsen. Endeavour 13:215–221.

———. 1954b. The origin of lizards. Endeavour 13:213–219.

———. 1959a. Hand und Fuss von *Tanystropheus longobardicus* (Bassani). Eclogae geologicae Helvetiae 52:921–941.

———. 1959b. Über das Gebiss von *Cyamodus*. Vierteljahrsschrift der Naturforschenden Gesellschaft in Zürich 104:174–188.

———. 1959c. Ein neuer Pachypleurosaurier von der Stulseralp bei Bergün (Kt. Graubünden, Schweiz). Eclogae geologicae Helvetiae 52:639–658.

———. 1962a. Ein weiterer Schädel von *Macrocnemus bassanii* Nopcsa aus der anisischen Stufe der Trias des Monte San Giorgio (Kt. Tessin, Schweiz). Paläontologische Zeitschrift H. Schmidt-Festband:110–133.

———. 1962b. La position des nothosauridés dans le système des reptiles. Colloque International CNRS 104: 135–144.

———. 1963a. Wege der Reptiliensystematik. Paläontologische Zeitschrift 37:61–87.

———. 1963b. Paläontologie als stammesgeschichtliche Urkundenforschung; pp. 238–417 in G. Heberer (ed.), Die Evolution der Organismen, 3. Auflage. Gustav Fischer, Stuttgart.

———. 1964. Die Wirbeltierfauna der Trias der Tessiner Kalkalpen. Geologische Rundschau 53:393–412.

———. 1965. Sind die Reptilien stammesgeschichtlich eine Einheit. Umschau in Wissenschaft und Technik Heft 5:149–155.

———. 1966. Der Schädel von *Paranothosaurus amsleri* Peyer aus dem Grenzbitumenhorizont der anisisch-ladinischen Stufe der Trias des Monte San Giorgio (Kt. Tessin, Schweiz). Eclogae geologicae Helvetiae 59:517–540.

———. 1967. Das Problem der Euryapsida. Colloques Internationaux du Centre National de la Recherche Scientifique 163:335–348.

———. 1968. Alles Lebendige meinet den Menschen. Die Grabungen am Monte San Giorgio im Tessin. Reprint, Schweizer Spiegel, Heft Nr. 12, no pagination.

———. 1971. Über einen Schädel von *Askeptosaurus italicus* Nopcsa aus der mittleren Trias des Monte San Giorgio (Kt. Tessin, Schweiz). Abhandlungen des hessischen Landesamtes für Bodenforschung zu Wiesbaden 60:89–98.

———. 1974. Die Triasfauna der Tessiner Kalkalpen. Neujahrsblatt der Naturforschenden Gesellschaft in Zürich, 176. Stück. Leemann AG, Zurich, 119 pp.

———. 1980. Observations on temporal openings of reptilian skulls and the classification of reptiles; pp. 153–175 in L. Jacobs (ed.), Aspects of Vertebrate History. Museum of Northern Arizona Press, Flagstaff.

———. 1987. Die Triasfauna der Tessiner Kalkalpen. XXVI. *Lariosaurus lavizzarii* n.sp. (Reptilia, Sauropterygia). Abhandlungen der Schweizerischen Paläontologischen Gesellschaft 110:1–24.

———. 1988. Bemerkungen zur Ordnung der Thalattosauria (Reptilia). Eclogae geologicae Helvetiae 81:879-886.

———. 1989a. The relationships of the Placodontia. Neues Jahrbuch für Geologie und Paläontologie, Monatshefte 1989:17–22.

———. 1989b. Sauropterygia und Placodontia (Reptilia). Eclogae geologicae Helvetiae 82:1049–1052.

———. 1990. Über Nothosauria (Sauropterygia, Reptilia)— ein Diskussionsbeitrag. Paläontologische Zeitschrift 64:313–316.

———. 1994. Bemerkungen über Pachypleurosaurier aus der Mittleren Trias des Monte San Giorgio, Schweiz. Eclogae geologicae Helvetiae 87:1023–1027.

Kuhn-Schnyder, E., and H. Rieber. 1984. Ziele und Grenzen der Paläontologie. Die Naturwissenschaften 71:199–205.

Kummer, B. 1975. Biomechanik fossiler und rezenter Wirbeltiere. Natur und Museum 105:156–167.

Lane, J. A. 2010. The Morphology and Relationships of the Hybodont Shark *Tribodus limae*, with a Phylogenetic Analysis of Hybodont Sharks (Chondrichthyes, Hybodontiformes). PhD Dissertation, City University of New York.

Lauder, G. V. 1985. Aquatic feeding in lower vertebrates; pp. 210–229 in M. Hildebrandt, D. M. Bramble, K. F. Liem, and D. B. Wake, D. B. (eds.), Functional Vertebrate Morphology. Belknap Press of Harvard University Press, Cambridge, Massachusetts.

Lauder, G.. V., and T. Pendergast, T. 1992. Kinematics of aquatic prey capture in the snapping turtle, *Chelydra serpentina*. Journal of Experimental Biology 164:55–78.

Lauder, G. V., and H. B. Schaffer. 1986. Functional design of the feeding mechanism in lower vertebrates: unidirectional and bidirectional flow systems in the tiger salamander. Zoological Journal of the Linnean Society 88:277–290.

Laurin, M., and R. R. Reisz. 1995. A reevaluation of early amniote phylogeny. Zoological Journal of the Linnean Society 113:165–223.

Lautenschlager S., and J. B. Desojo. 2011. Reassessment of the Middle Triassic rauisuchian archosaurs *Ticinosuchus ferox* and *Stagonosuchus nyassicus*. Paläontologische Zeitschrift 85:357–381.

Lee, M. S. Y. 1993. The origin of the turtle body plan: bridging a famous morphological gap. Science 261:1716–1720.

———. 1996. Correlated progression and the origin of turtles. Nature 379:811–815.

Leidy, J. 1868. Notice of some reptilian remains from Nevada. Proceedings of the Academy of Natural Sciences Philadelphia 20:177–178.

Li, C. 2003. First record of a protorosaurid reptile (Order Protorosauria) from the Middle Triassic of China. Acta Geologica Sinica 77:419–423.

———. 1999. Ichthyosaur from Guizhou, China. Chinese Science Bulletin 44:1329–1333.

———. 2000. Placodont (Reptilia: Plcodontia) from the Upper Triassic of Guizhou, southwest China. Vertebrata PalAsiatica 38:314–317.

———. 2007. A juvenile *Tanystropheus* sp. (Protorosauria, Tanystropheidae) from the Middle Triassic of Guizhou, China. Vertebrata PalAsiatica 45:37–42.

———. 2010. Amazing reptile fossils from the marine Triassic of China. Bulletin of the Chinese Academy of Sciences 24:80–82.

Li., C., N. C. Fraser, O, Rieppel, and X.-C. Wu. 2018. A Triassic stem turtle with an edentulous beak. Nature 560: 476-479.

Li, C., D.-Y. Jiang, L. Cheng, X.-C. Wu, and O. Rieppel. 2014. A new species of *Largocephalosaurus* (Diapsida: Saurosphargidae), with implications for the morphological diversity and phylogeny of the group. *Geological Magazine* 151: 100–120.

Li, C., and O. Rieppel. 2002. A new cyamodontoid placodont from the Triassic of Guizhou, China. Chinese Science Bulletin 47:403–407.

Li, C., O. Rieppel, C. Cheng, and N. C. Fraser. 2016. The earliest herbivorous marine reptile and its remarkable jaw apparatus. Science Advances 2016;2:e1501659.

Li, C., O. Rieppel, and M. C. LaBarbera. 2004. A Triassic aquatic protorosaur with an extremely long neck. Science 305:1931.

Li, C., O. Rieppel, X.-C. Wu, L.-J. Zhao, and L.-T. Wang. 2011. A new Triassic marine reptile from southwestern China. Journal of Vertebrate Paleontology 31:303–312.

Li, C., X.-C. Wu, Y.-N. Cheng, T. Sato, and L.-T. Wang. 2006. An unusual archosaurian from the marine Triassic of China. Die Naturwissenschaften 93:200–206.

Li, C., X.-C. Wu, O. Rieppel, L.-T. Wang, and L.-J. Zhao. 2008. An ancestral turtle from the Late Triassic of southwestern China. Nature 456:497–501.

Li, C., X.-C. Wu, L.-J. Zhao, S. J. Nesbitt, M. R. Stocker, and L.-T. Wang. 2016. A new armored archosauriform (Diapsida: Archosauromorpha) from the marine Middle Triassic of China, with implications for the diversity of life styles of archosauriforms prior to the diversification of Archosauria. Science of Nature—Die Naturwissenschaften 103:95; DOI: 10.1007/s00114-016-1418-4.

Li, C., and H.-L. You. 2002. *Cymbospondylus* from the Upper Triassic of Guizhou, China. Vertebrata PalAsiatica 40:9–16.

Li, C., L.-J. Zhao, and L.-T. Wang. 2007. New species of *Macrocnemus* (Reptilia: Protorosauria) from the Middle Triassic of southwestern China and its paleogeographic implication. Science in China, Series D (Earth Sciences) 50:1601–1605.

Li, J.-L. 2006. A brief summary of the Triassic marine reptiles of China. Vertebrata PalAsiatica 44:99–108.

Li, J.-L., J. Liu, and O. Rieppel. 2002. A new species of *Lariosaurus* (Sauropterygia: Nothosauridae) from Triassic of Guizhou, Southwest China. Vertebrata PalAsiatica 40:114–126.

Li, J.-L., and O. Rieppel. 2004. A new nothosaur from Middle Triassic of Guizhou, China. Vertebrata PalAsiatica 42: 1–12.

Li, Z.-G., D.-J. Jiang, O. Rieppel, R. Motani, A. Tintori, Z.-Y. Sun, and C. Ji. 2016. A new species of *Xinpusaurus* (Reptilia, Thalattosauria) from the Ladinian (Middle Triassic) of Xingyi, Guizhou, southwestern China. Journal of Vertebrate Paleontology 36:4, e1218340; DOI: 10.1080/02724634.2016.1218340.

Lin, K., and O. Rieppel. 1998. Functional morphology and ontogeny of *Keichousaurus hui* (Reptilia, Sauropterygia). Fieldiana (Geology) N.S. 39:1–35.

Liu, G.-B, and G.-Z. Yin. 2008. Preliminary researches for mixosaurid fossils from Middle Triassic Guanling Formation in Panxian of Guizhou. Acta Palaeontologica Sinica 47:73-90. (In Chinese with English abstract)

Liu, G.-B., G.-Z. Yin, X.-H. Wang, S.-Y. Wang, and L.-Z. Huang. 2002. On a new marine reptile from Middle Triassic Yangliujing Formation of Guizhou, China. Geological Journal of China Universities 8:220–226.

Liu, J. 1999. New discovery of sauropterygian from Triassic of Guizhou, China. Chinese Science Bulletin 44:1312–1315.

———. 2001. Postcranial skeleton of *Xinpusaurus*; pp. 1–7 in T. Deng and Y. Wang (eds.), Proceedings of the Eighth Annual Meeting of the Chinese Society of Vertebrate Paleontology. Ocean Press, Beijing.

———. 2007. A juvenile specimen of *Anshunsaurus* (Reptilia: Thalattosauria). American Museum Noviates 3582:1–9.

———. 2013. On the taxonomy of *Xinpusaurus* (Reptilia: Thalattosauria). Vertebrata PalAsiatica 51:17–23.

Liu, J., S.-X. Hu, O. Rieppel, D.-Y. Jiang, M. Benton, N. P. Kelley, J. C. Aitchison, C.-Y. Zhou, W. Wen, J.-Y. Huang, T. Xie, and T. Lv. 2014. A gigantic nothosaur (Reptilia, Sauropterygia) from the Middle Triassic of SW China and its implications for the Triassic biotic recovery. Scientific Reports 4:7142; DOI: 10.1038/srep07142.

Liu, J., R. Motani, D.-Y. Jiang, S.-X. Hu, J. C. Aitchison, O. Rieppel, M. J. Benton, Q.-Y. Zhang, and C.-Y. Zhou. 2013. The first specimen of the Middle Triassic *Phalarodon atavus* (Ichthyosauria: Mixosauridae) from south China, showing postcranial anatomy and peri-Tethyan distribution. Palaeontology 56:849-866.

Liu, J., C. L. Organ, M. J. Benton, M. C. Brandley, and J. C. Aitchison. 2017. Live birth in an archosauromorph reptile. Nature Communications 8:14445; DOI: 10.1038/ncomms14445.

Liu J., and O. Rieppel. 2001. The second thalattosaur from the Triassic of Guizhou, China. Vertebrata PalAsiatica 39: 77–87.

———. 2005. Restudy of *Anshunsaurus huangguoshuensis* (Reptilia: Thalattosauria) from the Middle Triassic of Guizhou, China. American Museum Novitates 3488:1–34.

Liu, J., O. Rieppel, D.-Y. Jiang, J. C. Aitchison, R. Motani, Q.-Y. Zhang, C.-Y. Zhou, and Y.-Y. Sun. 2011. A new pachypleurosaur (Reptilia, Sauropterygia) from the lower Middle Triassic of southwestern China and the phylogenetic relationships of Chinese pachypleurosaurs. Journal of Vertebrate Paleontology 31:292–302.

Liu, J., L.-J. Zhao, C. Li, and T. He. 2013. Osteology of *Concavispina biseridens* (Reptla, Thalattosauria) from the Xiaowa Formation (Carnian), Guanling, Guizhou, China. Journal of Paleontology 87:341–350.

Liu, X.-Q., W.-B. Lin, O. Rieppel, Z.-Y. Sun, Z.-G. Li, H. Lu, and D.-Y. Jiang. 2015. A new specimen of *Diandongosaurus acutidentatus* (Sauropterygia) from the Middle Triassic of Yunnan, China. Vertebrata PalAsiatica 53:281–290.

Lombardo, C., and A. Tintori. 2005. Feeding specializations in Norian fishes. Annali dell' Università degli Studi di Ferrara, Museologia Scientifica e Naturalistica, volume speciale:1–9.

López-Arbarello, A., T. Bürgin, H. Furrer, and R. Stockar. 2016. New holostean fishes (Actinopterygii: Neopterygii) from the Middle Triassic of the Monte San Giorgio (Canton Ticino, Switzerland). PeerJ; DOI: 10.7717/peerj.2234.

Lovelace, D. M., and A. C. Doebbert. 2015. A new age constraint for the Early Triassic Alcova Limestone (Chugwater Group), Wyoming. Paleogeography, Paleoclimatology, Paleoecology 424:1–5.

Luo, Y.-M., and Y.-Y. Yu. 2002. The restudy on the skull and mandible of *Xinpusaurus suni*. Guizhou Geology 19:71–75.

Lurie, E. 1988. Louis Agassiz: A Life in Science. Johns Hopkins University Press, Baltimore, Maryland, 504 pp.

Lydekker, R. 1888. Catalogue of the Fossil Reptilia and Amphibia in the British Museum (Natural History), Part I. British Museum (Natural History), London, 309 pp.

———. 1889. Catalogue of the Fossil Reptilia and Amphibia in the British Museum (Natural History), Part II. British Museum (Natural History), London, 307 pp.

———. 1890. Catalogue of the Fossil Reptilia and Amphibia in the British Museum (Natural History), Part IV. British Museum, London, 295 pp.

Lyell, C. 1855. A Manual of Elementary Geology, or the Ancient Changes of the Earth and Its Inhabitants as Illustrated by Geological Monuments. Reprinted from the Fifth Edition, Greatly Enlarged. A. Appleton, New York, 647 pp.

Ma, L.-T., D.-Y. Jiang, O. Rieppel, R. Motani, and A. Tintori. 2015. A new pistosauroid (Reptilia, Sauropterygia) from the late Ladinian Xingyi marine reptile level, southwestern China. Journal of Vertebrate Paleontology 35:1, e881832.1-6; DOI: 10.1080/02724634.2014.881832.

Maisch, M. W. 2010. Phylogeny, systematics, and origin of the Ichthyosauria—the state of the art. Plaeodiversity 3:151–214.

———. 2014. On the morphology and taxonomic status of *Xinpusaurus kohi* JIANG et al., 2004 (Diapsida: Thalattosauria) from the Upper Triassic of China. Palaeodiversity 7:47–59.

Maisch, M. W., D.-Y. Jiang, W.-C. Hao, Y.-L. Sun, Z.-Y. Sun, and H. Stöhr. 2008. A well preserved skull of *Qianichthyosaurus zhoui* Li, 1999 (Reptilia: Ichthyosauria) from the Upper Triassic of China and the phylogenetic position of the Toretocnemidae. Neues Jahrbuch für Geologie und Paläontologie, Abhandlungen 248:257–266.

Maisch, M. W., and A. T. Matzke. 1997a. Observations of Triassic ichthyosaurs. Part I: Structure of the palate and mode of tooth implantation in *Mixosaurus cornalianus* (BASSANI, 1886). Neues Jahrbuch für Geologie und Paläontologie, Monatshefte 1997:717–732.

———.1997b. *Mikadocephalus gracilirostris* n. gen., n. sp., a new ichthyosaur from the Grenzbitumenzone (Anisian-Ladinian) of Monte San Giorgio (Switzerland). Paläontologische Zeitschrift 71:267–289.

———. 1998a. Observations on Triassic Ichthyosaurs. Part II: A new ichthyosaur with palatal teeth from Monte San Giorgio. Neues Jahrbuch für Geologie und Paläontologie, Monatshefte 1998:26–41.

———. 1998b. Observations on Triassic ichthyosaurs. Part IV: On the forelimb of *Mixosaurus* BAUR, 1887. Neues Jahrbuch für Geologie und Paläontologie, Abhandlungen 209:247–272.

———. 1999. Observations on Triassic ichthyosaurs. Part V: The skulls of *Mikadocephalus* and *Wimanius* reconstructed.

Neues Jahrbuch für Geologie und Paläontologie, Monatshefte 1999:345–356.

———. 2000. The Ichthyosauria. Stuttgarter Beiträge zur Naturkunde Serie B (Geologie und Paläontologie) 298:1–159.

Maisch, M. W., X.-R. Pan, Z.-Y. Sun, T. Cai, D.-P. Zhang, and L. Xie. 2010. Cranial osteology of *Guizhouichthyosaurus tangae* (Reptilia: Ichthyosauria) from the Upper Triassic of China. Journal of Vertebrate Paleontology 26:588–597.

Maisey, J. G. 1977. The fossil selachian fishes *Palaeospinax* Egerton, 1872, and *Nemacanthus* Agassiz, 1837. Zoological Journal of the Linnean Society 60:259–273.

———. 1978. Growth and form of finspines in hybodont sharks. Palaeontology 21:657–666.

———. 1979. Finspine morphogenesis in squalid and heterodontid sharks. Zoological Journal of the Linnean Society 66:161–183.

———. 1982. The anatomy and interrelationships of Mesozoic hybodont sharks. American Museum Novitates 2724:1–48.

———. 2008. The postorbital palatoquadrate articulation in elasmobranchs. Journal of Morphology 269:1022–1040.

———. 2011. The braincase of the Middle Triassic shark *Acronemus tuberculatus* (Bassani, 1866). Palaeontology 5:417–428.

———. 2012. What is an "elasmobranch"? The impact of paleontology in understanding elasmobranch phylogeny and evolution. Journal of Fish Biology 80:918–951.

Maisey, J. G., G. J. P. Naylor, and D. J. Ward. 2004. Mesozoic elasmobranchs, neoselachian phylogeny and the rise of modern elasmobranch diversity; pp. 17–56 in G. Arratia and A. Tintori (eds.), Mesozoic Fishes 3—Systematics, Paleoenvironments and Biodiversity. Friedrich Pfeil, Munich.

Mariani, E. 1923. Su un nuovo esemplare di *Lariosaurus balsami* Cur. trovato negli scisti di Perledo sopra Varenna (Lago di Como). Atti della Società Italiana di Scienze Naturali e del Museo Civico di Storia Naturale di Milano 62:218–225.

Mateer, N. J. 1976. On two new specimens of *Pachypleurosaurus* (Reptilia: Nothosauria). Bulletin of the Geological Institutions of the University of Uppsala N.S. 6:107–123.

Maxwell, E. E., T. Argyriou, R. Stockar, and H. Furrer. 2017. Re-evaluation of the ontogeny and reproductive biology of *Saurichthys* (Actinopterygii). Society of Vertebrate Paleontology 77th Annual Meeting Program and Abstracts, p. 160.

Maxwell, E. E., and M. W. Caldwell. 2016. First record of live birth in Cretaceous ichthyosaurs: closing an 80 million year gap. Proceedings of the Royal Society of London B (Suppl.) 270:S104–S107.

Maxwell, E. E., H. Furrer, and M. R. Sánchez-Villagra. 3013. Exceptional fossil preservation demonstrates a new mode of axial skeleton elongation in early ray-finned fishes. Nature Communications 4:2570; DOI: 10.1038/ncomms3570 |www.nature.com/naturecommunications.

Maxwell, E. E., C. Romano, F. Wu, and H. Furrer. 2015. Two new species of *Saurichthys* (Actinopterygii: Saurichthyidae) from the Middle Triassic of Monte San Giorgio, Switzerland, with implications for character evolution in the genus. Zoological Journal of the Linnean Society 173:887–912.

Mazin, J. M. 1982. Affinités et phylogénie des Ichthyopterygia. Geobios 6:85–98.

Mazin, J. M., and G. Pinna. 1993. Paleoecology of the armoured placodonts. Paleontologia Lombarda N.S. 2:83–91.

McGowan, C., and R. Motani, R. 2003. Ichthyopterygia. Handbook of Paleoherpetology, Part 8. Friedrich Pfeil, Munich, 175 pp.

Merriam, J. C. 1895. On some reptilian remains from the Triassic of northern California. American Journal of Science, Series 3, 50:55–57.

———. 1904. A new marine reptile from the Triassic of California. University of California Publications, Bulletin of the Department of Geology 3:419–421.

———. 1905. The Thalattosauria, a group of marine reptiles from the Triassic of California. Memoirs of the California Academy of Sciences 5:1–38.

———. 1908. Triassic Ichthyosauria with special reference to the American forms. Memoirs of the University of California 1:1–196.

———. 1910. The skull and dentition of a primitive ichthyosaurian from the Middle Triassic. University of California Publications, Bulletin of the Department of Geology 5:381–390.

Meyer, H. v. 1830. *Protorosaurus*. Isis von Oken 1830:517–519.

———. 1832. Palaeologica zur Geschichte der Erde und ihrer Geschöpfe. Siegmund Schmerber, Frankfurt a.M., 560 pp.

———. 1842. *Simosaurus*, die Stumpfschnauze, ein Saurier aus dem Muschelkalke von Lunéville. Neues Jahrbuch für Mineralogie, Geognosie, Geologie und Petrefakten-Kunde 1842:184–197.

———. 1847–1855. Zur Fauna der Vorwelt. Die Saurier des Muschelkalkes mit Rücksicht auf die Saurier aus buntem Sandstein und Keuper. Heinrich Keller, Frankfurt a.M., 165 pp.

———. 1849. Fische, Crustaceen, Echinodermen und andere Versteinerungen aus dem Muschelkalk Oberschlesiens. Palaeontographica 1:216–242.

———. 1861. Reptilien aus dem Stubensandstein des oberen Keupers. Palaeontographica 7:253–346.

Miles, R. S. 1971. Palaeozoic Fishes. Chapman and Hall, London, 259 pp.

Modesto, S. P., and H.-D. Sues. 2004. The skull of the Early Triassic archosauromorph reptile *Prolacerta broomi* and its phylogenetic significance. Zoological Journal of the Linnean Society 140:335–351.

Motani, R. 1999a. The skull and taxonomy of *Mixosaurus* (Ichthyopterygia). Journal of Paleontology 73:924–935.

———. 1999b. Phylogeny of the Ichthyopterygia. Journal of Vertebrate Paleontology 19:473–496.

———. 1999c. On the evolution and homologies of ichthyopterygian forefins. Journal of Vertebrate Paleontology 19:28–41.

———. 2005a. Detailed tooth morphology in a durophagous ichthyosaur captured by 3D laser scanner. Journal of Vertebrate Paleontology 25:462–465.

———. 2005b. Evolution of fish-shaped reptiles (Reptilia: Ichthyopterygia) in their physical environment and constraints. Annual Review of Earth and Planetary Sciences 33:395–420.

Motani, R., X.-H. Chen, D.-Y. Jiang, L. Cheng, A. Tintori, and O. Rieppel. 2015. Lunge feeding in early marine reptiles and first evolution of marine tetrapod feeding guilds. Scientific Reports 5:8900; DOI: 10.1038/srep08900.

Motani, R., D.-Y. Jiang, G.-B. Chen, A. Tintori, O. Rieppel, C. Ji, and J.-D. Huang. 2015. A basal ichthyosauriform with a short snout from the Lower Triassic of China. Nature 517:485–488.

Motani, R., D.-Y. Jiang, A. Tintori, Y.-L. Sun, W.-C. Hao, A. Boyd, S. Hinic-Frlog, L. Schmitz, J. Shin, and Z.-Y. Sun. 2008. Faunal assemblages of Anisian (Middle Triassic) marine reptiles from Panxian, Guizhou, China. Journal of Vertebrate Paleontology 28:900–903.

Motani, R., D.-Y. Jiang, A. Tintori, O. Rieppel, and G. Chen. 2014. Terrestrial origin of viviparity in Mesozoic marine reptiles. Triassic embryonic fossils. PLoS One DOI: 10.1371/journal.pone.0088640.

Motani, R., D.-Y. Jiang, A. Tintori, O. Rieppel, G.-B. Chen, and H. You. 2015a. Status of *Chaohusaurus chaoxianensis* (Chen, 1985). Journal of Vertebrate Paleontology 35:1, e892011.1-8; DOI: 10.1080/02724634.2014.892011.

Müller, J. 2003. Early loss and multiple return of the lower temporal arcade in diapsid reptiles. Die Naturwissenschaften 90:473–476.

———. 2005. The anatomy of *Askeptosaurus italicus* from the Middle Triassic of Monte San Giorgio and the interrelationships of thalattosaurs (Reptilia, Diapsida). Canadian Journal of Earth Sciences 42:1347–1367.

———. 2007. First record of a thalattosaur from the Upper Triassic of Austria. Journal of Vertebrate Paleontology 27:236–240.

Müller, J., S. Renesto, and S. E. Evans. 2005. The marine diapsid reptile *Endennasaurus* from the Upper Triassic of Italy. Palaeontology, 48: 15–30.

Münster, G. 1830. Über einige ausgezeichnete fossile Fischzähne aus dem Muschelkalk bei Bayreuth. Birner, Bayreuth, 4 pp.

———. 1834. Vorläufige Nachricht über einige neue Reptilien im Muschelkalke von Baiern. Neues Jahrbuch für Mineralogie, Geognosie, Geologie und Petrefaktenkunde 1834:521–527.

———. Ueber einige merkwürdige Fische aus dem Kupferschiefer und dem Muschelkalk. Beiträge zur Petrefacten-Kunde 1:114–121.

Mutter, R. J. 1998. Zur systematischen Stellung einiger Bezahnungsreste von *Acrodus georgii* sp. nov. (Selachii, Hybodontoidea) aus der Grenzbitumenzone (Mittlere Trias) des Monte San Giorgio (Kanton Tessin, Schweiz). Eclogae geologicae Helvetiae 91:513–519.

———. 2003. Case 3230. *Colobodus* Agassiz, 1844 (Osteichthyes, Perleidiformes): proposed designation of *C. bassanii* de Alessandri, 1910 as the type species, with designation of a neotype. Bulletin of Zoological Nomenclature 60:135–137.

———. 2004. The "perleidiform" family Colobodontidae: A review; pp. 197–208 in G. Arratia and A. Tintori (eds.), Mesozoic Fishes—Systematics, Paleoenvironments and Biodiversity. Friedrich Pfeil, Munich.

Mutter, R. J., J. Cartanyà, and S. A. U. Basaraba. 2008. New evidence of *Saurichthys* from the Lower Triassic with an evaluation of early saurichthyid diversity; pp. 103–127 in G. Arratia, H.-P. Schultze, and M. V. H. Wilson (eds.), Mesozoic Fishes 4—Homology and Phylogeny. Friedrich Pfeil, Munich.

Mutter, R. J., and A. Herzog. 2004. A new genus of Triassic actinopterygian with an evaluation of deepened flank scales in fusiform fossil fishes. Journal of Vertebrate Paleontology 24:794–801.

Neenan, J. M., N. Klein, and T. M. Scheyer. 2013. European origin of placodont marine reptiles and the evolution of crushing dentition in Placodontia. Nature Communications 4:1621; DOI: 10.1038/ncomms2633.

Neenan, J., C. Li, O. Rieppel, and T. M. Scheyer. 2015. The cranial anatomy of Chinese placodonts and the phylogeny of Placodontia (Diapsida: Sauropterygia). Zoological Journal of the Linnean Society 175:415–428.

Nelson, G. J. 1969. Gill arches and the phylogeny of fishes, with notes on the classification of vertebrates. Bulletin of the American Museum of Natural History 141:475–552.

Nelson, J. S. 2006. Fishes of the World. 4th ed. John Wiley, New York, 624 pp.

Nesbitt, S. J., 2011. The early evolution of archosaurs: relationships and the origin of major clades. Bulletin of the American Museum of Natural History 325:1–292.

Nesbitt, S. J., S. L. Brusatte, J. B. Desojo, A. Liparini, M. A. G. de França, J. C. Weinbaum, and D. J. Gower. 2013. Rauisuchia. Geological Society of London Special Publication 379:241–274.

Nesbitt, S. J., R. J. Butler, M. D. Ezcurra, P. M. Barett, M. R. Stocker, K. D. Angielczyk, R. M. H. Smith, C. A. Sidor, G. Niedzwiedzki, A. G. Sennikov, and A. J. Charig. 2017. The earliest bird-line archosaurs and the assembly of the dinosaur body plan. Nature 544:484–487.

Nicholls, E. L. 1999. A reexamination of *Thalattosaurus* and *Nectosaurus* and the relationships of the Thalattosauria (Reptilia: Diapsida). PaleoBios 19:1–29.

Nicholls, E. L., and D. B. Brinkman. 1993a. A new specimen of *Utatsusaurus* (Reptilia: Ichthyosauria) from the Lower Triassic Sulphur Mountain Formation of British Columbia. Canadian Journal of Earth Sciences 30:486–490.

———. 1993b. New thalattosaurs (Reptilia: Diapsida) from the Triassic Sulphur Mountain Formation of Wapiti Lake, British Columbia. Journal of Paleontology 67:263–278.

Nicholls, E. L., D. B. Brinkman, and C. M. Callaway. 1999. New material of *Phalarodon* (Reptilia: Ichthyopterygia) from the Triassic of British Columbia and its bearing on the interrelationships of mixosaurs. Palaeontographica A 252:1–22.

Nicholls, E. L., and C. M. Callaway. 1997. Preface; pp. xvii–xviii in C. M. Callaway and E. L. Nicholls (eds.), Ancient Marine Reptiles. Academic Press, San Diego, California.

Nicholls, E. L., and M. Manabe. 2004. Giant ichthyosaurs of the Triassic—a new species of *Shonisaurus* from the Pardonet Formation (Norian: Late Triassic) of British Columbia. Journal of Vertebrate Paleontology 24:838–849.

Nicholls, E. L., C. Wei, and M. Manabe. 2002. New material of *Qianichthyosaurus* Li, 1999 (Reptilia, Ichthyosauria) from the Late Triassic of Southern China, and implications for the distribution of Triassic Ichthyosaurs. Journal of Vertebrate Paleontology 22:759–765.

Nopcsa, F. 1923. Neubeschreibung des Trias-Pterosauriers. Palaeontologische Zeitschrift 5:160–181.

———. 1925. *Askeptosaurus*, ein neues Reptil der Trias von Besano. Centralblatt für Mineralogie, Geologie, und Paläontologie B 1925:265–267.

———. 1928. Palaeontological notes on reptiles. Geologica Hungaria, Ser. Palaeontologica 1:3–84.

———. 1930. Notizen über *Macrochnemus Bassanii* nov. gen. et spec. Centralblatt für Mineralogie, Geologie und Paläontologie B 1930:252–255.

———. 1931. *Macrocnemus*, nicht *Macrochnemus*. Centralblatt für Mineralogie, Geologie und Paläontologie B 1931:655–656.

Nosotti, S. 2007. *Tanystropheus longobardicus* (Reptilia, Protorosauria): re-interpretations of the anatomy based on new specimens from the Middle Triassic of Besano (Lombardy, northern Italy). Memorie della Società Italiana di Scienze Naturali e del Museo Civico di Storia Naturale di Milano 35 (3):1–88.

Nosotti, S., and O. Rieppel. 2003. *Eusaurosphargis dalsassoi* n. g. n. sp., a new, unusual diapsid reptile from the Middle Triassic of Besano (Lombardy, Northern Italy). Memorie della Società Italiana di Scienze Naturali e del Museo Civico di Storia Naturale di Milano 31:1–33.

Nosotti, S., and G. Teruzzi. 2008. I rettili di Besano-Monte San Giorgio. Natura, Rivista di Scienze Naturali 98:1–99.

O'Keefe, F. R. 2001. A cladistic analysis and taxonomic revision of the Plesiosauria (Reptilia: Sauropterygia). Acta Zoologica Fennica 213:1–63.

O'Keefe, F. R., and L. M. Chiappe. 2011. Viviparity and k-selected life history in a Mesozoic marine plesiosaur (Reptilia, Sauropterygia). Science 333:870–873.

O'Keefe, F. R., O. Rieppel, and P. M. Sander. 1999. Shape disassociation and inferred heterochrony in a clade of pachypleurosaurs. Paleobiology 25:504–517.

O'Keefe, F. R., and P. M. Sander. 1999. Paleontological paradigms and inferences of phylogenetic pattern: a case study. Paleobiology 25:518–533.

Olsen, P. E. 1979. A new eosuchian from the Newark Supergroup (Late Triassic—Early Jurassic) of North Carolina and Virginia. Postilla 176:1–14.

Ortlam, D. 1967. Fossile Böden als Leithorizonte für die Gliederung des Höheren Buntsandsteins im nördlichen Schwarzwald und südlichen Odenwald. Geologisches Jahrbuch 84:485–590.

Osborn, H. F. 1903. On the primary division of the Reptilia into two subclasses, Synapsida and Diapsida. Science 17:275–276.

Owen, R. 1841. Description of parts of the skeleton and teeth of five species of the genus *Labyrinthodon* (*Lab. leptognathus, Lab. pachygnathus,* and *Lab. ventricosus*, from the Coton-end and Cubbington Quarries of the lower Warwick Sandstone; *Lab. jaegeri*, from Guy's Cliff, Warwick; and *Lab. scutulatus*, from Leamington); with remarks on the probable identity of *Cheirotherium* with this genus of extinct Batrachians. Transactions of the Geological Society of London (2) 6:515–543.

———. 1842. Report on British Fossil Reptiles, Part II. Report of the Eleventh Meeting of the British Association for the Advancement of Science, 1842 (1841): 60–204.

———. 1858. Description of the skull and teeth of the *Placodus laticeps* OWEN, with indications of other new species of *Placodus*, and evidence of the saurian nature of that genus. Philosophical Transactions of the Royal Society of London 148:169–184.

———. 1860. Palaeontology; or, a systematic summary of extinct animals and their geologic remains. Adam and Charles Black, Edinburgh, 420 pp.

———. 1879. On the enthiodont Reptilia, with evidence of the species *Enthiodon uniseries* Ow. Quarterly Journal of the Geological Society of London 35:557–564.

Parrington, F. R. 1935. On *Prolacerta broomi* gen. et sp. n., and the origin of lizards. Annals and Magazine of Natural History (10) 16:197–205.

Patterson, C. 1973. Interrelationships of holosteans; pp. 233–305 in P. H. Greenwood, R. S. Miles, and C. Patterson (eds.), Interrelationships of Fishes. Academic Press, London.

Payne, J. L., and B. van de Schootbrugge. 2007. Life in Triassic oceans: links between planktonic and bethic recovery and radiation; pp. 165–189 in P. G. Falkowski and A. H. Knoll (eds.), Evolution of Primary Producers in the Sea. Elsevier, Amsterdam.

Peters, D. 2000. A reexamination of four prolacertiforms with implications for pterosaur phylogenetics. Rivista Italiana di Paleontologia e Stratigrafia 106:293–336.

Peyer, B. 1931a. Die Triasfauna der Tessiner Kalkalpen. IV. *Ceresiosaurus calcagnii* nov. gen. nov. spec. Abhandlungen der schweizerischen Paläontologischen Gesellschaft 51:1–68.

———. 1931b. Die Triasfauna der Tessiner Kalkalpen. I. Einleitung. Abhandlungen der schweizerischen Paläontologischen Gesellschaft 50:1–5.

———. 1931c. Die Triasfauna der Tessiner Kalkalpen. II. *Tanystropheus longobardicus* Bass sp. Abhandlungen der schweizerischen Paläontologischen Gesellschaft 50:7–110.

———. 1931d. Die Triasfauna der Tessiner Kalkalpen. III. Placodontia. Abhandlungen der schweizerischen Paläontologischen Gesellschaft 51:1–125.

———. 1931e. *Paraplacodus broilii* nov. gen. nov. sp., ein neuer Placodontier aus der Tessiner Trias. Vorläufige Mitteilung. Centralblatt für Mineralogie, Geologie und Paläontologie B 1931:190–192.

———. 1931f. *Macrocnemus*, nicht *Macrochnemus*. Centralblatt für Mineralogie, Geologie und Paläontologie B 1931:190–192.

———. 1932. Die Triasfauna der Tessiner Kalkalpen. IV. *Pachypleurosaurus edwardsii* Corn. spec. Abhandlungen der schweizerischen Paläontologischen Gesellschaft 52:1–18.

———. 1933. Die Triasfauna der Tessiner Kalkalpen. VII. Neubeschreibung der Saurier von Perledo. Abhandlungen der schweizerischen Paläontologischen Gesellschaft 53–54:1–130.

———. 1935. Die Triasfauna der Tessiner Kalkalpen. VIII. Weitere Placodontierfunde. Abhandlungen der schweizerischen Paläontologischen Gesellschaft 55:1–26.

———. 1936a. Die Triasfauna der Tessiner Kalkalpen. X. *Clarazia schinzi* nov. gen. nov. spec. Abhandlungen der schweizerischen Paläontologischen Gesellschaft 57:1–61.

———. 1936b. Die Triasfauna der Tessiner Kalkalpen. XI. *Hescheleria rübeli* nov. gen. nov. spec. Abhandlungen der schweizerischen paläontologischen Gesellschaft 58:1–48.

———. 1937. Die Triasfauna der Tessiner Kalkalpen. XII. *Macrocnemus bassanii* Nopcsa. Abhandlungen der schweizerischen Paläontologischen Gesellschaft 59:1–140.

———. 1939a. Die Triasfauna der Tessiner Kalkalpen. XIV. *Paranothosaurus amsleri* nov. gen. nov. spec. Abhandlungen der schweizerischen Paläontologischen Gesellschaft 62:1–87.

———. 1939b. Über die Rekonstruktion des Skelettes von *Tanystropheus*. Eclogae geologicae Helvetiae 32:203–209.

———. 1944. 1924–1944: Die Repilien vom Monte San Giorgio. Gebr. Fritz, Zurich, 95 pp.

———. 1946. Die schweizerischen Funde von *Asteracanthus* (*Strophodus*). Schweizerische Paläontologische Abhandlungen 64:1–101.

————. 1955. Die Triasfauna der Tessiner Kalkalpen. XVIII. *Helveticosaurus zollingeri* n. g. n. sp. Abhandlungen der schweizerischen Paläontologischen Gesellschaft 72: 1–50.

————. 1956. Über Zähne von Haramiyden, von Triconodonten und von wahrscheinlich synapsiden Reptilien aus dem Rhät von Hallau, Kt. Schaffhausen. Abhandlungen der schweizerischen Paläontologischen Gesellschaft 72:1–72.

————. 1957. Über die morphologische Deutung der Flossenstacheln einiger Haifische. Mitteilungen der naturforschenden Gesellschaft Bern N.F. 14:159–176.

Peyer, B., and E. Kuhn-Schnyder. 1955a. Placodontia; pp. 459–486 in J. Piveteau (ed.), Traîté de Paléontologie, Vol. 5. Masson, Paris.

————. 1955b. Squamates du Trias; pp. 578–605 in J. Piveteau (ed.), Traîté de Paléontologie, Vol. 5. Masson, Paris.

Peyer, H. C. 1963. Prof. Dr. Bernhard Peyer, 1885–1963; Publikationen bearbeitet von Prof. Dr. E. Kuhn-Schnyder, Zürich. Verhandlungen der Schweizerischen Naturforschenden Gesellschaft 143:242–249.

Pinna, G. 1967. La collezione di rettili triasici di Besano (Varese) del Museo Civico di Storia Naturale di Milano. Natura, Rivista di Scienze Naturali 58:178–192.

————. 1976. Osteologia del cranio del rettile placodonte *Placocheylanus stoppanii* (Osswald, 1930) basta su un nuovo esemplare del Retico Lombardo. Atti della Società Italiana di Scienze Naturali e del Museo Civico di Storia Naturale di Milano 117:3–45.

————.1980. *Drepanosaurus unguicaudatus*, nuovo genere nuova specie di Lepidosauro del Trias Alpino (Reptilia). Atti della Societá Italiana di Scienze Naturali e del Museo Civico di Storia Naturale di Milano 121:181–192.

————. 1984. Osteologia di *Drepanosaurus unguicaudatus*, Lepidosauro triasico del Sottordine Lacertilia. Memorie della Societá Italiana di Scienze Naturali e del Museo Civico di Storia Naturale di Milano 24:7–28.

————. 1989. Sulla regione temporo-jugale dei rettili placodonti e sulle relazioni fra placodonti e ittiotterigi. Atti della Società Italiana di Scienze Naturali e del Museo Civico di Storia Naturale di Milano 130:149–158.

————. 1990. Notes on the stratigraphy and geographical distribution of placodonts. Atti della Società Italiana di Scienze Naturali e del Museo Civico di Storia Naturale di Milano 131:145–156.

————. 1992. *Cyamodus hildegardis* Peyer, 1931 (Reptilia, Placodontia). Memorie della Società Italiana di Scienze Naturali e del Museo Civico di Storia Naturale di Milano 26:1–21.

————. 1993. The Norian reptiles of Northern Italy. Paleontologia Lombarda N.S. 2:115–124.

————. 1999. Placodontia (Reptilia Triadica). Fossilium Catalogus Animalia, Pars. 136. Backhuys Publ., Leiden, 75 pp.

Pinna, G., and P. Arduini. 1978. Un nuovo esemplare di *Ticinosuchus ferox* Krebs, rinvenuto nel giacimento triassico di Besano in Lombardia. Natura, Rivista di Scienze Naturali 69:73-80.

Pinna, G., and S. Nosotti. 1989. Anatomia, morphologia funzionale e paleoecologia del rettile placodonte *Psephoderma alpinum* Meyer, 1858. Memorie della Societá Italiana di Scienze Naturali e del Museo Civico di Storia Naturale di Milano 25:1–50.

Pinna, G., and G. Teruzzi, G. 1991. Il giacimento paleontologico di Besano. Natura, Rivista di Scienze Naturali 82:1–55.

Piveteau, J. 1944–45. Paléontologie de Madagascar XXV—Les poissons du Trias inférieur, la famille des Saurichthyidés. Annales de Paléontologie 31:79–89.

Plieninger,T. 1847a. Über ein neues Sauriergenus und die Einreihung der Saurier mit flachen, zweischneidigen Zähnen in eine Familie. Jahreshefte des Vereins für vaterländische Naturkunde in Württemberg 2:148–154.

————. 1847b. Nachträgliche Bemerkungen zu dem Vortrage (S. 148 dieses Heftes) über ein neues Sauriergenus und die Einreihung der Saurier mit flachen, zweischneidigen Zähnen in eine Familie. Jahreshefte des Vereins für vaterländische Naturkunde in Württemberg 2:247–254.

Preto, N., E. Kustatscher, and P. B. Wignall. 2010. Triassic climates—state of the art and perspectives. Paleogeography, Paleoclimatology, Paleoecology 290:1–10.

Pritchard, A. C., and S. Nesbitt. 2017. A bird-like skull in a Triassic diapsid reptile increases heterogeneity of the morphological and phylogenetic radiation of Diapsida. Royal Society Open Science 4:170499

Pritchard, A. C., A. H. Turner, S. J. Nesbitt, R. B. Irmis, and N. D. Smith. 2015. Late Triassic tanystropheids (Reptilia, Archosauromorpha) from northern New Mexico (Petrified Forest Member, Chinley Formation) and the biogeography, functional morphology, and evolution of Tanystropheidae. Journal of Vertebrate Paleontology 35:2, e911186; DOI: 10.1080/02724634.2014.911186.

Quenstedt, W. 1963. Fossilium Catalogus I: Animalia. Pars 102: Clavis bibliographica. W. Junk, 's-Gravenhage, 118 pp.

Radinsky, L. 1961. Tooth histology as a taxonomic criterion for cartilaginous fishes. Journal of Morphology 109: 3–92.

Reif, W.-E. 1973. Morphologie und Ultrastruktur des Hai-"Schmelzes." Zoologica Scripta 2:231–250.

————. 1976. Morphogenesis, pattern formation and function of the dentition of *Heterodontus* (Selachii). Zoomorphologie 83:1-47.

————. 1978a. Types of morphogenesis of the dermal skeleton in fossil sharks. Paläontologische Zeitschrift 52:110–128.

————. 1978b. Flow pattern in the boundary layer of Recent fast swimming sharks. Neues Jahrbuch für Geologie und Paläontologie, Abhandlungen 157:136–138.

Reif, W.-E., and F. Stein. 1999. Morphology and function of the dentition of *Henodus chelyops* Huene, 1936 (Placodontia, Triassic). Neues Jahrbuch für Geologie und Paläontologie, Monatshefte 1999:65–80.

Reiff, W. 1942. Ergänzungen zum Panzerbau von *Henodus chelyops* v.Huene. Palaeontographica 94:31–42.

Renesto, S. 1984. A new lepidosaur (Reptilia) from the Norian beds of the Bergamo Prealps. Preliminary note. Rivista Italiana di Paleontologia e Stratigrafia 90:165–176.

————. 1991. The anatomy and relationships of *Endennasaurus acutirostris* (Reptilia, Neodiapsida), from the Norian (Late Triassic) of Lombardy. Rivista Italiana di Paleontologia e Stratigrafia 97:409–430.

————. 1993. A juvenile *Lariosaurus* (Reptilia, Sauropterygia) from the Kalkschieferzone (uppermost Ladinian) near Viggiù (Varese, Northern Italy). Rivista Italiana di Paleontologia e Stratigrafia 99:199–212.

————. 1994. A new prolacertiform reptile from the Late Triassic of Northern Italy. Rivista Italiana di Paleontologia e Stratigrafia 100:285–306.

————. 2000. Bird-like head on a chameleon body: a new specimen of the enigmatic diapsid reptile *Megalancosaurus* from the Late Triassic of Northern Italy. Rivista Italiana di Paleontologia e Stratigrafia 106:157–180.

————. 2005. A new specimen of *Tanystropheus* (Reptilia Protorosauria) from the Middle Triassic of Switzerland and the ecology of the genus. Rivista Italiana di Paleontologia e Stratigrafia 111:377–394.

Renesto, S., and M. Avanzini. 2002. Skin remains in juvenile *Macrocnemus bassanii* Nopcsa (Reptilia, Prolacertiformes) from the Middle Triassic of Northern Italy. Neues Jahrbuch für Geologie und Paläontologie, Abhandlungen 224:31–48.

Renesto, S., G. Binelli, and H. Hagdorn. 2014. A new pachypleurosaur from the Middle Triassic Besano Formation of Northern Italy. Neues Jahrbuch für Geologie und Paläontologie, Abhandlungen 271:151–168.

Renesto, S., F. Confortini, E. Gozzi, M. Malzanni, and A. Paganoni. 2003. A possible rauisuchid (Reptilia, Archosauria) tooth from the Carnian (Late Triassic) of Lombardy (Italy). Rivista del Museo civico di Scienze Naturali "Enrico Caffi" Bergamo 22:109–114.

Renesto, S., and M. Felber. 2007. Un Pachipleurosauridae dai livelli centrali del Calcare di Meride in Canton Ticino (Svizzera). Geologia Insubrica 10:9–12.

Renesto, S., C. Lombardo, A. Tintori, and G. Danini. 2003. Nothosaurid embryos from the Middle Triassic of northern Italy: an insight into the viviparity of nothosaurs? Journal of Vertebrate Paleontology 23:957–960.

Renesto, S., J. A. Spielmann, S. G. Lucas, and F. T. Spagnoli. 2010. The taxonomy and paleobiology of the Late Triassic (Carnian—Norian: Adamanian—Apachean) drepanosaurs (Diapsida: Archosauromorpha: Drepanosauromorpha). New Mexico Museum of Natural History & Science, Bulletin 46:1–81.

Renesto, S., and R. Stockar. 2009. Exceptional preservation of embryos in the actinopterygian *Saurichthys* from the Middle Triassic of Monte San Giorgio, Switzerland. Swiss Journal of Geosciences 102:323–330.

————. . 2015. Prey content in a *Saurichthys* reveals the presence of advanced halecomorph fishes in the Middle Triassic of Monte San Giorgio. Neues Jahrbuch für Geologie und Paläontologie, Abhandlungen 278:95–107.

Renesto, S., and A. Tintori. 1995. Functional morphology and mode of life of the Late Triassic placodont *Psephoderma alpinum* Meyer from the Calcare di Zorzino (Lombardy, N Italy). Rivista Italiana di Paleontologia e Stratigrafia 101:37–48.

Repossi, E. 1902. Il Mixosauro degli strati triasici di Besano in Lombardia. Atti della Società Italiana di Scienze Naturali 41:361–372.

Rieber, H. 1968. Die Artengruppe der *Daonella elongata* Mojs. aus der Grenzbitumenzone der Mittleren Trias des Monte San Giorgio (Kt. Tessin, Schweiz). Paläontologische Zeitschrift 42:33–61.

————. 1969. Daonellen aus der Grenzbitumenzone der Mittleren Trias des Monte San Giorgio (Kt. Tessin, Schweiz). Eclogae geologicae Helvetiae 62:657–683.

————. 1973a. Cephalopoden aus der Grenzbitumenzone (Mittlere Trias) des Monte San Giorgio (Kt. Tessin, Schweiz). Schweizerische Paläontologische Abhandlungen 93:1–96.

————. 1973b. Ergebnisse paläontologisch-stratigraphischer Untersuchungen in der Grenzbitumenzone (Mittlere Trias) des Monte San Giorgio (Kanton Tessin, Schweiz). Eclogae geologicae Helvetiae 66:667–685.

————. 1975. Der Posidonienschiefer (oberer Lias) von Holzmaden und die Grenzbitumenzone (mittlere Trias) des Monte San Giorgio: Ein Vergleich zweier Lagerstätten fossiler Wirbeltiere. Jahreshefte der Gesellschaft für Naturkunde Württemberg 130:163–190.

————. 1982. The formation of the bituminous layers of the Middle Triassic of Ticino (Switzerland); pp. 527 in G. Einsele and A. Seilacher (eds.), Cyclic and Event Stratification. Springer, Berlin, [abstract].

————. 1995. Emil Kuhn-Schnyder † 29.4.1905–30.7.1994. Paläontologische Zeitschrift 69:313–320.

Rieber, H., and H. Lanz, H. 1999. Geschichte der Grabungstätigkeit am Monte San Giorgio; pp. 77–85 in H. Rieber (ed.), Paläontologie in Zürich. Fossilien und ihre Erforschung in Geschichte und Gegenwart. Zoologisches Museum der Universität, Zurich.

Rieber, H., and L. Sorbini. 1983. Middle Triassic bituminous shales of Monte San Giorgio (Tessin, Switzerland); pp. 1–17 in H. Rieber and L. Sorbini (eds.), First International Congress on Paleoecology, Excursion 11A, Guidebook.

Rieppel, O. 1980a. Additional specimens of *Saurichthys madagascariensis* Piveteau from the Eotrias of Madagascar. Neues Jahrbuch für Geologie und Paläontologie, Monatshefte 1980:43–51.

————. 1980b. A new coelacanth from the Middle Triassic of Monte San Giorgio, Switzerland. Eclogae geologicae Helvetiae 73:921–939.

————. 1981. The hybodontiform sharks from the Middle Triassic of Mte. San Giorgio, Switzerland. Neues Jahrbuch für Geologie und Paläontologie, Abhandlungen 161:324–353.

————. 1982. A new genus of shark from the Middle Triassic of Monte San Giorgio, Switzerland. Palaeontology 25:399–412.

————. 1985a. Die Triasfauna der Tessiner Kalkalpen. XXV. Die Gattung *Saurichthys* (Pisces, Actinopterygii) aus der mittleren Trias des Monte San Giorgio, Kanton Tessin. Schweizerische Paläontologische Abhandlungen 108:1–81.

————. 1985b. A second actinistian from the Middle Triassic of Monte San Giorgio, Kanton Tessin, Switzerland. Eclogae geologicae Helvetiae 78:707–713.

————. 1987a. The Pachypleurosauridae: an annotated bibliography. With comments on some lariosaurs. Eclogae geologicae Helvetiae 80:1105–1118.

————. 1987b. *Clarazia* and *Hescheleria:* a re-investigation of two problematical reptiles from the Middle Triassic of Monte San Giorgio, Switzerland. Palaeontographica A 195:101–129.

————. 1989a. *Helveticosaurus zollingeri* Peyer (Reptilia, Diapsida): skeletal paedomorphosis, functional anatomy and systematic affinities. Palaeontographica A 208:123–152.

————. 1989b. A new pachypleurosaur (Reptilia: Sauropterygia) from the Middle Triassic of Monte San Giorgio, Switzerland. Philosophical Transactions of the Royal Society of London B 323:1–73.

————. 1989c. The hindlimb of *Macrocnemus bassanii* Nopcsa (Reptilia: Prolacertiformes): development and functional anatomy. Journal of Vertebrate Paleontology 9:373–387.

————. 1992. New species of the genus *Saurichthys* (Pisces: Actinopterygii) from the Middle Triassic of Monte San Giorgio (Switzerland), with comments on the phylogenetic interrelationships of the genus. Palaeontographica A 221:63–94.

———. 1994a. Osteology of *Simosaurus gaillardoti*, and the phylogenetic interrelationships of stem-group Sauropterygia. Fieldiana (Geology) N.S. 28:1–85.

———. 1994b. Middle Triassic reptiles from Monte San Giorgio: recent results and future potential for analysis. Paleontologia Lombarda N.S. 2:131–144.

———. 1998a. *Corosaurus alcovensis* Case and the phylogenetic interrelationships of Triassic stem-group Sauropterygia. Zoological Journal of the Linnean Society 124:1–41.

———. 1998b. The status of the sauropterygian reptile genera *Ceresiosaurus*, *Lariosaurus* and *Silvestrosaurus* from the Middle Triassic of Europe. Fieldiana (Geology) N.S. 38:1–46.

———. 1998c. The systematic status of *Hanosaurus hupehensis* (Reptilia, Sauropterygia) from the Triassic of China. Journal of Vertebrate Paleontology, 18:545–557.

———. 1998d. Ichthyosaur remains (Reptilia, Ichthyosauria) from the Middle Triassic of Makhtesh Ramon, Negev, Israel. Neues Jahrbuch für Geologie und Paläontologie, Monatshefte 1998:537–544.

———. 1999a. Phylogeny and Paleobiogeography of Triassic Sauropterygia: problems resolved and unresolved. Paleogeography, Paleoclimatology, Paleoecology 153:1–15.

———. 1999b. The sauropterygian genera *Chinchenia*, *Kwangsisaurus*, and *Sanchiaosaurus* from the Lower and Middle Triassic of China. Journal of Vertebrate Paleontology 19:321–337.

———. 2000a. Sauropterygia I: Placodontia, Pachypleurosauria, Nothosauroidea, Pistosauroidea. Encyclopedia of Paleoherpetology, Part 12A. Friedrich Pfeil, Munich, 134 pp.

———. 2000b. *Paraplacodus* and the phylogeny of the Placodontia (Reptilia: Sauropterygia). Zoological Journal of the Linnean Society 130:635–659.

———. 2001a. The cranial anatomy of *Placochelys placodonta* Jaekel, 1902, and a review of the Cyamodontoidea (Reptilia, Placodontia). Fieldiana (Geology) N.S. 45:1-104.

———. 2001b. A new species of *Nothosaurus* (Reptilia: Sauropterygia) from the Upper Muschelkalk (Lower Ladinian) of southwestern Germany. Palaeontographica A 263:137–161.

———. 2001c. A new species of *Tanystropheus* (Reptilia: Protorosauria) from the Middle Triassic of Makhtesh Ramon, Israel. Neues Jahrbuch für Geologie und Paläontologie, Abhandlungen 221:271–287.

———. 2002a. Feeding mechanics in Triassic stem-group sauropterygians: the anatomy of a successful invasion of Mesozoic seas. Zoological Journal of the Linnean Society 135:33–63.

———. 2002b. The dermal armor of cyamodontoid placodonts (Reptilia, Sauropterygia): morphology and systematic value. Fieldiana (Geology) N.S. 46:1–41.

Rieppel, O., N. C. Fraser, and S. Nosotti. 2003. The monophyly of Protorosauria (Reptilia, Archosauromorpha): a preliminary analysis. Atti della Società Italiana di Scienze Naturali e del Museo Civico di Storia Naturale di Milano 144:359–382.

Rieppel, O., and R. W. Gronowski. 1981. The loss of the lower temporal arcade in diapsid reptiles. Zoological Journal of the Linnean Society 72:203–217.

Rieppel, O., and H. Hagdorn. 1997. Paleobiogeography of Middle Triassic Sauropterygia in Central and Western Europe; pp. 121–144 in J. M. Callaway and E. L. Nicholls (eds.), Sea Reptiles of the Past. Academic Press, San Diego,California.

Rieppel, O., D.-Y. Jiang, N. C. Fraser, W.-C. Hao, R. Motani, Y.-L. Sun, and Z.-Y. Sun. 2010. *Tanystropheus* cf. *T. longobardicus* from the early Late Triassic of Guizhou Province, southwestern China. Journal of Vertebrate Paleontology 30:1082–1089.

Rieppel, O., C. Li, and N. C. Fraser. 2008. The skeletal anatomy of the Triassic protorosaur *Dinocephalosaurus orientalis* Li, from the Middle Triassic of Guizhou Province, southern China. Journal of Vertebrate Paleontology 28:95–110.

Rieppel, O., J.-L. Li, and J. Liu. 2003. *Lariosaurus xingyiensis* (Reptilia, Sauropterygia) from the Triassic of China. Canadian Journal of Earth Sciences 40:621–634.

Rieppel, O., and K. Lin. 1995. Pachypleurosaurs (Reptilia: Sauropterygia) from the Lower Muschelkalk, and a review of the Pachypleurosauroidea. Fieldiana (Geology) N.S. 32:1–44.

Rieppel, O., and J. Liu. 2006. On *Xinpusaurus* (Reptilia: Thalattosauria). Journal of Vertebrate Paleontology 26:200–204.

Rieppel, O., J. Liu, and H. Bucher. 2000. The first record of a thalattosaur reptile from the Late Triassic of Southern China (Guizhou Province, PR China). Journal of Vertebrate Paleontology 20:507–514.

Rieppel, O., J. Liu, and C. Li. 2006. A new species of the thlattosaur genus *Anshunsaurus* (Reptilia: Thalattosauria) from the Middle Triassic of Guizhou Province, southwestern China. Vertebrata PalAsiatica 44:285–296.

Rieppel, O., J.-M. Mazin, and E. Tchernov. 1997. Speciation along rifting continental margins: a new nothosaur from the Negev (Israel). Comptes Rendus de l'Académie des Sciences (2) 325:991–997.

Rieppel, O., J.-M. Mazin, and E. Tchernov. 1999. Sauropterygia from the Middle Triassic of Makhtesh Ramon, Negev, Israel. Fieldiana (Geology) N.S. 40:1–85.

Rieppel, O., J. Müller, and J. Liu. 2005. Rostral structure in Thalattosauria (Reptilia, Diapsida). Canadian Journal of Earth Sciences 42:2081–2086.

Rieppel, O., P. M. Sander, and G. W. Storrs. 2002. The skull of the pistosaur *Augustasaurus* from the Middle Triassic of northwestern Nevada. Journal of Vertebrate Paleontology 22:577–592.

Rieppel, O., and R. Wild. 1996. A revision of the genus *Nothosaurus* (Reptilia, Sauropterygia) from the Germanic Triassic, with comments on the status of *Conchiosaurus clavatus*. Fieldiana (Geology) N.S. 34:1–82.

Röhl, H. J., A. Schmid-Röhl, H. Furrer, A. Frimmel, A. Oschmann, and L. Schwark. 2001. Microfacies, geochemistry and paleoecology of the Middle Triassic Grenzbitumenzone from Monte San Giorgio (Canton Ticino, Switzerland). Geologia Insubrica 6:1–13.

Romano, C., and W. Brinkmann. 2009. Reappraisal of the lower actinopterygian *Birgeria stensiöi* ALDINGER, 1931 (Osteichthyes; Birgeriidae) from the Middle Triassic of Monte San Giorgio. Neues Jahrbuch für Geologie und Paläontologie, Abhandlungen 252:17–31.

Romano, C., I. Jogan, J. Jenks, I. Jerjen, and W. Brinkmann. 2012. *Saurichthys* and other fossil fishes from the late Smithian (Early Triassic) of Bear Lake County (Idaho, USA), with a discussion of saurichthyid paleogeography and evolution. Bulletin of Geosciences 87:543–570.

Romer, A. S. 1933. Vertebrate Paleontology. University of Chicago Press, Chicago, 491 pp.

———. 1945. Vertebrate Paleontology. 2nd ed. University of Chicago Press, Chicago, 687 pp.

———. 1956. Osteology of the Reptiles. University of Chicago Press, Chicago, 772 pp.

———. 1966. Vertebrate Paleontology. 3rd ed. University of Chicago Press, Chicago, 468 pp.

———. 1968. Notes and Comments on Vertebrate Paleontology. University of Chicago Press, Chicago, 304 pp.

———. 1971. Unorthodoxies in reptilian phylogeny. Evolution 25:103–112.

Rusconi, M. R., C. Lombardo, and A. Tintori. 2007. Colobodontidae from the Upper Triassic (Carnian) of Friuli Venezia Giulia (Udine, NE Italy). Gortania—Atti del Museo Friulano di Storia Naturale 28:59–72.

Saller, F., S. Renesto, and F. Dalla Vecchia. 2013. First record of *Langobardisaurus* (Diapsida, Protorosauria) from the Norian (Late Triassic) of Austria, and a revision of the genus. Neues Jahrbuch für Geologie und Paläontologie, Abhandlungen 268:83–95.

Sander, P. M. 1988. A fossil reptile embryo from the Middle Triassic of the Alps. Science 239:780–783.

———. 1989a. The pachypleurosaurids (Reptilia: Nothosauria) from the Middle Triassic of Monte San Giorgio (Switzerland), with the description of a new species. Philosophical Transactions of the Royal Society of London B 325:561–670.

———. 1989b. The large ichthyosaur *Cymbospondylus buchseri*, sp. nov., from the Middle Triassic of Monte San Giorgio (Switzerland), with a survey of the genus in Europe. Journal of Vertebrate Paleontology 9:163–173.

———. 1992. *Cymbospondylus* (Shastasauridae: Ichthyosauria) from the Middle Triassic of Spitsbergen: filling a biogeographic gap. Journal of Paleontology 66:332–337.

———. 2000. Ichthyosauria: their diversity, distribution, and phylogeny. Paläontologische Zeitschrift 74:1–35.

Sander, P. M., X. Chen, L. Cheng, and X. Wang. 2011. Short-snouted toothless ichthyosaur from China suggests Late Triassic diversification of suction feeding ichthyosaurs. PLoS ONE 6:e19480; DOI: 10.1371/journal.pone.0019480.

Sander, P. M., and C. Faber. 1998. New finds of *Omphalosaurus* and a review of Triassic ichthyosaur paleobiology. Paläontologische Zeitschrift 72:149–162.

Sander, P. M., N. Klein, P. C. H. Albers, C. Bickelmann, and H. Winkelhorst. 2014. Postcranial morphology of a basal Pistosauroidea (Sauropterygia) from the Lower Muschelkalk of Winterswijk, the Netherlands. Paläontologische Zeitschrift 88:55–71.

Sander, P. M., O. Rieppel, and H. Bucher. 1997. A new pistosaurid (Reptilia: Sauropterygia) from the Middle Triassic of Nevada and its implications for the origin of plesiosaurs. Journal of Vertebrate Paleotology 17:526–533.

Sanz, J. L. 1983. Los Nothosaurios (Reptilia, Sauropterygia) espanoles. Estudios geológicos 39:193–215.

Sato, T., Y.-N. Cheng, X.-C. Wu, and C. Li. 2010. Osteology of *Yunguisaurus* Cheng et al., 2006 (Reptilia; Sauropterygia), a Triassic pistosauroid from China. Journal of Vertebrate Paleontology 14:179–195.

———. 2014. A new specimen of the Triassic pistosauroid *Yunguisaurus*, with implications for the origin of Plesiosauria (Reptilia, Sauropterygia). Palaeontology 57:55–76.

Sato, T., Y.-N. Cheng, X.-C. Wu, and H.-Y. Shan. 2014. *Diadongosaurus acutidentatus* Shang, Wu & Li, 2011 (Diapsid: Sauropterygia) and the relationships of Chinese eosauropterygians. Geological Magazine 151:121–133.

Schatz, W. 2001. Taxonomic significance of biometric characters and the consequences for classification and biostratigraphy, exemplified through moussoneliform daonellas (*Daonella*, Bivalvia; Triassic). Paläontologische Zeitschrift 75:51–70.

Scheyer, T. M., and J. B. Desojo. 2011. Paleohistology and external microanatomy of rauisuchian osteoderms (Archosauria: Pseudosuchia). Palaeontology 54:1289–1302.

Scheyer, T. M., J. M. Neenan, T. Bodogan, H. Furrer, C. Obrist, and M. Plamondon. 2017. A new, exceptionally preserved juvenile specimen of *Eusaurosphargis dalsassoi* (Diapsida) and implications for Mesozoic marine diapsid phylogeny. Scientific Reports 7:4406; DOI: 10.1038/s41598-017-04514-x.

Scheyer, T. M., C. Romano, J. Jenks, and H. Bucher. 2014. Early Triassic marine biotic recovery: the predator's perspective. PLoS ONE 9:e88987; DOI: 10.1371/journal.pone.0088987.

Schinz, H., and A. Wolfer. 1931. Notizen zur schweizerischen Kulturgeschichte. Vierteljahrsschrift der Naturforschenden Gesellschaft in Zürich 76: 435–532.

Schmid, L., and M. R. Sánchez-Villagra. 2010. Potential genetic bases of morphological evolution in the Triassic fish *Saurichthys*. Journal of Experimental Zoology (Molecular and Developmental Evolution) 314B:519–526.

Schmidt, M. 1928. Die Lebewelt unserer Trias. Ferdinand Rau, Öhringen, 461 pp.

Schoch, R., and H.-D. Sues. 2015. A Middle Triassic stem-turtle and the evolution of the turtle body plan. Nature 523:584–587.

Schroeder, H. 1914. Wirbeltiere der Rüdersdorfer Trias. Abhandlungen der Königlich Preussischen Geologischen Landesanstalt, Neue Folge 65:1–98.

Schwarz, W. 1970. Die Triasfauna der Tessiner Kalkalpen XX. *Birgeria stensiöi* Aldinger. Schweizerische Paläontologische Abhandlungen 89:1–93.

Seeley, H. G. 1882. On *Neusticosaurus pusillus* (Fraas), an amphibious reptile having affinities with terrestrial Nothosauria and with marine Plesiosauria. Quarterly Journal of the Geological Society of London 38:350–366.

———. 1889. Researches on the structure, organization, and classification of the fossil Reptilia. VI. On the anomodont Reptilia and their allies. Philosophical Transactions of the Royal Society of London B 180:215–296.

Seilacher, A., W.-E. Reif, and F. Westphal. 1985. Sedimentological, ecological and temporal patters of fossil Lagerstätten. Philosophical Transactions of the Royal Society of London B 311:5–23.

Sennikov, A. G. 2011. New tanystropheids (Reptilia: Archosauromorpha) from the Triassic of Europe. Paleontological Journal 45:90–104.

Seymour, R. S. 1982. Physiological adaptations to aquatic life; pp. 1–51 in C. Gans and F. H. Pough (eds.), Biology of the Reptilia, Vol. 13, Physiology D. Academic Press, London.

Shang, Q.-H. 2006. A new species of *Nothosaurus* from the early Middle Triassic of Guizhou, China. Vertebrata PalAsiatica 44:237–249.

Shang, Q.-H., and C. Li. 2009. On the occurrence of the ichthyosaur *Shastasaurus* in the Guanling bota (Late Triassic), Guizhou, China. Vertebrata PalAsiatica 47:178–193.

———. 2013. The sexual dimorphism of *Shastasaurus tangae* (Reptilia: Ichthyosauria) from the Triassic Guanling biota, China. Vertebrata PalAsiatica 51:253–264.

———. 2015. A new small-sized eosauropterygian (Diapsida: Sauropterygia) from the Middle Triassic of Luoping, Yunnan, southwestern China. Vertebrata PalAsiatica 53:265–280.

Shang, Q.-H., C. Li, and X.-C. Wu. 2017. New information on *Dianmeisaurus gracilis* Shang & Li, 2015. Vertebrata PalAsiatica 55:145–161.

Shang, Q.-H., X.-C. Wu, and C. Li. 2011. A new eosauropterygian from Middle Triasssic of eastern Yunnan Province, southwestern China. Vertebrata PalAsiatica 4:155–171.

Sickler, F. K. L. 1835. Sendschreiben an J. F. Blumenbach über die höchst merkwürdigen, vor einigen Monaten erst entdeckten Reliefs der Fährten urweltlicher grosser und unbekannter Thiere in den Hessberger Sandsteinbrüchen bei der Stadt Hildburghausen. Jahrbuch für Mineralogie, Geognosie, Geologie und Petrefaktenkunde 1835:230–234.

Simpson, G. G. 1953. The Major Features of Evolution. Columbia University Press, New York, 434 pp.

———. 1961. Principles of Animal Taxonomy. Columbia University Press, New York, 247 pp.

Snyder, R. C. 1954. The anatomy and function of the pelvic girdle and hind limb in lizard locomotion. American Journal of Anatomy 95:1–45.

———. 1962. Adaptations for bipedal locomotion of lizards. American Zoologist 2:191–203.

Sobral, G., R. B. Sookias, B. A. S. Bhullar, R. Smith, R. J. Butler, and J. Müller. 2016. New information on the briancase and inner ear of *Euparkeria capensis* Broom: implications for diapsid and archosaur evolution. Royal Society Open Science 3:160072; DOI: 10.1098/rsos.160072.

Soergel, W. 1925. Die Fährten der Chirotheria. Gustav Fischer, Jena, 92 pp.

Steel, R. 1973. Crocodylia. Encyclopedia of Paleoherpetology, Part 16. Gustav Fischer, Stuttgart, 116 pp.

Stefani, M., P. Arduini, A. Garassino, G. Pinna, G. Teruzzi, and G. L. Trombetta. 1992. Paleoenvironment of extraordinary fossil biota from the Upper Triasssic of Italy. Atti della Società Italiana di Scienze Naurali e del Museo Civico di Storia Naturale di Milano 132:309–335.

Stensiö, A. E. 1916. Uber einige Fischreste aus der Cava Tre Fontane, Tessin. Bulletin of the Geological Institution of the University of Upsala 15:13–34.

———. 1919. Einige Bemerkungen über die systematische Stellung von *Saurichthys mougeoti* Agassiz. Senckenbergiana 1:177–181.

———. 1925. Triassic fishes from Spitzbergen, Part II. Kungl. Svenska Vetenskaps Academiens Handlingar (3) 2:1–261.

Stockar, R. 2010. Facies, depositional environment, and paleoecology of the Middle Triassic Cassina beds (Meride Limestone, Monte San Giorgio, Switzerland). Swiss Journal of Geosciences 103:101–119.

Stockar, R., and S. Rensto. 2011. Co-occurrence of *Neusticosaurus edwardsii* and *N. peyeri* (Reptilia) in the Lower Meride Limestone (Middle Triassic, Monte San Giorgio). Swiss Journal of Geosciences 104 (Suppl. 1): S167–S178.

Stone, R. 2010. Excavation yields tantalizing hints of earliest marine reptiles. Science 330:1164–1165.

Stoppani, A. 1860–1865. Géologie et Paléontologie des Couches a *Avicula contorta* en Lombardie. Bernardoni, Milan, 267 pp.

———. 1863. Sulle ricerche fatte a spese della Società nelle Palafitte del Lago di Varese e negli schist bituminosi di Besano. Atti della Società Italiana di Scienze Naturali 5:423–434.

Storrs, G. W. 1991a. Anatomy and relationships of *Corosaurus alcovensis* (Diapsida: Sauropterygia) and the Triassic Alcova Limestone of Wyoming. Bulletin of the Peabody Museum of Natural History 44:1–151.

———. 1991b. Note on a second occurrence of thalattosaur remains (Reptilia: Neodiapsida) in British Columbia. Canadian Journal of Earth Sciences 28:2065–2068.

———. 1993a. Function and phylogeny in sauropterygian (Diapsida) evolution. American Journal of Science 293A:63–90.

———. 1993b. The systematic position of *Silvestrosaurus* and a classification of Triassic sauropterygians (Neodiapsida). Paläontologische Zeitschrift 67:177–191.

Strohl, J. 1940. Karl Hescheler, 1868–1940. Verhandlungen der Schweizerischen Naturforschenden Gesellschaft 120:445–450.

Sues, H.-D. 1987a. On the skull of *Placodus gigas* and the relationships of the Placodontia. Journal of Vertebrate Paleontology 7:138–144.

———. 1987b. Postcranial skeleton of *Pistosaurus* and interrelationships of the Sauropterygia (Diapsida). Zoological Journal of the Linnean Society 90:109–131.

Sun, Z.-Y, W.-C. Hao, Y.-L. Sun, and D.-Y. Jiang. 2009. The conodont genus *Nicoraella* and a new species from the Anisian of Guizhou, South China. Neues Jahrbuch für Geologie und Paläontologie, Abhandlungen 252:227–235.

Sun, Z.-Y, Y.-L. Sun, W.-C. Hao, and D.-Y. Jiang. 2006. Conodont evidence for the age of the Triassic Panxian Fauna, Guizhou, China. Acta Geologica Sinica 80:621–630.

Sun, Z.-Y., A. Tintori, C. Lombardo, D.-Y. Jiang, W.-C. Hao, Y.-L. Sun, F. Wu, and M. Rusconi. 2008. A new species of the genus *Colobodus* Agassiz, 1844 (Osteichthyes, Actinopyerygii) from the Pelsonian (Anisian, Middle Triassic) of Guizhou, South China. Rivista Italiana di Paleontologia e Stratigrafia 114:363–376.

Taylor, M. A. 1987. How tetrapods feed in water: a functional analysis by paradigm. Zoological Journal of the Linnean Society 91:171–195.

———. 1989. Neck and neck. Nature 341:688–689.

———. 1993. Stomach stones for feeding or buoyancy? The occurrence and function of gastroliths in marine tetrapods. Philosophical Transactions of the Royal Society of London B 341:163–175.

———. 1997. Before the dinosaur: the historical significance of the fossil marine reptiles; pp. xix–xlvi in J. M. Callaway and E. L. Nicholls (eds.), Ancient Marine Reptiles. Academic Press, San Diego, California.

———. 2000. Functional significance of bone ballast in the evolution of buoyancy control strategies by aquatic tetrapods. Historical Biology 14:15–31.

Tintori, A. 1990. The vertebral column of the Triassic fish *Saurichthys* (Actinopterygii) and its stratigraphical significance. Rivista Italiana di Paleotologia e Stratigrafia 96:93–102.

———. 1992. Fish taphonomy and Triassic anoxic basins from the Alps: a case study. Rivista Italiana di Paleotologia e Stratigrafia 97:393–408.

———. 2013. A new speccies of *Saurichthys* (Actinopterygii) from the Middle Triassic (early Ladinian) of the northern Grigna Mounain (Lombardy, Italy). Rivista Italiana di Paleontologia e Stratigrafia 119:287–302.

Tintori, A., and M. Felber. 2015. I vertebrati marini del Triassico medio nel XXI secolo: dal Monte San Giorgio alla Cina. Geologia Insubrica 11:63–80.

Tintori, A., T. Hitij, D.-Y. Jiang, C. Lombardo, and Z.-Y. Sun. 2014. Triassic actinopterygian fishes: the recovery after the end-Permian crisis. Integrative Zoology 9:394–411.

Tintori, A., and C. Lombardo. 1999. Late Ladinian fish faunas from Lombardy (North Italy): stratigraphy and paleobiology; pp. 495–504 in G. Arratia and G. Viohl (eds.), Mesozoic Fishes—Systematics and Paleoecology. Friedrich Pfeil, Munich.

Tintori, A., G. Muscio, and S. Nardon. 1985. The Triassic fossil fishes localities in Italy. Rivista Italiana di Paleontologia e Stratigrafia 91:197–210.

Tintori, A., and S. Renesto. 1990. A new *Lariosaurus* from the Kalkschieferzone (uppermost Ladinian) of Valceresio (Varese, N. Italy). Bollettino della Società Paleontologica Italiana 29:309–319.

Tschanz, K. 1986. Funktionelle Anatomie der Halswirbelsäule von *Tanystropheus longobardicus* (Bassani) aus der Trias (Anis/Ladin) des Monte San Girgio (Tessin) auf der Basis vergleichend morphologischer Untersuchungen an der Halsmuskulatur der rezenten Echsen. PhD Dissertation, University of Zurich.

———. 1988. Allometry and heterochrony in the growth of the neck of Triassic prolaceriform reptiles. Palaeontology 31: 997–1011.

———. 1989. *Lariosaurus buzzii* n. sp. from the Middle Triassic of Monte San Giorgio (Switzerland), with comments on the classification of nothosaurs. Palaeontographica A 208:153–179.

Turner, S., C. J. Burrow, H.-P. Schultze, A. Blieck, W.-E. Reif, C. B. Rexroad, P. Bultynck, and G. S. Nowlan. 2010. False teeth: conodont-vertebrate phylogenetic relationships revisited. Geodiversitas 32:545–594.

Vaughn, P. P. 1955. The Permian reptile *Araeoscelis* restudied. Bulletin of the Museum of Comprative Zoology 113:305–467.

Vickers-Rich, P. C., T. H. Rich, O. Rieppel, R. A. Thulborn, and H. A. McClure (†). 1999. A Middle Triassic vertebrate fauna from the Jilh Formation, Saudi Arabia. Neues Jahrbuch für Geologie und Paläontologie, Abhandlungen 213:201–232.

Wang, K.-M. 1959. Discovery of a new reptile fossil from Hubei, China. Acta Palaeontologica Sinica 7:315–356. (in Chinese with English abstract)

Wang, X.-F., G. H. Bachmann, H. D. Hans, P. M. Sander, G. Cuny, X.-H. Chen, C. Wang, L.-D. Chen, L. Cheng, F.-S. Meng, and G. Xu. 2008. The Late Triassic black shales of the Guanling area, Guizhou Province, south-west China: a unique marine reptile and pelagic crinoid fossil Lagerstätte. Palaeontology 51:27–61.

Watson, D. M. S. 1914. *Pleurosaurus* and the homologies of the bones in the temporal region of the lizard's skull. Annals and Magazine of Natural History (8) 14:84–95.

Watson, T. 2017. How giant marine reptiles terrorized the ancient seas. Nature 543:603–607.

Weishampel, D. B., and O. Kerscher. 2013. Franz Baron Nopcsa. Historical Biology 25:391–544.

Weiss, G. 1983. Bayreuth als Stätte alter erdgeschichtlicher Entdeckungen. Druckerei Ellwanger, Bayreuth, 70 pp.

Werth, A. 2000. Feeding in marine mammals; pp. 487–526 in K. Schwenk (ed.), Feeding. Form, Function, and Evolution in Tetrapod Vertebrates. Academic Press, San Diego, California.

Wild, R. 1973. Die Triasfauna der Tessiner Kalkalpen. XXIII. *Tanystropheus longobardicus* (Bassani) (Neue Ergebnisse). Schweizerische Paläontologische Abhandlungen 95:1–162.

———. 1975. *Tanystropheus* H. v. Meyer, 1855 (Reptilia): request for conservation under the plenary powers. Z.N. (S.) 2084. Bulletin of Zoological Nomenclature 32:124–126.

———. 1976. *Tanystropheus* H. von Meyer, [1852] (Reptilia): revised request for conservation under the plenary powers. Z.N. (S.) 2084. Bulletin of Zoological Nomenclature 33:124–126.

———. 1980a. *Tanystropheus* (Reptilia: Squamata) and its importance for stratigraphy. Mémoires de la Société Géologique de France N.S. 139:201–206.

———. 1980b. Die Triasfauna der Tessiner Kalkalpen. XXIV. Neue Funde von *Tanystropheus* (Reptilia, Squamata). Schweizerische Paläontologische Abhandlungen 102:1–31.

———. 1987. An example of biological reasons for extinction: *Tanystropheus* (Reptilia: Squamata). Mémoires de la Société Géologique de France N.S. 150:37–44.

Wild, R., and H. Oosterink, H. 1984. *Tanystropheus* (Reptilia: Squamata) aus dem Unteren Muschelkalk von Winterswijk, Holland. Grondbor en Hamer 5:142–148.

Williston, S. W. 1910, New Permian reptiles; rhachitomous vertebrae. Journal of Geology 18:585–600.

———. 1914a. The osteology of some American Permian vertebrates. Journal of Geology 22:364–419.

———. 1914b. Water Reptiles of the Past and Present. University of Chicago Press, Chicago, 251 pp.

———. 1917. The phylogeny and classification of reptiles. Journal of Geology 25:411–421.

———. 1925 The Osteology of the Reptiles. Harvard University Press, Cambridge, Massachusetts, 304 pp.

Wiman, C. 1910. Ichthyosaurier aus der Trias Spitzbergens. Bulletin of the Geological Institution of the University of Upsala 10:124–148.

Winsor, M. P. 1976. Starfish, Jellyfish, and the Order of Life. Issues in Nineteenth-Century Science. Yale University Press, New Haven, Connecticut, 228 pp.

Woodward, A. S. 1888. On some remains of the extinct selachian genus *Asteracanthus* from the Oxford Clay of Peterborough, preserved in the collection of Alfred N. Leeds, Esq., of Eyebury. Annals and Magazine of Natural History (6) 2:336–342.

———. 1889. Catalogue of the Fossil Fishes in the British Museum (Natural History). Part I. British Museum, London, 474 pp.

———. 1916. The Fossil Fishes of the English Wealden and Purbeck Formations. Monograph of the Palaeontological Society London 69:1–48.

Wu, F., Y.-L. Sun, G. Xu, W.-C. Hao, G.-Y. Jiang, and Z.-Y. Sun. 2011. New saurichthyid actinopterygian fishes from the Anisian (Middle Triassic) of southwestern China. Acta Palaeontologica Polonica 56:581–614.

Wu, X.-C., Y.-N. Cheng, C. Li, L.-J. Zhao, and T. Sato. 2011. New information on *Wumengosaurus delicatomandibularis* Jiang et al., 2008 (Dipsida: Sauropterygia), with revision of the osteology and phylogeny of the taxon. Journal of Vertebrate Paleontology 31:70–83.

Wu, X.-C., Y.-N. Cheng, Y. Sato, and H.-Y. Shan. 2009. *Miodentosaurus brevis* Cheng et al., 2007 (Diapsida;

Thalattosauria): its postcranial skeleton and phylogenetic relationships. Vertebrata PalAsiatica 47:1–20.

Xu, G-H., K.-Q. Gao, and M. I. Coates. 2015. Taxonomic revision of *Plesiofuro mingshuica* from the Lower Triassic of northern Gansu, China, and the relationships of early neopterygian clades. Journal of Vertebrate Paleontology 35:6, e1001515; DOI: 10.1080/02724634.2014.1001515.

Xue, Y., D.-Y. Jiang, R. Motani, O. Rieppel, Y.-L. Sun, Z.-Y. Sun, C. Ji, and P. Yang. 2015. New information on skeletal dimorphism and allometric growth in *Keichousaurus hui*, a pachypleurosaur from the Middle Triassic of Guizhou, South China. Acta Palaeontologica Polonica 60:681–687.

Yang, P., C. Ji, D.-Y. Jiang, R. Motani, A. Tintori, Y. Aun, and Z.-Y. Sun. 2013. A new species of *Qianichthyosaurus* (Reptilia: Ichthyosauria) from Xingyi Fauna (Ladinian, Middle Triassic) of Guizhou. Acta Scientiarum Naturalium Universitatis Pekinensis 49:1002–1008.

Yin, G.-Z., X. Zhou, Y. Cao, Y. Yu, and Y. Luo. 2000. A preliminary study on the early Late Triassic marine reptiles from Guanling, Guizhou, China. Geology-Geochemistry 28:1–23.

Young, C.-C. 1958. On the new Pachypleurosauroidea from Keichow, South-West China. Vertebrata PalAsiatica 2:69–81.

———. 1959. On the new Nothosauria from the Lower Triassic beds of Kwangsi, South-West China. Vertebrata PalAsiatica 3:73–78.

———. 1965. On the new nothosaurs from Hupeh and Kweichou, China. Vertebrata PalAsiatica 9:315–356.

———. 1972. A marine lizard from Nanchang, Hupeh Province. Memoirs of the Institute of Vertebrate Paleontology and Paleoanthropology, Academia Sinica A 9:17–27.

Young, C.-C., and Z.-M. Dong. 1972. *Chaohusaurus geishanensis* from Anhui Province. Aquatic Reptiles from the Triassic of China. Academia Sinica, Institute of Vertebrate Paleontology and Paleoanthropology, Memoir 9:11–14. (In Chinese)

Young, M. T., S. L. Brusatte, M. Ruta, and M. B. Andrade. 2010. The evolution of Metriorhynchoidea (Mesoeucrocodilia, Thalattosuchia): an integrated approach using geometric morphometrics, analysis of disparity, and biomechanics. Zoological Journal of the Linnean Society 158:801–859.

Zangerl, R. 1935. Die Triasfauna der Tessiner Kalkalpen. IX. *Pachypleurosaurus edwardsi* Cornalia sp., Osteologie-Variationsbreite-Biologie. Abhandlungen der schweizerischen Paläontologischen Gesellschaft 56:1–80.

Zanon, R. T. 1989. *Paraplacodus* and the diapsid origin of Placodontia. Journal of Vertebrate Paleontology 9:47A.

Zhang, Q.-Y., C.-Y. Zhou, T. Lü, and J.-K. Bai. 2010. Discovery of Middle Triassic *Saurichthys* in the Luoping area, Yunnan, China. Geological Bulletin of China 29:26–30. (in Chinese, with English abstract)

Zhang, Q.-Y., C.-Y. Zhou, T. Lü, T. Xie, X.-Y. Lou, W. Liu, Y.-Y. Sun, J.-Y. Huang, and L.-S. Zhao. 2009. A conodont-based Middle Triassic age assignment for the Luoping Biota of Yunnan, China. Science in China Series D-Earth Sciences 52:1673–1678.

Zhang, Q.-Y., C.-Y. Zhou, T. Lü, T. Xie, X.-Y. Lou, W. Liu, Y.-Y. Sun, and X.-S. Jiang. 2008. Discovery and significance of the Middle Triassic Anisian biota from Luoping, Yunnan Province. Geological Review 54:145–149.

Zhao, L.-J., C. Li, J. Liu, and T. He 2008. A new armored placodont from the Middle Triassic of Yunnan Province, Southwestern China. Vertebrata PalAsiatica 43:171–177.

Zhao, L.-J., J. Liu, C. Li, and T. He. 2013. A new thalattosaur, *Concavispina biseridens* gen. et sp. nov. from Guanling, Guizhou, China. Vertebrata PalAsiatica 51:24–28.

Zhao, L.-J., T. Sato, J. Liu, C. Li, and X.-C. Wu. 2010. A new skeleton of *Miodentosaurus brevis* (Diapsida: Thalattosuria) with further study of the taxon. Vertebrata PalAsiatica 48:1–10.

Zhao, L.-J., L.-T. Wang, and C. Li. 2008. Studies of the Triassic marine reptiles of China: A review. Acta Palaeontologica Sinica 47:232–239.

Ziegler, B. 1975. Forscher und akademischer Lehrer. Zum 70. Geburtstag des Paläontologen Emil Kuhn-Schnyder (29 April). Neue Zürcher Zeitung, nr. 98, 29. April 1975.

Zittel, K. A. v. 1887–1890. Handbuch der Palaeontologie, I. Abtheilung. Palaeozoologie, III. Band. Vertebrata (Pisces, Amphibia, Reptilia, Aves). R. Oldenbourg, Munich, 900 pp.

Zorn, H. 1971. Paläontologische, stratigraphische und sedimentologische Untersuchungen des Savatoredolomits (Mitteltrias) der Tessiner Kalkalpen. Schweizerische Paläontologische Abhandlungen 91:1–90.

Index

Acipenser, 57, 59
Acipenseriformes, 54
Acqua del Ghiffo, *2, 3, 7, 11*; *Ceresiosaurus* from, 141, 142; pachypleurosaurs from, 127, 135
Acqua Ferruginosa, *Ceresiosaurus* from, *138*, 142; *Habroichthys* from, *67*; pachypleurosaurs from, 127
acrodin, 54, 56
Acrodus, 37–39, 41, *42*, 46
Acrodus anningiae, 41
Acrodus bicarinatus, 47, 49, 50
Acrodus georgii, 37, *38, 39, 40*
Acrodus lateralis, 41; tooth histology of, 46
Acrodus tuberculatus, 49
Acrodus undulatus, 41
Acronemus, 49–51
Acronemus tuberculatus, 47, *48, 49, 50*, 51
Acrosauria, 107
Actinistia, 37
actinopterygian fishes from Mt. San Giorgio, 52–69
Actinopterygii, *30*, 34, 35
actinotrichia, 35
adaptive zone, 108
Aetheodontus, 66–68
Aetosauria, 76
Agassiz, Louis, 37, 41, 42; on *Colobodus*, 55; on *Placodus*, 106; on *Saurichthys*, 52, 53, 57; on *Strophodus*, 43
Agkistrognathus, 164
Alaska, 82
Aldinger, Hermann, 33
Alessandri, Giulio de, 32; on *Belonorhynchus*, 60; on *Colobodus*, 55; on Triassic sharks, 47
Alla Cascina, 128, 134; *Ceresiosaurus* from, *142*; pachypleurosaurs from, 135
Allolepidotus, 66
Alpine Triassic, 27, 130; *Birgeria* from, 52; *Cymbospondylus* from, 99; *Lariosaurus* from, 140; paleobiogeography of, 199, 203; *Prohalecites* from, 33; *Tanystropheus* from, 182, 184; *Ticinosuchus* from, 194
Altisolepis, 66
alveolus, 191
Amblyrhynchus, 187

Amblyrhynchus cristatus, 84
ambush predator, 58
ammonoids, 22
Amniota, 72
Amotosaurus rotfeldensis, 171, 181
amphistylic jaw suspension, 38, 47
Amsler, Alfred J., 147
anaerobic metabolism, 85
anagenetic evolution, 128, 134
anagenetic lineage, 133
anagenetic transition, 133
Anapsida, 72, 73, 107
Anarosaurus, 107, 129
Anarosaurus-Dactylosaurus clade, 199, 203
Anarosaurus heterodontus, 129
angiosperms, 192
anguilliform locomotion, 86, 101
Anhui Province, 78, 79, 83, 89, 91, 200
Anning, Mary, 41, 105
anoxic, 27, 28
Anshunsaurus, 156, 206; paleobiogeography of, 212; relationships of, 165, 210; rostrum of, 165
Anshunsaurus huangguoshuensis, 164, 206, 209
Anshunsaurus huangnihensis, 206
Anshunsaurus wushaensis, 206
Antarctica, 75, 169, 193, 222
Antrimpos, 27
apex predator, 101, 150, 191
aquatic adaptations in reptiles, 84–89
Araeoscelidia, 72, 74, 107
Araeoscelis, 72, 107, 169, 172; skull of, 170, 173
Archaeosemionotus, 66
archipterygium, 36
Archosauria, 75, 76, 207
Archosauriformes, 76, 170
Archosauromorpha, 72, 75, 163
Arganasuchus dutuiti, 194
Argentina, 159; *Chirotherium* tracks from, 195; early dinosaurs from, 77; rauisuchians from, 193
Asteracanthus, 42, 43, 46
Asteracanthus cf. *reticulatus*, 37, *44*, 45
Asteracanthus semiverrucosus, 44, *45*
Askeptosauroidea, 158; relationships of, 210
Askeptosaurus, 17, 18, 154–156, 158, 159, 169, 206, 216; paleobiogeography of, 212;

relationships of, 163, 164, 165, 173, 174, 210; rostrum of, 157, 158, 165
Askeptosaurus italicus, *152*, 155, *156, 157*, 158
Atopodentatus unicus, 88, 204
Augustasaurus, 212
Augustasaurus hagdorni, 81, 84
Avemetatarsalia, 76
Aves, 72

bacterial mat, 28
Balsamo–Crivelli, Guiseppe, 31, 105, 123
Barth, Carl, 195
Bassani, Francesco, 15, 32, 47, 93; on *Belonorhynchus*, 53, 60; on *Ichthyosaurus*, 92; on *Macrocnemus*, 171, 172; on *Tribelesodon*, 179; on *Undina*, 70
Bayreuth, 16, 52, 53, 57; *Placodus* from, 105, 109; *Tanystropheus* from, 178, 181
Bdellodus, 45
belemnites, 22, *25*, 26
belemnite phragmocone, 38
Bellotti, Cristoforo, 32, 47; on *Belonorhynchus*, 60; on *Ichthyorhynchus*, 60
Belonorhynchus, 53
Belonorhynchus macrocephalus, 60
Belonorhynchus robustus, 60
Bender-collection, 33, 93
Bergamasque Prealps, 164, 171
Bergün, 125
Berlichingen, 150
Berlin-Ichthyosaur State Park, 84
Besania, 66
Besano, 7, 8, 15, 22, 26, 32, 119, 155, 216, 217; actinopterygians from, 136; *Lariosaurus* from, 140; pachypleurosaurs from, 134
Besanosaurus, 92, 95, 101, 103; relationships of, 211
Besanosaurus leptorhynchus, 94, *102*
biotic crisis, 193
Birgeria, 52–54, 57, 58, 65
Birgeria acuminata, 53
Birgeria mougeoti, 53
Birgeria stensioei, 52, *53*, 215
bivalves, 22, 25, 28, 208
Blumenbach, Johann Friedrich, 195
Bobasatrania, 66
Bobasatrania ceresiensis, 54, *55*, 68
bone ballast, 84–86, 121, 176

bone density analysis, 188
Borromeo, Vitaliano, 123
Botta, Mario, 7
bowfin, 52
brachiopods, 108
Brazil, 192, 193, 196
Brinkmann, Winand, 94, 97; on
  Shastasauridae indet., 103
British Columbia, 53, 54, 78, 82–84, 91,
  92, 102, 164, 166, 208
Broili, Ferdinand, 4, 7, *16*, *18*, 110,
  115, 116; on *Lariosaurus*, 124;
  on *Pachypleurosaurus*, 124; on
  Tanystropheus, 178
Bronn, Heinrich Georg, 195
Brough, James, 33
Broughia, 66
Buchser, Fritz, *24*, 99, *100*
Buckland, William, 105
buoyancy, 37, 84, 121, 128, 176
buoyancy control, 85
Bürgin, Toni, 33, 67
Burgundian gate, 80
Buzzi, Guiseppe, *2*, *11*, *18*, *24*, 146

Ca' del Frate, 22, 25; fossil fishes from,
  32, 33; *Lariosaurus* from, 89, 123,
  140, 145; *Neusticosaurus* from,
  134–136
Caelatichthys, 66
Calcagni, Emilio, 3; and Acqua del
  Ghiffo, 142
California, 53, 82–84, 87, 92,93, 102,
  153, 164–166, 210–211
Callawaya wolonggangensis, 101, 102,
  108; relationships of, 211
Callaway, Jack, 93, 97, 101, 208, 216
cannibalism, 103; intraspecific, 63
Captorhinomorpha, 72, 74
carapace, 79, 118, 120, 121;
  in hupehsuchians, 200;
  in *Sinocyamodus*, 209; in
  Sinosaurosphargis, 204
carnivorous habits, 177, 191
Carpathian gate, 80, 199
Carroll, Robert L., 125, 134, 135, 216
Cartorhynchus lenticarpus, 201
Cascinello, 127
caudal autotomy, 185
caudal peak, 97, 102
Cava Tre Fontane, 9, 10, 22; fossil
  fishes from, 32, 33; *Acrodus* from,
  39; *Askeptosaurus* from, 158,
  Birgeria from, 53; *Cymbospondylus*
  from, 98, 99, *100*; *Helveticosaurus*
  from, *104*, 110; *Hybodus* from,
  42; *Macrocnemus* from, *168*, 172,
  173; pachypleurosaurs from, 126;
  Paranothosaurus from, 147, *148*;
  Paraplacodus from, 115, *116*;
  Ticinosuchus from, 193
cephalic spines, *38*, 39, 51
cephalopods, 22, *25*, 26; and
  Ceresiosaurus, 143; and

Cymbospondylus, 101; and mixosaurs,
  94; and *Paranothosaurus*, 148;
  and *Tanystropheus*, 182, 183; and
  thalattosaurs, 82
Cephaloxenus, 66, 68
Ceratitidae, 26
ceratohyal, 50
ceratotrichia, 34
Ceresiosaurus, 3, 24, 142, 143, 147,
  215; possible synonymy with
  Lariosaurus, 144; sister taxon of
  Lariosaurus, 146
Ceresiosaurus calcagnii, 3, 24, *138*, 141,
  143, 144
Ceresiosaurus lanzi, *142*, 144
Chaohu City, 78, 79, 83, 89, 91, 98,
  200, 201, 211
Chaohu fauna, 200–201, 211
Chaohusaurus, 78, 89, 91, 98
Chaohusaurus chaoxianensis, 201
Chaohusaurus geishanensis, 200
Chaohusaurus zhangjiawanensis, 201
Chelydra serpentina, 131
chemoautotrophic bacteria, 27, 28
Chinchenia suni, 201
Chirosaurus, 195
Chirotherium, 216
Chirotherium barthi, 196
Chirotherium Barthii, 195
Chirotherium monument, 196, *197*
Chirotherium tracks 192, 194, *197*;
  British, 195; footprints, 196; history of
  discovery, 194–196
choanae, 37
Chondrichthyes, *30*, 34
Chondrostei, 36, 52
Choristodera, 75
Cimmeria, 213
clades, 108
Claraz, Georges, 159
Clarazia, 153–155, 160–162, 216;
  relationships of, 163–165; rostrum
  of, 166
Clarazia schizi, *159*, *160*
Claraziidae, 153, 154, 162
Clarazisauria, 154
cleithrum, 59
coelacanth, 33, 70, 215
coelacanths from Mt. San Giorgio,
  70, 71
Coleoidea, 26
Colobodus, 52, 56–57, 67
Colobodus baii, 56
Colobodus bassanii, 54, *55*, *56*, 57
Concavispina, 209; paleobiogeography
  of, 210; relationships of, 210
Concavispina biseridens, 206, 209
conodonts, 25
conservation deposit, 26
convergent evolution, 59, 66, 75, 76,
  88; in *Dinocephalosaurus*, 203; in
  Saurichthys, 65; of salt glands, 78, 85
Conybeare, William Daniel, 105
Cope, Edward Drinker, 106

coprolite, 142
Cornalia, Emilio, 8, 31, 123, 124, 135
Corosaurus, 212
Corosaurus alcovensis, 81, 84
Costasaurichthys Group, 62
Crenilepis, 56
Crocodylia, 72
Crocodylus porosus, 85
Crossopterygii, 33, 36, 37
crurotarsal joint, 193
crustacean, 25
Cryptocleidoidea, 79
Ctenognathichthys, 66, 67
Ctenognathichthys bellottii, 54, 66,
  67, *68*
Curioni, Giulio, 31, 32; on *Lariosaurus*,
  123, 139, 141, 142
Cyamodontoidea, 79, 86, 118
Cyamodus, 110
Cyamodus hildegardis, 109, 117, *119*,
  120, 215; as stomach content, 121,
  146
Cyamodus rostratus, 109
Cyanobacteria, 27, 28
Cymatosaurus, 81, 87
Cymbospondylidae, 92, 98
Cymbospondylus, 92, 99, 103; from Mt.
  San Giorgio, 98–101
Cymbospondylus asiaticus, 208
Cymbospondylus buchseri, 95, *96*, 98,
  99, *100*, 101
Cymbospondylus nicholsi, 98, 99
Cymbospondylus petrinus, 101
Cymbospondylus piscosus, 98, 99, 101

Dactylosaurus, 129, 205
Dadocrinus beds, 181
Dal Sasso, Cristiano, 112
Daninia, 66
Daonella, 22, 25, 28
Dawazisaurus brevis, 203
Deecke, Wilhelm, 32; on
  Belonorhynchus, 60
De la Beche, Henry, 105
developmental constraints, 188
Diandongosaurus acutidentatus, 202;
  relationships of, 203
Dianmeisaurus gracilis, 202
Dianopachysaurus, 203
Dianopachysaurus dingi, 202
diapophysis, 101
Diapsida, 72, 74, 106, 107
diastema, 117, 120; in thalattosaurs,
  158, 165, 166
digitigrade stance, 177, 196
Dingxiaosaurus luyinensis, 206
Dinocephalosaurus, 188, 203;
  paleobiogeography of, 212
Dinocephalosaurus orientalis, 89; cervical
  vertebrae of, 188, 203
Dinosauria, 72
dinosaurs, 77
Dipnoi, 36, 37
Dipteronotus, 66

discovery of black shales at Besano / Mt. San Giorgio, 8

discovery of fossil fishes at Besano / Mt. San Giorgio, 31–33

dispersalist scenario, 213

Drepanosaurus unguicaudatus, 75

Ducanichthys, 66

durophagy, 45, 79; in ichthyosaurs, 91; in mixosaurs, 94; in thalattosaurs, 82 153, 161, 163; in *Ticinolepis*, 69

eastern Alps, 20

eastern Swiss Alps, 83, 113, 114

eastern Pacific faunal province, 80, 83, 212; faunal relations to western Tethys, 167, 211; macropredators in, 202; thalattosaurs from, 81

eastern Tethyan faunal province, 83, 203; close faunal relations to western Tethys, 177, 211, 212, 217; early occurrence of fossil reptiles, 150, 199, 201; macropredators in, 202

Elasmosauridae, 79

endemism, 200, 211, 212, 217; in the Panxian-Luoping fauna, 201, 203, 204

Endennasaurus, 165, 210

Endennasaurus acutirostris, 164

endobenthic, 87, 88

endoskeleton, 34, 36

endothermic, 86

end-Permian mass extinction, 73, 202, 217

Eoeugnathus, 66

Eohupehsuchus, 200

Eoprotrachyceras curionii, 8

Eoraptor, 77

Eorhynchochelys sinensis, 210

Eosaurichthys, 57

Eosauropterygia, 109

Eosemionotus, 66

Eosuchia, 153, 154, 158

epaxial musculature, 51

epibenthic, 46

Equisetites, 192

erect gait, 192

Eretmorhipis, 200

Etter, Walter, 28

Euparkeria, 75, 76, 170, 196, 197

Eureptilia, 72, 73

Euryapsida, 107, 169

Eusaurosphargis, 83, 105, *112, 113, 114,* 115

Eusaurosphargis dalsassoi, 111

Euselachii, 51

Eusuchia, 77

evolutionary reversal, 134

evolutionary trends, 65

exoskeletal, 34

feeding, 85, 86; filter-feeding, 88, 204; lunge feeding, 200; raptorial feeding, 87; suction feeding, 86, 87, 208

Felber, Markus, 19

fifth metatarsal, in *Askeptosaurus*, 159; in *Helveticosaurus*, 111; in *Macocnemus*, 177; in *Prestosuchus*, 196; in *Tanystropheus*, 171, 186

fin spines, 32, 35, *38*, 39, 41; costate, *42*, 43, 44; tuberculated, 43, 44, *45*; of *Palaeobates*, 46, *47*; of *Acronemus*, *47, 48,* 49, *50,* 51

footprints, 194–196

France, 41, 55, 195

freshwater, 171

fulcral scales, 59, 62, 65

Furo, 66

Furo trotii, 31

Furrer, Heinz, 22, 25, 27; on Ca' del Fate, 136

Fusea (Udine), 149

Fuyuan Country, 207

ganoid scales, 56, 59

ganoin ornamentation, 56, 57

gars, 52

gas exchange, 84, 85

gastroliths, 84, 121; in *Guizhouichthyosaurus*, 208; in the Nile crocodile, 121

gastropods, 25

Germanic basin, 55, 80; nothosaur taxonomy, 140, 141, 149, 150; pachypleurosaurs from, 199, 203; *Tanystropheus* from, 180–182

Germanosaurus, 81, 87

ghost lineage, 133

Glyphoderma, 205

Glyphoderma kangi, 205

Gondwanaland, 19, 169

gonopodium, 63; in *Peltopleurus*, 68, 69; in *Saurichthys*, *64*, 65

Gracilignathichthys, 66

grades, 108

Grigna mountains, 31, 62; *Lariosaurus* from, 123, 126

Grippia, 91, 115

Grosser Pachypleurosaurus, 125, 126, 132, 134

Guangxi Autonomous Region, 200

Guanling, 205, 207

Guanling biota, 92, 164, 165, 207–210, 211, 217; ichthyosaurs from, 208; placodonts from, 209; thalattosaurs from, 209, 210

Guanlingsaurus, 101, 102; relationships of, 211

Guanlingsaurus liangae, 208

Guizhouichthyosaurus, 101; relationships of, 211

Guizhouichthyosaurus tangae, 208

Guizhou Province, 56, 57, 81–83, 89, 92, 101, 102 114, 150, 164, 166, 198, 201, 205, 207, 208, 211, 217

Gyrolepis, 66

habitat partitioning, 58; in actinopterygians, 67; in ichthyosaurs, 79; in mixosaurs, 94; in nothosaurs, 81; in *Saurichthys*, 60, 67; in *Ticinolepis*, 68; in *Xinpusaurus*, 82

Habroichthys, 66, 68

Habroichthys minimus, *67*

Haeckel, Ernst, 4, 162

Hanosaurus, 203

Hanosaurus hupehensis, 200

head spines, 39, of *Palaeobates*, 46

Helveticosaurus, 105, 110–114

Helveticosaurus zollingeri, *104,* 110

hemal arches, 59

hemitrichia, 35

Henodus chelyops, 86, 88

Heptonema paradoxa, 32

herbivores, 74, 76, 77, 88, 153, 193, 204

Herrerasaurus, 77

Hertwig, Richard, 4

Hescheler, Karl, 4, 9, *10,* 12, 161

Hescheleria, 87, 153–155, 161, *162,* 216; relationships of, 163–165; rostrum of, 165, 166

Hescheleria ruebeli, 161

heterocercal caudal fin, 35, 36, *38,* 51, 54, 56, 58

heterochronic shift, 133, 134

heterodont dentition, 41, 54; in *Mixosaurus*, 92–93; in nothosaurs, 139; in *Tanystropheus*, 187

Heterodontiformes, 40

Heterodontus, 35, 40, 41

heterodonty, 44, 46, 50

heterotopic ossifications, 186

Hildburghausen, 194–196, *197*

Holostei, 52

Holzmaden, 38, 39, 78, 89, 216

homocercal caudal fin, 35, 36, 58

homodont dentition, 112, 115; in *Askeptosaurus*, 158; in *Macrocnemus*, 176; in pachypleurosaurs, 139

Hubei Province, 83, 200, 201

Huene, Friedrich von, 155; on *Prestosuchus*, 192, 196; on *Rauisuchus*, 192; on *Tanystropheus*, 178, 179, 181; on *Zanclodon*, 179

Hupehsuchus, 200

Huxley, Julian Sorell, 108

Huxley, Thomas Henry, 169

Hybodontiformes, 34

hybodontiform sharks, 37–49

Hybodus, 38, 40, *42*, 46

Hybodus plicatilis, *42*; *Hybodus* cf. *plicatilis*, 37, *43*

hyoid apparatus, 208

hyomandibula, 50

hyostylic jaw suspension, 47

hypermorphosis, 134

hyperphalangy, 81, 86; in *Besanosaurus*, 102; in *Ceresiosaurus*, 143, 144; in *Lariosaurus*, 140; in *Yunguisaurus*, 206

hyposphene-hypantrum articulation, 117

Ichthyopterygia, 78, 111, 113
Ichthyorhynchus [Saurichthys] curionii, 32, 60
Ichthyosauria, 78, 106, 107, 109
Ichthyosauromorpha, 78
ichthyosaurs, 78–79, 91–92
Ichthyosaurus, 92
Ichthyosaurus Cornalianus, 92
Idaho, 91
India, 193
insectivorous habits, 74; in *Macrocnemus*, 177, 191; in *Tanystropheus*, 179, 183, 187
internal fertilization, 39, 58, 63
interspecific competition, 68
isodont dentition, 94, 165
Israel, 81, 83, 118, 201

Jaekel, Otto, 41, 42, 46; on *Anarosaurus*, 107
Japan, 76, 84, 91
Jehol biota, 77
Jordan, 83

Kaup, Johann Jakob, 195
Keichousaurus, 89, 131, 202; paleobiology of, 205; relationships of, 203
Keichousaurus hui, 79, 205
Keichousaurus yunnanensis, 200
Konservat Lagerstätte, 26, 41, 47, 215
Kuehneosauridae, 75
Kuhn-Schnyder, Emil, 5, 10, 12, *13*, *14*, 16–19, 22, 27, 215; on *Acrodus*, 40, 49; on *Askeptosaurus*, 154–157; on *Birgeria*, 52; on *Cyamodus*, 119; on *Cymbospondylus*, 99; on *Helveticosaurus*, 110–111; on *Lariosaurus*, 145–147; on *Macrocnemus*, 173, 174; on *Mixosaurus*, 93; on *Neusticosaurus*, 134; on pachypleurosaurs, 125; on *Paranothosaurus*, 149; on *Paraplacodus*, 115, 117; on polyphyly, 108; on protorosaurs, 173; on Shastasauridae indet., 103; on *Simosaurus*, 107; on thalattosaurs, 164
Kwangsisaurus orientalis, 200

labial cartilages, 50
labyrinthodont amphibians, 195
Lacertilia, 163
Lago Ceresio, 142
Lago di Lario, 142
Lake Como, 31, 105, 123, 142
Lake Lecco, 31
Lake Lugano, 19, *20*, 126, 142
Lang, Arnold, 4
Langobardisaurus, 181
Langobardisaurus pandolfii, 171
Lanz, Heinz, *14*, 144
Largocephalosaurus, 114, 165, 203, 204; paleobiogeography of, 210

Largocephalosaurus polycarpon, 204
Largocephalosaurus qianensis, 204
Lariosauridae, 139, 146, 147
Lariosaurus, 80, 81, 89, 123, 124, 127, 136, 139, 147, 216; differs from *Nothosaurus*, 140; from Val Mare, 145; named after Lago di Lario, 142; possible synonymy with *Ceresiosaurus*, 144; sister taxon of *Ceresiosaurus*, 146; systematics of, 141, 205;
Lariosaurus balsami, 31, 123, 124, 126, 139, 140; differs from *Ceresiosaurus*, 141; from Val Mare, 145
Lariosaurus buzzi, 121, *146*, 147
Lariosaurus hongguoensis, 141
Lariosaurus lavizzarii, 145
Lariosaurus valceresii, 145
Lariosaurus vosseveldensis, 141
Laiosaurus xingyiensis, 141, 205
Lariosaurus hongguoensis, 202
Latimeria, 37
Latimeriidae, 70
Laurasia, 19, 169
Lavizzari, Luigi, 145
Legnonotus, 66
Lepidosauria, 153, 154
Lepidosauromorpha, 72, 75, 163
lepidotrichia, 34, 35, 36; in *Saurichthys*, 61, 65
Lepidotus Trotti, 31
Leptacanthus cornaliae, 32
Liaoning Province, 77
Limax cinereoniger, 5
Litorosuchus somnii, 207
locomotion, 38, 85, 86; bipedal, 177; digitigrade, 194; in *Askeptosaurus*, 156; in *Ceresiosaurus*, 143; in *Macrocnemus*, 174, 177; in *Paranothosaurus*, 148, 150; in plesiosaurs, 143, 144; in *Qianosuchus*, 204; in *Ticinosuchus*, 192, 196
longevity, 132
Lugano, 7, 8, 19, 95
Luganoia, 66
Luoping County, 62, 83, 86, 88, 201–203, 205
Lyell, Charles, 195

Macrochemus, 172
Macrochemus bassanii, 171, 172, 207
Macrocnemus, 17, 18, 24, 170, 185, 216; from China, 207; paleobiogeography of, 177, 212; relationships of, 171, 173, 181
Macrocnemus bassanii, *168*, *174*; from the Black Forest, 181; insectivorous, 177, 191; juvenile specimen of, *175*; terrestrial, 177
Macrocnemus fuyuanensis, 177, 207; *Macrocnemus* aff. *M. fuyuanensis*, 177
Macrocnemus obristi, 177
Macromerosaurus, 123
Macromirosarus Plinj, 31, 123, 124, 139

macropredators, 201, 202
Macroscelosaurus, 177, 178
Madagascar, 53, 54, 59
Majiashan quarry, 200, 201
Majiashanosaurus discocoracoidis, 201
Makhtesh Ramon, 81, 83, 201, 234
Mammalia, 72
mammal-like reptiles, 72
Maniraptora, 77
Meckel's cartilage, 34, 46, 50
Megalancosaurus preonensis, 75
Meride, *6–8*, *9*, 11, 15, 19
Meridensia, 66, 67
Mesosauria, 72
Mesosaurus, 107
Meyer, Hermann von, 16, 46, 105, 106; on *Nothosaurus*, 149, 178; on *Teratosaurus*, 192
Mikadocephalus, 95, 103
Mikadocephalus gracilirostris, 94
Milan, 8, 15
Milleretta, 72
Miodentosaurus, 166
Miodentosaurus brevis, 209; relationships of, 210
Mixosauridae, 92, 93
mixosaurs from Mt. San Giorgio, 92–98
Mixosaurus, 22, 91, 92, 95, 155
Mixosaurus cornalianus, 33, *90*, 93–95, *96*, 97, *98*, *103*; 'neoholotype' of, 93, 95; neotype of, 93; morphotypes of, 95; relationships of, 211
Mixosaurus guanlingensis, 208
Mixosaurus kuhnschnyderi, 94–96; relationships of, 211
Mixosaurus nordenskioeldii, 94, 95
Mixosaurus panxianensis, 202; paleobiogeography of, 211
monimostylic quadrate, 164
monocuspid teeth, 183
monophyly, 73; of Chondrostei, 36; Kuhn-Schnyder's critique of, 147; of *Lariosaurus*, 139, 144; of mixosaurs, 96; of *Nothosaurus*, 139; of pachypleurosaurs, 80, 128, 200, 206; of Protorosauria, 75, 169, 170; of rauisuchians, 76, 193; of Sauropterygia, 105, 109; of Thalattosauria, 153; principle of, 108
Morocco, 193, 194
morphometric analysis, 134
mortality rate, 188
Motani, Ryosuke, 95
Münster, Georg Ludwig, 109; on *Macroscelosaurus*, 178

Nanchangosaurus, 200
Nautilus, 26
Nectosaurus, 87, 153, 154; paleobiogeography of, 211; relationships of, 164, 165, 210; rostrum of, 166
Nectosaurus halius, 153
negative allometric growth, 135

Negev, 81, 83, 201

nektonic, 27

Nemacanthus tuberculatus, 47, 49, 51; riddle of, 215

Neodiapsida, 72, 74, 75

Neopterygii, 36

Neoselachii, 35

neural arches, 59

Neusticosaurus, 80, 89, 123–126, 129, 131, 136, 205; in *Ceresiosaurus* coprolite, 142

Neusticosaurus edwardsii, 8, 108, *122*, 124–126, 128, 129, 133; longevity, 132; sexual maturity, 132; type specimen of, 134–137

Neusticosaurus peyeri, 127, 128, *129*, 130, 133–135; embryo of, 131; longevity, 132; sexual maturity, 132

Neusticosaurus pusillus, 24, 125, *127*, 128, 130, 133, 134, *138*, 142, *143*; longevity, 132; sexual maturity, 132

Nevada, 81–84, 91, 92, 98, 99, 101, 102, 166, 202

niche partitioning, 211

Nopcsa, Franz von, 15, 16; on *Askeptosaurus*, 155; on *Macrochemus*, 171, 177; on *Pachypleurosaurus*, 124; on protorosaurs, 172

Nothosauridae, 139

nothosaurs, 80–81; from Mt. San Giorgio, 141–150

Nothosaurus, 72, 80, 81, 87, 106, 149, 216; differs from *Lariosaurus*, 140; taxonomy of, 141, 205

Nothosaurus giganteus, 148, 149, 202

Nothosaurus jagisteus, 150

Nothosaurus juvenilis, 141

Nothosaurus marchicus, 147

Nothosaurus mirabilis, 149, 178

Nothosaurus raabi, 147

Nothosaurus winkelhorsti, 141

Nothosaurus yangjuanensis, 202

Nothosaurus youngi, 141, 205

Nothosaurus zhangi, 202

notochord, 59, 65

notochordal sheath, 51

obturator foramen, 131

Odoiporosaurus, 199

Odoiporosaurus-Serpianosaurus-Neusticosaurus clade, 203

Odoiporosaurus teruzzii, 129, *130*

Odontochelys semitestacea, 210

opercular apparatus, 58

opercular bone, 54; in *Saurichthys*, 60

Ophidia, 163

Ophiopsis, 66

Ornithischia, 77

orthodentine, 46

Osteichthyes, *30*, 34, 35

osteodentine, 40; in *Acrodus*, 45, 46; in *Asteracanthus*, 45

osteoderms, 179; in aetosaurs, 77; in *Cyamodus hildegardis*, 118–121; in *Eusaurosphargis*, 112; in hupehsuchians, 200; in *Paraplacodus*, 116; in *Placodus*, 117; in *Psephochelys*, 209; in rauisuchians, 193; in *Saurosphargis*, 82, 204; in *Sinocyamodus*, 209; in *Sinosaurosphargis*, 86, 204; in *Ticinosuchus*, 192–194

ovoviviparity, 88

Owen, Richard, 109; on *Placodus*, 106; on *Chirotherium*, 195

*P*achypleura, 124, 139

Pachypleura Edwardsii, 8, 123, 135

Pachypleurosauroidea, 206

pachypleurosaurs, 80; anatomy, 130–131; different from *Lariosaurus*, 139–140; evolution, 133–134; history of taxonomy, 123–129, 134–137; life history, 131–133

Pachypleurosaurus, 124, 139

Pachypleurosaurus edwardsii, *122*, 124–126, 128, 129; longevity, 132; type specimen of, 134–137; sexual maturity, 132

Pachypleurosaurus staubi, 125, 126

paddle-fish, 52

pachyostosis, 85, 121, 127; in *Ceresiosaurus*, 144; in hupehsuchians, 200; in *Lariosaurus*, 140; in *Neusticosaurus*, 128, 130; in *Nothosaurus*, 149; in *Paranothosaurus*, 149; in *Tanystropheus*, 185

Palaeobates, 45; tooth histology of, 46

Palaeobates angustissimus, 37, *47*

Palaeoniscus, 31

Palatodonta bleekeri, 79, 115

palatoquadrate, 34, 46; cleaver-shaped, 49; postorbital process of, 50

paleoecology of Mt. San Giorgio basin: 26–28

paleobiography, 83–84, 210–213

Paleotethys, 199

pallial dentine, 46

Pangea, 19, 20

Panjiangsaurus epicharis, 208

pan-Tethyan faunal element, 182

Panthalassic Ocean, 20

Panxian County, 82, 83, 114, 141, 201

Panxian-Luoping fauna, 56, 201–205, 210–212

Paraceratites trinodosus, 7

Parahupehsuchus, 200

Paralonectes, 164; rostrum of, 166

Paranothosaurus, 149; apex predator, 150

Paranothosaurus amsleri, 147, *148*, 149

Paraplacodus, 110, 112, 117, 118, 215

Paraplacodus broilii, 115, *116*

parapophysis, 101

Parapsida, 107

Pareiasaurus, 72

paraphyletic, 33, 52, 74, 158; pachypleurosaurs, 203

Parareptilia, 72, 74

patagium, 75

pelagic, 26, 79, 81, 165, 207; *Birgeria* 53–54; *Cymbospondylus*, 101, 102; ichthyosaurs, 78, 83, 91, 211; *Neusticosaurus edwardsii*, 133

Peltoperleidus, 66–68

Peltopleurus, 66, 68

Peltopleurus lissocephalus, 68, *69*

Pelycosauria, 106

Peripeltopleurus, 66, 68

Perledo, 31, 32, 105, 123, 124; actinopterigians from, 136; *Lariosaurus* from, 140

Perleididae, 67

Perleidus, 66, 67

Pessosaurus, 95

Petrolacosaurus, 72

Peyer, Bernhard, 3, 4, *5*, 7–9, *10*, 12, 13, *16*, 17, *18*, 19, 22, 24, 215; on *Askeptosaurus*, 155, 156; on *Asteracanthus*, 43; on *Ceresiosaurus*, 141, 142; on *Clarazia*, 161; on Claraziidae, 153, 162; on *Cyamodus*, 109, 121; on *Helveticosaurus*, 111; on *Hescheleria*, 161; on *Lariosaurus*, 123, 124, 136; on *Macrocnemus*, 171, 172; on *Mixosaurus*, 91; on pachypleurosaurs, 128, 139; on *Paranothosaurus*, 147, 149; on *Paraplacodus*, 115, 116; on protorosaurs, 173; on *Saurosphargis*, 82, 110; on *Tanystropheus*, 177, 179, 180, 186; on *Tocosauria*, 160, 162–163

Phalarodon, 93

Phalarodon atavus, 202

Phalarodon cf. *P. fraasi*, 202

Pholidophoridae, 36

Pholidopleurus, 66

Phygosaurus, 126

Phygosaurus perledicus, 127

phylogeny of fishes, 33–37

Phytosauria, 76

Pinna, Giovanni, 93, 95; on placodont relationships, 108, 115, 118

piscivory, 191

pistosaurs, 81

Pistosaurus, 81, 87, 106

Placochelys, 205, 209

placodont relationships, 105–109

Placodontia, 79, 105–107, 111

Placodontoidea, 79

Placodus, 106, 109, 112, 114, 116, 117; paleobiogeography of, 212

Placodus gigas, 109

Placodus inexpectatus, 202

placoid sales, 34, 35, 44; of *Acronemus*, 51; of *Asteracanthus*, 45–46

Placopleurus, 66

planktonic, 27

plastron, 79, 120
Platysiagum, 66
pleromin, 45
Plesiosauria, 79
Plesiosauroidea, 79
Plesiosaurus, 79, 105
Plesiosaurus dolichodeirus, 31
plicidentine, 97
Pliosauridae, 79
Pliosauroidea, 79
Pogliana, 22, 26; *Cyamodus* from, 119, 120, *121*
Point 902, 13, *14*, 22, *23*, 27, 215; *Acrodus* from, *39*; *Acronemus* from, *50*; cf. *Undina* from, *71*; *Ctenognathichthys* from, *68*; *Cyamodus* from, 119, *120*; *Hybodus* from, *42*; *Lariosaurus* from, *146*; *Mixosaurus* from, *98*; pachypleurosaurs from, *126*; *Palaeobates* from, *47*; *Peltopleurus* from, *69*; *Phalarodon* from, *94*; *Saurichthys* from, 62, *63*; *Serpianosaurus* from, *126, 132*; *Tanystropheus* from, *184*; *Ticinepomis* from, *70*; *Ticinolepis* from, *69*
polyphyletic, 107, 108
Porto Ceresio, 3, 7, 19, 32, 142
positive allometric growth, 182, 188
postembryonic growth, 121
Prestosuchus, 192, 196
Procolophon, 72
Proganosauria, 106
Prohalecites, 33, 66
Prolacerta, 75, 107, 155, 157, 174; skull of, 170, 173
Prolacerta broomi, 169
Prolacertiformes, 169, 170
Prolacertilia, 174
Protanystropheus antiquus, 182
Proterochampsidae, 76
Proterosuchus, 75, 170
Protorosauria, 75, 107, 163, 169, 170, 173, 188
protorosaurs, 75; from Mt. San Giorgio, 171–188; history of taxonomy, 169–170
Protorosaurus, 107, 170, 172, 173
Protorosaurus speneri, 169
Psephochelys polyosteoderma, 205, 209
Psephoderma, 87, 199, 209
Psephoderma alpinum, 199
Pseudosuchia, 76, 196
Pterosauria, 179
pterosaurs, 77
Ptycholepis, 66

*Q*ianichthyosaurus xingyiensis, 205; relationships of, 211
Qianichthyosaurus zhoui, 208
Qianosuchus mixtus, 204
Qianxisaurus chajiangensis, 206
Qinglong County, 207

*r*adials, 35, 36, 51
Rauisuchia, 76
rauisuchians, 76, 192–194
Rauisuchidae, 192
Rauisuchinae, 192
Rauisuchus, 76, 192, 193
Reptilia, 72, 106, 108
Rhipidistia, 37
Rhomaleosauridae, 79
Rhynchocephalia, 75, 153–155, 163
Rhynchosauria, 75
rhynchosaurs, 76
Rio Vallone, 129
Rüdersdorf (Berlin), 147, 181, 182
Russia, 193

*s*alinity tolerance, 28
salt glands, 85; in *Macrocnemus*, 176
Sanchiaosaurus dengi, 201
Sander, Paul Martin, 95, 135; on pachypleurosaurs, 126–128
Sangiorgiosaurus, 94, 95
Sangiorgiosaurus kuhnschnyderi, 94
Saphaeosaurus, 107
Sarcopterygii, *30*, 34, 35
Saudi Arabia, 83
Sauria, 74, 75, 105, 163
Saurichthyidae, 57
Saurichthys, 52–54, 57–59, 61, 62, 215, 217; embryos of, *64*, 65; gonopodium of, *64*, 65
Saurichthys acuminatus, 53
Saurichthys apicalis, 57
Saurichthys breviabdominalis, 61
Saurichthys costasquamosus, 61, 62
Saurichthys curionii, *58*, *59*, *61*, 63, *64*, 65
Saurichthys deperditus, 57
Saurichthys grignae, 62
Saurichthys macrocephalus, *58*, 60, *61*, 62, 63
Saurichthys madagascariensis, 59
Saurichthys (Sinosaurichtys) minuta, 57
Saurichthys paucitrichus, 61, *62*
Saurichthys rieppeli, 62, *63*
Saurichthys striolatus, 57
Saurichthys yunnanensis, 62
Saurischia, 77
Sauropodomorpha, 77
Saaauropterygia, 72, 105–107, 109
Saurorhynchus, 57
Saurosphargidae, 78, 82
saurosphargids, 82–83
Saurosphargis, 110, 112, 114, 118, 204; paleobiogeography of, 210
Saurosphargis voltzi, 82, 118, 204
sauropterygians, 79–81
Saurosuchus, 196
scapulocoracoid, 36, 46, 51
scleral ossicles, 131, 158
Scleromochlus, 77
Scotland, 77, 180, 222
semi-durophagous, 52, 95
Serpiano, 3, 9, 11, 19, 32, 141
Serpianosaurus, 129, 131, 133, 134, 205

Serpianosaurus-Neusticosaurus clade, 133, 134, 199
Serpianosaurus mirigiolensis, *126*, 127, 129, *132*, 133, 134; longevity, 132; sexual maturity, 132
sexual dimorphism, 80; in *Mixosaurus*, 97; in pachypleurosaurs, 128, 131; in *Keichousaurus*, 131, 205; in *Neusticosaurus*, 131, 205; in *Serpianosaurus*, 131, 205; in *Tanystropheus*, 186; in *Tanytrachelos*, 186
sexual maturity, 132
Shastasauridae, 92, 101
Shastasauridae indet., *103*
shastasaurids from Mt. San Giorgio, 101–103
Shastasaurus, 101–103, 208; relationships of, 211
Shastasaurus tangae, 208
Shonisaurus, 101, 102; relationships of, 211
Sickler, Friedrich, 194
Silvestrosaurus, 144, 147
Silvestrosaurus buzzii, 147
Simosaurus, 81, 87, 106, 107
Simosaurus gaillardoti, 125
Simosaurus pusillus, 125
Simpson, George Gaylord, 107
Sinocyamodus xinpuensis, 209
Sinosaurosphargis, 83, 113, 165, 203; paleobiogeography of, 212
Sinosaurosphargis yunguiensis, 86, 204
skeletal paedomorphosis, 131; in *Macrocnemus*, 186; in *Tanystropheus*, 186
skeletochronological studies, 132
Slovenia, 199
Smilodon laevis, 179
sodium balance, 84, 85
Soergel, Wolfgang, 196
Somerset, 105
South Africa, 36, 75, 76, 107, 169, 193
South China Block, 20, 83
southern Alps, 20, 118, 125, 140, 207; speciation center, 200
Spain, 83, 118, 182, 195
spatium interosseum, 92, 97, 111
speciation center, 200
Sphenodon, 153, 170, 172, 216
Sphenodon punctatus, 17, 73, 75
Sphenodontida, 72
Spinirolo, 8, *9*
Spitsbergen (Svalbard), 33, 53, 84, 91, 92, 94, 95
Squamata, 72, 75, 106, 107, 153, 154, 163, 169, 172
stagnation deposit, 26
Stagonolepidae, 193
Stagonosuchus 193
Steinsfurt, 109
Stenopterygius, 89
Stensiö, Erik A:son, 33, on *Birgeria*, 52, 53
Stoppani, Antonio, 8, 53
stratigraphic overlap, 134

244   Index

stratigraphic succession, 133, 134
stratigraphy at Mt. San Giorgio, *21*, 22–25
streptostylic quadrate, 172; in *Macrocnemus*, 173, 176
Strophodus, 43
Strophodus reticulatus, 44
sturgeon, 52, 54, 57
Subholostei, 52
suboscillatory undulation, 54
subthecodont, 110
sympatric speciation, 68
Synapsida, 72, 73, 106, 107
Synaptosauria, 106, 107
synchronomorial denticles, 35

tailbend, 78, 96, 97, 99, 101
Tanystropheus, 17, 18, 24, 105, 170, 172, 176, 177, 180, 184, 203, 216; amphibious, 188; axial-subundulatory swimmer, 187; bone histology of, 187; from China, 207; heterotopic ossifications, 186; history of taxonomy, 177–182; juvenile specimens of, 183, 187; paleobiogeography of, 177, 182, 212; paleoecology, 186–188; relationships of, 171, 173, 174, 181; stomach content of, 183, 186; terrestrial, 179, 181, 182, 186, 187
Tanystropheus antiquus, 181, 182
Tanystropheus conspicuus, 16, 178–180, 182
Tanystropheus fossai, 184
Tanystropheus longobardicus, 15, 179, *180*, *182*, *183*, *184*, *185*; mortality rate, 188; cervical vertebrae of, 188, 203; developmental constraints, 188; ecology of, 186; from the Black Forest, 181; paleobiogeography of, 212
Tanystropheus cf. *T. longobardicus*, 207
Tanystropheus meridensis, 183, 184
Tanytrachelos ahynis, 171; heterotopic ossifications, 186
Tanzania, 193
Teleostei, 52
Teratosaurus suevicus, 192
Tethys Ocean (Tethys Sea), 20, 27, 28, 140, 199
teutophagous, 103
Texas, 106, 169
Thalattoarchon, 91; apex predator, 101
Thalattosauria, 78, 81, 153–155, 160, 163, 164
Thalattosauridae, 153; relationships of, 210
Thalattosauriformes, 81, 164
Thalattosauroidea, 158
thalattosaurs, 81–82; from Mt. San Giorgio, 155–163; history of taxonomy, 153–155; interrelationships of, 163–167
Thalattosaurus, 153–155, 160; relationships of, 163–165; rostrum of, 157, 158, 166

Thalattosaurus alexandrae, 153, 155
thecodont, 194
Therapsida, 106
Theromorpha, 106
thyroid fenestra, 111; in *Eusaurosphargis,* 112; in pachypleurosaurs, 131; in *Clarazia*, 159; in *Macrocnemus*, 176; in *Tanystropheus*, 186
Tibet, 83, 84
Ticinepomis, 215
Ticinepomis peyeri, *70*
Ticinolepis, 66
Ticinolepis crassidens, 68, 69
Ticinolepis longaeva, 68
Ticinosuchus, 76; erect gait, 192; terrestrial, 191
Ticinosuchus ferox, *190*, *192*, 193–194; apex predator, 191; carnivorous, 191; crurotarsal joint, 193; locomotion, 196; piscivorous, 191; terrestrial, 194
Tintori, Andrea, 19, 27, 52, 136
Tocosauria, 160, 162, 163
Toretocnemus, 211
Trachelosaurus, 172
trans-Pacific distribution, 211
trans-Pacific relationships, 165
Transylvania, 81
Tribelesodon, 16, 171, 177, 179, 215
Tribelesodon longobardicus, 15, 179
tricuspid teeth, 179, 182, 183, 187
Trilophosauria, 75
trophic competition, 92
trophic pyramid, 150
trophic specialization, 211, 216
Trotti, Lodovico, 31
Tschanz, Karl, 146, 147, 187
tuatara, 17, 73, 75, 153, 170
Tunisia, 83
Turkey, 83
Typicusichthyosaurus tsaihuae, 208

Undina picnea, 70, 71; cf. *Undina picnea*, *71*
Upper Silesia, 46; *Tanystropheus* from, 181–182; *Saurosophargis* from, 82, 110, 113, 204

Val Brembana, 184
Valle Stelle; *Acrodus* from, 39, *42*; *Askeptosaurus* from, 156
Vallone mine, 157
Val Porina, 9, 10, *11*, 109, 110; *Acrodus* from, 39, *40*; *Asteracanthus* from, 43, *44*, 45; *Birgeria* from, 52, *53*; *Clarazia* from, *159*; *Cyamodus* from, 109, 118, *119*, 120, 121; *Hescheleria* from, *162*; *Macrocnemus* from, 172; pachypleurosaurs from, 124, 126–128, 135; *Palaeobates* from, 46; *Paraplacodus* from, 110, 115; *Tanystropheus* from, 16, 179, *180*; *Ticinosuchus* from, *190*, 193
Val Serrata, 127; pachypleurosaurs from, 128

Varenna, 31, 105, 142
Vipera aspis, 4
viviparity, 58, 78, 88, 89; in *Saurichthys*, 63; in *Birgeria*, 65; in *Mixosaurus*, 96, *98*; in *Keichousaurus*, 132, 205; in lariosaurs, 148; in *Neusticosaurus*, 132; in nothosaurs, 148; in pachypleurosaurs, 128; in *Peltopleurus*, 68; in protorosaurs, 174
Voltzia, 27

*W*angosaurus brevirostris, 80, 206
water balance, 84–85
western Tethyan faunal province, 65, 83, 167, 199, 202; *Colobodus* from, 55–56; faunal relations to eastern Tethys 177, 200, 201, 211, 212, 217; nothosaur taxonomy, 80, 81, 150; placodont origins, 79
Wild, Rupert, 179; on *Tanystropheus*, 181, 186, 187; on juvenile *Tanystropheus*, 183
Wiman, Carl, 32
Wimanius, 95
Wimanius odontopalatus, 94
Winterswijk, 79; ancestral placodont from, 115; *Lariosaurus* from, 141, 150, 182; *Nothosaurus* from, 150; saurosphargid from 83, 112; *Tanystropheus* from, 182
Woodward, Arthur Smith, 41, 43
Wumengosaurus, 202; relationships of, 203
Wumengosaurus delicatomandibularis, 202
Wuming District, 200
Wusha District, 80, 205–207
Wyoming, 81, 83, 84, 228, 236

*X*ingyi City, 81, 205, 206
Xingyi fauna, 81, 83, 92, 141, 205–207, 210–212
Xinminosaurus catactes, 202
Xinpu District, 205
Xinpusaurus, 82, 206, 208, 209; paleobiogeography of, 210, 211; relationships of, 165, 210; rostrum of, 166
Xinpusaurus bamaolinensis, 209
Xinpusaurus kohi, 209
Xinpusaurus suni, 164, 209
Xinpusaurus xingyiensis, 206

*Y*angtze Plate, 83
Yunguisaurus, 81
Yunguisaurus liae, 26
Yunnan Province, 62, 83, 86, 88, 201, 202, 205, 207

*Z*anclodon, 177, 179
Zanclodon laevis, 179
Zangerl, Rainer, 124, 125, 128
Zogno, 164

OLIVIER RIEPPEL is Rowe Family Curator of Evolutionary Biology at the Field Museum in Chicago. He is the author of *Turtles as Hopeful Monsters: Origins and Evolution*. His main current research interests focus on Triassic marine reptiles from southern China. He has also contributed extensively to the comparative anatomy and evolution of modern reptiles, most notably the evolutionary origin of turtles and snakes. He has published widely in the history and philosophy of comparative biology on topics as diverse as species concepts, mid-eighteenth-century French biology, and the history of phylogenetic systematics. Rieppel is on the editorial board of several peer-reviewed scientific journals and has published eight books and more than 350 scientific papers.